The Child Who Plays With Time

The Child Who Plays With Time

Sylvana Caridi & Thierry Babot

GravityPro Consulting, LLC

Published, edited and distributed by
Gravity Pro Consulting, LLC
United States

Library of Congress Control Number: Applied for.

First edition: 2025
ISBN: 978-1-970408-01-0

Printed by various publishing and printing partners

This book is dedicated to every reader,

May you feel within it a new breath, an inspiration—perhaps even a hope—in the face of the challenges of our time, and may you each, in your own way, contribute to the evolutionary leap that our era is calling for.

Love and Truth

Table of Contents

PART I

CHAPTER 1

CHAPTER 2

CHAPTER 3

PART II

CHAPTER 4

CHAPTER 5

CHAPTER 6

PART III

CHAPTER 7

PART IV

CHAPTER 8

CHAPTER 9

CHAPTER 10

CHAPTER 11

CHAPTER 12

CHAPTER 13

PART V

CHAPTER 14

CHAPTER 15

CHAPTER 16

CHAPTER 17

PART VI

CHAPTER 18

CHAPTER 19

CHAPTER 20

CHAPTER 21

PART VII

CHAPTER 22

Acknowledgments

Bringing a new thought into existence is never an easy task.

It demands courage, patience, and above all, the trust of those who know how to recognize, behind the unknown, the promise of a possibility.

Many welcomed this approach with benevolence and genuine enthusiasm, discerning in it an unexpected coherence and a fertile potential for science and consciousness brought together.

But the formalism of institutions—designed to validate acquired knowledge—still leaves little room for the refoundation of fundamental science.

This is why we wish to express our profound gratitude to those who, despite everything, believed in this research.

To the scientists, researchers, engineers, philosophers, and friends who supported, advised, or simply encouraged Thierry throughout this journey: your trust was an essential support.

You offered what every discovery first requires—not resources, but sincere listening and an open gaze.

We also express our gratitude to those who, before us, opened the great paths of scientific and philosophical thought.

From Bohr to Einstein, from Schrödinger to Heisenberg, from de Broglie to Lemaître, and many others besides—each of them dared to think beyond the frame, between intuition and rigor, between reason and wonder.

Their work has shaped the world as we understand it today, and continues to inspire all those who seek to unite truth and meaning, logic and life.

A very special acknowledgment goes to Charleen Coche, whose editorial work and English translation have been far more than a technical task: a work of understanding and accuracy. Through her precision, her patience, and her profound sensitivity to the text, she refined its clarity, coherence, and musicality—in French as in English.

Our thanks also go to Gabriel Uribe, befuse.com, whose artistic sensibility and rigor gave form to the visual universe of the project. From the book cover to the design of the GravityPro.com website, he captured the spirit of the discourse, revealing through image what the words suggested: a unity between science, art, and consciousness. His gaze, in resonance with our vision, made it possible to express its full depth.

Finally, our gratitude goes to all those who join us in this adventure: readers, researchers, students, dreamers, and builders of the future.

Your curiosity and openness bring this endeavor to life.

May you find in it not certainties, but paths—not closed answers, but surges of intelligence and trust toward the real.

May this book be, for you as for us, a living bridge between knowledge and consciousness, an invitation to think, feel, and co-create the world differently.

For science has meaning only if it illuminates, connects, and elevates.

Prologue

Thierry

There comes a moment when curiosity ceases to be a mere game of the intellect and becomes a demand of the heart.

It is there, at that intimate border, that true science is born: no longer as an exercise in analysis, but as an act of love toward the real—the meeting point between the desire to understand and the need to love what one understands.

Since childhood, I have never been able to accept the idea of a fragmented world, without connection or depth.

Everything in the texture of the real—from the beat of a cell to the course of galaxies—hinted at a silent coherence: a living, underlying, intelligent order, woven from a memory that science had not yet named.

Very early on, I felt that the key to this coherence lay in Time.

Not the time of clocks, but living Time, that profound flow that does not merely accompany phenomena, but engenders them, orders them, and binds them together.

It does not measure change: it creates it. It does not follow matter: it precedes it, structures it, and guides it.

For years, I searched within the foundations of classical science for traces of this originating Time.

But the more I explored established models, the more I perceived a silent fracture: physics described space, matter, energy—but it reduced Time to a mere parameter, a passive axis along which events follow one another, without acknowledging its creative role.

It was in observing this absence that a clear truth emerged: the Universe is not a fixed mechanism, but a dynamic of living Time.

There, Time is not a backdrop, but the very substance of the real—the breath that engenders energy, organizes matter, weaves memory, and renders consciousness possible.

Every particle, every being, every thought is a modulation of Time: a wave of existence within a single field that transforms itself, remembers, and continually reinvents itself.

In other words, there exists only one Reality: the universal Life of Time, which unfolds, reflects itself, and discovers itself through us.

We are not its observers, but its local expressions, the sensitive instruments of its own becoming.

It is from this understanding that Absolute Relativity was born—an attempt at unification between physics and consciousness, between the visible and the invisible, between the human being and the Universe.

Not to confuse science and spirituality, but to rediscover their common source: the coherence of the real, that silent intelligence that connects everything to everything, and of which Time is the primordial language.

Sylvana

Through this vision, GravityPro found a new momentum.

Already engaged in informational systems and the architecture of complex structures, the company naturally extended its field of research toward the foundations of information itself—all the way to its quantum dimension, where physical, energetic, and cognitive processes converge into a single dynamic: that of living Time.

For us, to understand no longer means accumulating data or building models; it means entering into resonance with the real, listening to it, translating it, and circulating it.

True understanding is a living act: it is shared, embodied, and transmitted like a wave of coherence through human society.

We live in a time when everything changes very quickly—sometimes more quickly than our capacity to give meaning.

Disciplines are fragmenting, discoveries are piling up, innovations are following one after another, but the guiding thread is fading: knowledge is multiplying, but wisdom is becoming scarce.

It is in the face of this dispersion that GravityPro chose another path: to rediscover the principles of organization of the real, those unifying laws that connect matter, energy, space, and consciousness, in order to apply them in the domains where the future of humanity is being decided:

The way we produce energy, explore the Universe, exchange information, and understand life.

Our objective is clear: to build a new way of co-creating with the Universe, founded on an understanding of Time and of the laws that connect matter, information, and life.

This approach does not seek to dominate nature, but to engage in dialogue with it—to harmonize with the logic of the Cosmos so as to unfold its creative power in sustainable and conscious forms.

We wish to share the philosophy of this science, not through closed equations or technical abstractions, but through the inner journey that made it possible: the intuitions, the thought experiments, and the moments of revelation that led to this vision of the world.

It is in this spirit that we have conceived this book.

The choice of dialogue was the obvious one: it makes it possible to follow, step by step, the genesis of an idea, to reveal its connections, its hesitations, and its discoveries, while making the reflection accessible.

Over the course of these exchanges, we revisited the foundations of current science—its successes, its paradoxes, and its frontiers—to show how Absolute Relativity is positioned as a coherent extension of scientific knowledge: a path to go beyond its limits without betraying its rigor.

This book does not seek to convince, but to make one feel a coherence—the coherence that links Time, matter, and consciousness within a single living fabric.

It invites another gaze: a way of thinking the real in which understanding becomes an act of conscious participation, a way of taking part in the creation of the world, no longer as a spectator, but as a partner of the universal Life.

Part I

Foundations of a New Paradigm

~

The Time, the Meaning, and the Self

Chapter 1

The Awakening of a Consciousness

~

From Chaos to the Call of the Cosmos

1. The Call of the Stars — The Beginning of a Journey toward the Infinite

Sylvana

Before we begin to explore your theory, I would like us to take a moment to go a little further back. Not to the origin of the Universe… but to the origin of your gaze upon the world.

Because, you see, listening to you, it quickly becomes clear that nothing you propose stems from abstract speculation. Everything was born from you. From a lived experience. From an intimate encounter. From a very particular relationship with the real.

And it is precisely for that reason that I want us to begin here. I believe that what you went through in your childhood—your solitude, your pain, but also your extraordinary sensitivity to the living—did not merely forge your way of perceiving. It oriented your entire destiny.

It was in that silence, in that overflow of sensations, that you began to seek coherence. Not to do science. But to stand upright. To survive, even, in a world that made no room for you. And at its core, it is this visceral quest for meaning that drove you to question reality down to its most profound foundations.

And what I admire is that this search did not remain internal. You did not remain in observation or reverie. You acted, you built, you experimented.

Your bond with nature, your passion for the Cosmos, your need to understand, to structure everything, to see things through to the end… all of that is the living root of your science.

Today, you are developing technologies to explore space—not as an escape, but as a path of exploration and emancipation. And when one knows that… one senses that each of your childhood memories already carried, in subtle threads, a share of what you would like to pass on one day.

That is why, if I had to choose a starting point, it would be this one. Because it is here that we truly begin to feel what animates you—what drove you to search. What, perhaps, already touches us.

So, if you do not mind… tell me.

Where are you from?

How did it all begin for you?

Thierry
I was born in Strasbourg in 1969, in France, to a German mother and a Martinican father. From a very young age, my life was marked by difficulties adapting to my environment because of my autism, but also by profound losses.

My parents were separated at my birth. My mother struggled to take care of me, because she had health problems. My father, for his part, traveled a lot for work, and he eventually sent me to live with his parents in Martinique.

Sylvana
That is heartbreaking… you had barely arrived in this world, and already, everything was changing around you.

Thierry
Yes. My mother eventually died when I was four, and, unfortunately, my father died in turn two years later.

That is how I remained in Martinique, where it was my paternal grandparents who raised me.

Sylvana
You were so little… I suppose you have no memory of them?

Thierry

No, I have no memory of my parents—neither their faces nor their voices. I have always considered my grandparents to be my parents.

Sylvana

They were there for you, then… truly present?

Thierry

My grandparents were extraordinary. They offered me a stable home, filled with love. Their affection, though discreet, enveloped me and allowed me to grow up in a reassuring environment.

Their house was simple, modest—but it was surrounded by incredible nature, vibrant, alive. Very quickly, nature became my refuge. Then my escape. And, over time, my very first teacher.

Sylvana

Your very first teacher… that is beautiful, said like that.

But how can one learn from a tree, from a wind, from an insect? What were you grasping from the nature around you?

Thierry

Nature fascinated me with its harmony, its silent order, and its raw beauty. It was there that, over the years, I began to observe, to reflect, and to seek answers.

Those first years of my life forged my unique way of perceiving the world, and they laid the first stones of my quest to understand the Universe—and my place within it.

Sylvana

Can you take me with you for a moment—describe what your world in Martinique looked like when you were little?

And above all… at what moment was your difference, in some way, recognized or put into words?

Thierry

One must remember that, when I arrived on the island, everything there was still very simple, almost raw. Martinique, a French island located in the Caribbean, then lived at a peaceful rhythm, far from the agitation of the great cities of the world. Modernity had not yet reached every corner.

In the countryside where I lived, we still lit our nights with kerosene lamps. We drew water, or collected it when it rained. The roads were narrow, sometimes impassable, and everything seemed to operate at a much slower rhythm than one can imagine today. In that context, nature occupied a central place.

My autism was diagnosed very late, around the age of nine, when my grandfather, aware of my difficulties, made the firm decision to help me. He found an American specialist who had settled in Martinique. She was a physician-psychotherapist, specialized in the treatment of autism—a neurodevelopmental disorder that was better understood in her country.

Every Wednesday, I went to a room arranged for children in difficulty, within a medical clinic. Her name was Ruth, and she took full charge of me, because my case was, according to her, very specific and rare.

She looked after me until I was thirteen, with extraordinary patience and devotion. I wish to bless her through this exercise of remembrance.

. Ruth and the Embracing of the World — When One Encounter Changes a Life

Sylvana
Your grandfather must have sensed very early on that you had a relationship to the world… most particular.

But what do you think set him on that path? What pushed him to seek outside help at that moment?

Thierry
I must say that I did not manage to pronounce my first word before the age of seven. I expressed myself through drawings that my beloved grandmother had the patience and wisdom to interpret to ease my worries. I literally could not communicate with others.

At school, from my earliest years, because of my difference, it was agreed that it would be more appropriate to try to educate me at home.

Until the age of nine, I therefore lived in a very closed world, mainly centered on myself. My autism made me extremely sensitive to what was happening around me, but I did not have the tools to understand it or explain it.

My universe was profoundly interior, where every sensation was amplified. I perceived nature and its workings, but I could not grasp the totality of its functioning—which distressed me profoundly. I saw elements, interactions, but there was always a common thread missing, a logic—and that unsettled me.

Sylvana
Ruth truly managed to reach you.

What did she know how to do, what did she understand… that no one had been able to touch before her?

Thierry

My sessions with Ruth helped me enormously—to the point that they marked a decisive turning point in my life. They allowed me to bring order to the chaos of my thoughts, to connect objects to one another and to associate words with things. She was manifestly very gifted: she succeeded in projecting my inner world into the external reality outside of me.

She had prepared an entire universe for me in that hospital room—on the floor, with figurines such as humans, animals, houses, fences, etc. She invited me to have fun with all those toys, she observed what I did, tried to understand my reasoning. She wanted to pierce my inner logic, so as to guide me ever better in my self-learning.

She knew that I had the natural faculty to learn by myself. "He is a born autodidact," she would say to my grandparents. She was amazed, astonished to see my ability to reorganize everything logically, to want to optimize everything, with an intelligence beyond the norm for my young age. "He has the soul of an inventor," she would conclude.

All those exercises gave me the tools I needed to understand and apprehend the world around me more clearly.

She had very quickly understood something essential: my only true organ of perception—capable of learning—was my eyes. That is why she presented the reality of the world to me in the form of a miniature model, so that I could grasp the global vision.

She was the genius in this story. An exceptional woman. The one who saved me from a fate far more limited.

Sylvana

When you say that your eyes were your only true organ of learning… do you mean that everything had to pass through image? Through the global vision? I am trying to imagine what that represents, in daily life… How did you manage to understand what surrounded you?

Thierry

Yes, that is a good question… and to answer you, I need to explain to you how I experienced things—sensorily.

I must say that my relationship to my sensory stimulations was very complicated. I had such sensitivity that a simple caress on the cheek triggered a reaction of repulsion. It was the same when I touched an object by accident, or because of the unpleasant sensation of clothing—but I adapted to it, because I hated being naked.

In another context, when people showed me numbers, letters, or said words to me, I understood nothing, and that tended to stir a powerful anger in me. I saw only intangible shapes, with no correspondence to external reality—an aberration, pure and simple.

So my fairy, Ruth, had me work, for years, with method and delicacy, to structure my thoughts and forge links between what I saw and what it meant. She had conceptualized exercises suited to my way of perceiving the world.

Without that encounter, my life would have been entirely different. She was a light in the darkness of my inner life, and it is thanks to her that I was able to begin building bridges between my inner world and external reality—luminous, full of richness and meaning.

Sylvana
And after those years of learning with Ruth… did you attempt school again? I tell myself that perhaps, with everything you had integrated, someone might have accompanied you differently?

Thierry
Yes. After finishing my sessions with Ruth, around the age of fourteen, my grandfather enrolled me in school again. It was a vocational high school, in industrial electricity. On the practical side, I was exceptional. Anything that required building, handling objects, observing them—I did it with great ease.

But on the theoretical side… it was still the same. I did not understand what I was being taught and, since I could not follow classes in the usual way, I eventually fell behind. After a few months, it was another failure, and I had to stop.

Sylvana
So, even after that remarkable journey with Ruth, even after learning to connect your inner world to external reality… school remained a dead end.

But did it nevertheless change the way you looked at nature? Did that bridge you had built with her allow you to perceive differently what was around you?

Thierry
Yes. It was in that context that my questions evolved. I had always been fascinated by nature, but through those exercises, I began to connect my perceptions to concrete concepts. That allowed me to enter a more structured

phase of reflection, where I could finally put words and shapes to what I felt and observed.

However, that quest for understanding remained difficult. Even with that help, I was still missing part of the logic of the world—that perfect order I so longed for.

Sylvana
It is moving… One feels that you were beginning to carve out a path, step by step, between the chaos of sensations and that desire to understand the world.

You were already equipping yourself from within, with that very particular rigor that defines you… and meanwhile, your life was following its own rhythm, far from conventional school. You had your own world…

But in the end, what did your days look like?

Thierry
Until I was sixteen, my days were paced by an external routine that repeated itself tirelessly—day after day. But inwardly, I was living a thousand and one experiences.

From six in the morning until four in the afternoon, I moved through a kind of euphoria. Because even if I did not yet understand everything—even if the order of the world still eluded me at times—I was perceiving something. A hidden coherence, an underlying logic.

I did not yet have all the tools to grasp it, and that frustrated me profoundly. At times, it even made me angry. That feeling of being on the threshold of something, without yet being able to name it, burned me from within.

But despite it all, most often, that anger gave way to a greater joy: the joy of sensing that order existed—somewhere. And that I might, perhaps, end up understanding it. As though every detail of the real carried within it a discreet magic, a mystery I was called to unravel.

Every morning I woke up, and the first person I saw was my grandmother—an incredibly resilient woman who always found a task to keep me occupied. For instance, she made brooms that she sold at the market, and it was often my role to gather the dried palm leaves needed to make them.

But do not imagine a grand expedition into the forest! No—I merely carried the leaves from one pile to another. These small tasks, simple though they were, gave me the feeling of being useful—as though my contribution were essential.

Everything was a source of fascination for me. Each object became a riddle to solve. I would hold them very close to my eyes to examine the tiniest details: textures, colors, shapes, and above all movement. The dynamic aspect of things captivated me in particular; I could spend hours observing insects, animals, small and large.

Even during meals, this obsession was omnipresent. On my plate, each food had to be separated and carefully arranged, or else it would provoke immense frustration in me. Before I began to eat, I would spend a long time structuring my plate.

The first bite of each food was a true exploration: I scrutinized, I observed, and I took all the time I needed to understand its textures and flavors. Only once I was fully satisfied with my observations would I begin to eat—and then I devoured everything with an insatiable appetite, because I loved it.

It made my grandmother laugh—she so loved preparing all those little dishes for me.

Sylvana
And I bet that, knowing you… even gathering fruit became a high-precision operation! Every leaf inspected, every gesture measured to the millimeter… you never did things halfway—even dangling like an explorer in your coconut tree!

Thierry
You know me… Obviously it was not just climbing a tree to pick a fruit. Each branch was a trajectory to optimize, each fruit a calibrated mission. And then, beyond the strategy… I loved it. Climbing trees to pick and savor seasonal fruits. In Martinique, I was particularly spoiled: nature was brimming with delights, and each season offered its share of sweet treasures.

But this perception of order and harmony was not always easy to live with day to day. I was completely obsessed with order. I remember that if two flowers stood side by side and one of them leaned slightly, I could not stand it. I would hurry to fix the flower to a stick to straighten it—but the stick in turn became a discordant element. So I had to find another for the second flower, which did not need it, but it was more harmonious, more symmetrical to my eyes.

In my universe, I saw myself as the guardian of natural order, the one who had to ensure that everything remained perfectly organized, harmonious, in its proper place. For me, it was essential, visceral, almost vital.

Imagine the hell in my head: I was constantly spotting irregularities, at every step I took.

Sylvana

I can picture you… a meticulous little explorer, a kind of Sherlock Holmes of nature. Except your magnifying glass was your gaze. Nothing escaped you, and one can already sense that you were on a mission—a secret mission to understand the world.

But listening to you, I tell myself it must also have been exhausting. Because by seeing everything, feeling everything, wanting to rebalance everything… you could never truly rest, could you?

Thierry

It is true—everything had to have meaning and a well-defined place. And you know me… that has not really changed!

When I was not absorbed by this quest for order, I loved to draw for hours, completely immersed in the world I created on paper.

My grandmother, always attentive to nurturing my curiosity, had bought encyclopedias, dictionaries, and even an illustrated Bible. Although I could not yet read these books, I spent hours contemplating and reproducing the drawings they contained—only the ones that fascinated me.

For instance, I had a true fascination for mechanical objects and machines—the more complex they were, the better. I would reproduce their static images, then start imagining how they might work. And brace yourself… most of the time, I understood their purpose perfectly, which on several occasions surprised and astonished a family friend—a schoolteacher by profession.

My grandmother later told me how this friend explained to them that, for my age, what I was doing was normally impossible. What struck him most was not only the finesse of my drawings, but also their mechanical relevance. Every part, every step I depicted had meaning.

My creations, although imaginary, followed logical and technical principles, as though I intuitively understood the mechanics of objects I had never actually seen in real life. For me, drawing was far more than a simple pastime. It was a way to explore and structure my inner world, to give life to ideas, and to understand, in my own way, how things worked.

My curiosity did not stop at drawing or deciphering images on paper; insatiable, it drove me to go further— examine, explore, experiment with the real, as I still do today. That should have been fatal to me on several occasions—but fate, mysteriously, willed otherwise.

Sylvana
Wait… what do you mean by fatal?

Thierry
Well, to put it simply, when I ventured into the forest that surrounded our house, I would spend hours getting close to wild animals, observing them up close, sometimes even touching them.

Snakes, in particular, fascinated me. I felt a kind of compassion for them because of the way they slithered: it was as though they were disabled—deprived of arms and legs, forced to twist right and left in order to move forward… and yet they slid with impressive efficiency.

Sylvana
Well now, Thierry… I think you are breaking records! Feeling compassion for a snake—who would do that? I must admit, I had never imagined what it would be like to slither without arms or legs.

And you, instead of being afraid, watched them glide with fascination. Only you could see that as a feat—almost a dance. It is crazy—where everyone else sees danger, you see a masterpiece of natural mechanics…

But tell me… were you never afraid?

Thierry
I did not gauge the danger they represented at all. To me, it was an animal like any other, a creature of nature, no more and no less, with its own characteristics.

But I felt safe with all animals, which, paradoxically, was not the case with many of the humans around me.

I felt such communion with nature, so powerful was my desire to understand it and to serve it.

To return to the snakes, I liked to stroke the top of their heads.

Sylvana
Wait, wait… you really touched them? You stroked snakes as though it were the most natural thing in the world? Just thinking about it gives me chills…

And you—utterly calm—you were in profound communion with them. No, truly, you must already have been tuned to another frequency!

Thierry
I understand that it may be surprising. But for me, it was perfectly natural. These animals posed no danger, so long as I respected their rhythm and their space. I would spend hours observing them, understanding their movements,

their reactions.

I had noticed that certain signs, such as an S-shaped posture or hissing, indicated that they felt threatened. So I made sure never to provoke them: I would approach slowly, without any abrupt gestures, avoiding direct eye contact. It was a kind of silent dialogue, in which I had to listen to their signals. It was not recklessness, but a form of mutual respect.

I even remember a snake to which I would give milk to drink, because I loved bringing milk to wild animals—they adored it. It was only much later that I learned that giving milk to a snake made its venom even more potent.

I also used to witness violent fights between animals, especially between snakes and mongooses—and I hated that.

So I devised strategies to restore calm and serenity, and I often succeeded. One of the most effective—which will surely make you laugh—was to play them the accordion my grandfather had given me when I was a little older. You cannot imagine how much that can throw off animals caught in the tension of survival!

Of course, my grandparents, who were absorbed by their daily work as farmers, were completely unaware of my activities as peacemaker and benefactor among snakes, mongooses, cats, dogs, monkeys, etc.

Sylvana

After proclaiming yourself guardian of the natural order—now you were also playing diplomat between species! Honestly, it is wild… at that age, many would have run to hide at the sight of a mongoose or a snake. You—you observed, you were almost brokering non-aggression treaties. You never cease to astonish me!

Thierry

Yes, I had created an entire world, and I was proud to hold an important place in it, at the heart of nature and among the animals.

But one day, when I was about eleven, I learned that a neighbor had died, killed by a snakebite. It was an immense shock for me, in two ways: first through the loss of a kind neighbor, and second through the realization that a snake could kill a human being.

I suddenly realized that I had risked my life dozens of times without even knowing it. I could have died—and that realization shook me profoundly.

After that, I stopped approaching snakes and other venomous animals, such as tarantulas, for a time. Yet I never truly gave up that profound curiosity that drove me to examine everything around me—sometimes in defiance of danger.

But deep within, I had this indestructible assurance that we were all of the same nature and that I risked nothing. I had much love for nature and for animals, and they felt it.

Sylvana
You stopped… but only for a time? You mean that, despite that shock, and aware of the danger, you went back to touching snakes after that?

Thierry
Yes… I know it may seem strange, but I did not stay away for long.

My fascination with snakes eventually caught the attention of one of my father's cousins, who was a snake hunter. He had noticed my ability to interact with them, and rather than dissuade me, he decided to teach me his trade. He showed me how to capture live snakes carefully and bring them back to him. Their venom was collected to produce antivenoms, then they were released.

Each snake earned me the equivalent of two dollars—and with that money, I would run to buy all sorts of treats, and especially Malabars chewing gum, which I loved.

Of course, my grandparents knew nothing about this arrangement and would certainly not have allowed it. Much later, when they found out, they sternly reprimanded me.

Sylvana
You know, listening to you, I am thinking of that little boy whom others surely saw as "different," as though he did not have much to offer… when in truth, he inhabited a world of quiet, profound richness that others could not perceive. A structured world, full of observation, meaning, adventures, risks—a world he protected in his own way.

Thierry
What you say moves me… My autism is no longer really visible, but in social situations it is still there—discreet, yet very present. Everything goes so fast. Exchanges follow one another, words fly.

Me, I often need a little more time… not because I do not understand, but because I want to find the right words. I need that time to formulate an answer that truly reflects what I feel or think.

And sometimes, that slight lag is enough to make me disappear. People assume I have nothing to say. Yet inside, everything is alive. It thinks, it vibrates.

It is not an empty world, on the contrary: it is a world of excess, of density, of intensity.

Sylvana

So, in fact… it was not absence—it was too much of everything, all the time?

Thierry

Exactly. It is not the void—it is the extreme.

Sylvana

It is overwhelming to see how rich that inner world, which no one saw, truly was… but also fragile, perhaps? Because you seem to say that the slightest thing could make you falter.

Do you remember particularly difficult moments in your days? Instants when that balance would break?

Thierry

Indeed, that constant euphoria that inhabited me each day did not last the whole day. Each day, around four in the afternoon, my universe would tip into the horror of suffering. A dreadful migraine would descend upon me—like a shadow that paralyzed me. Unable to do anything, I had no choice but to lie down and wait for it to pass.

Those hours from four to eight were sheer torment for me—and for my grandmother too who, despite her herbalist's knowledge, could find no remedy for me.

My grandmother, in her wisdom and love, had arranged a space a little apart from the house. It was an open area, shaded by tall trees, where she had placed cushions that she stuffed herself.

She had wanted this place to be welcoming, not only for me, but also for the animals that liked to come near me. She knew how profoundly their presence comforted me.

Sylvana

You know… hearing you speak of those cushions, the animals around you, that inhabited silence… I cannot stop picturing this: a hidden sanctuary, in a battered childhood. As though, at the heart of suffering, you had known how to create a place of fragile but real peace. A bit like a clearing in the middle of the storm.

Thierry

Yes, that was exactly it. That resting corner was my inner clearing—a place where the pain could express itself without screaming, where something in me kept breathing, even when everything became too heavy.

I would lie there every day, with that pain consuming me. It was so hard to get there, as though I were walking each time toward a small purgatory. Yet that place, though steeped in suffering, possessed a certain beauty. The cushions, worn by time and by my many comings and goings, were arranged in a circle, like an invitation to comfort.

My pets often came to settle near me. Sometimes even the neighborhood dogs and cats would come lie down by my side. Those moments when I felt their warmth and their silent presence were almost magical. They did nothing extraordinary, but their simple nearness was soothing.

I considered them my friends, my brothers and sisters—companions who seemed to understand my pain without judgment, without questioning.

Then, at eight o'clock, the migraines vanished as suddenly as they had appeared. But that respite was short-lived.

Another ordeal began: an inner itching sensation, in my brain, almost impossible to describe, would invade my head. It was so unbearable that I had no choice but to swing my head from left to right, like a mechanical motion that was the only thing able to soothe me.

I would fall asleep like that, and my grandfather did everything he could to help me not move my head too forcefully, for they both knew it would shorten my life.

He had tried once to fix two wooden boards around my head, but it ended up hurting me. So the only solution was found by my grandmother: she made me a large U-shaped pillow, which limited the angle of rotation of my head.

Sylvana
The love of your grandparents, and especially that of your grandmother, shines through so profoundly in what you share. But apart from them, did you have interactions with other children or other people when you were a child?

Thierry
My interactions with others were very limited. I had tics, nervous gestures, which made most people uncomfortable. A standard school education was therefore out of the question, as we have already discussed. And, in my village, things were complicated. The children did not like me—they were afraid and kept their distance.

Four times a week, I had to go buy bread about a kilometer from the house.

I loved that moment, because I loved eating the bread just as it came out of the baker's wood-fired oven. My grandmother knew this and allowed me to buy an extra baguette so that I could eat it on the way back. It made me laugh and brought me a small, simple joy.

However, going to get the bread caused me a certain stress, because it often coincided with the time when the children left school to go have lunch. On several occasions, I had to endure stones being thrown at me, but I sometimes struck back in turn, when I had been hit and was bleeding.

I got into serious trouble every time I injured an assailant. The parents of those children hated me for it and sometimes threatened my grandfather, saying they would come after my life. My grandfather himself commanded respect. Threatening me was therefore not a good idea.

There was so much violence and injustice among humans that my heart bled to live through it. Not only was I attacked, but if I tried to defend myself, my life was threatened on top of it.

Sylvana
That is terrible… You were already vulnerable, and instead of being protected, you were exposed to even more violence. I understand better why you say that your heart bled.

Thierry
Yes. And my atypical behavior seemed to fuel their cruelty even more. Under stress or fear, I laughed nervously instead of crying. In their eyes, it was provocation. They believed I was making fun of them, which only fed their anger further.

I was also often the target of mockery because of the color of my skin. Since I was white and everyone knew that my mother was German, it set me apart from the others, on an island where the majority of Martinicans had much darker skin than mine.

In their eyes, that made me a foreigner, someone who did not belong among them. Between my appearance, my tics, and my behavior, I combined everything needed to be rejected outright. But my grandparents knew how to provide certain insights to those people and, over time, they came to simply ignore me.

But let us rather speak of positive things. Once a month, I went to the island's capital, Fort-de-France, with my grandmother. It was one of the rare moments I looked forward to.

We would get up very early, at four in the morning, and my grandmother would dress me well and do my hair—I was so proud. We took a taxi to get there, and there she sold the vegetables, fruit, and brooms that came from our farm.

Those days were very different from the ones I spent in the village. Fort-de-France was full of life, with very colorful houses close to one another; there were also many shops, etc., unlike our village where everything was far apart and isolated.

What I particularly loved was the port, and the old stone fort overlooking the harbor fascinated me. For me, it was as though I were watching a dizzying spectacle: I saw large ships, almost like floating cities, and many people, which made me a little anxious.

My grandmother, with the money she earned at the market, would take the opportunity to go shopping. I remember the fabrics she bought to make us clothes. She was wonderful, my grandmother; she knew how to do so many things, she was my hero.

I also remember that with every visit to Fort-de-France, my grandmother bought me zakari—this very hard white bread that I loved. She knew she had to buy a dozen, because I would already eat two whole loaves during the taxi ride back.

With each bite, she would look at me and gently shake her head, amused, and I would burst out laughing. Those moments of simplicity and delight were a small happiness, a bubble of lightness in a life often complicated.

She sometimes took advantage of being in Fort-de-France to visit her two friends, my "mammas." Those moments were magical for me.

Those two women always awaited me with great attentiveness and kindness. Each prepared something special: homemade cookies and a large glass of very fresh fruit juice.

They welcomed me like a prince, and, in those moments, I felt profoundly loved.

To me, they were like extra mothers. Their presence in Fort-de-France, even if it was rare, brought a precious light into a childhood often marked by solitude and rejection.

Those moments of tenderness and pride were unforgettable.

. THE THRESHOLD OF A MIRACLE — WHEN SUFFERING ENCOUNTERS THE UNEXPECTED

Sylvana

What you are describing is so vivid… One feels how much each memory matters, how every detail left a mark—the kind that forges a life, in silence.

You say the suffering lasted until your adolescence… until you were sixteen. What happened then? Was there a trigger, a change that overturned everything?

Thierry

Indeed, around the age of sixteen, I had grown a great deal—and with my height came a physical strength that made my problem of involuntary head movements at night increasingly dangerous. My grandparents were terribly worried. My grandmother took me to the hospital to consult a neurologist.

At that time, autism was still very poorly understood. The neurologist explained to my grandmother that, unfortunately, he had no solution for me and that, in the rare cases he had known, those children did not survive beyond twenty or twenty-one. He believed my condition would be fatal at that age. This announcement was a terrible shock for my grandparents.

My grandmother was an herbalist. She had tried every plant, every ointment, everything she knew, but nothing had succeeded in soothing my crises. Determined to find a solution, she decided to take me to see a shaman whom an acquaintance had recommended to her.

This shaman lived in Fort-de-France, so we visited him one day after the market.

I still remember my first meeting with this shaman very clearly. We entered a simple house, yet imbued with a mysterious aura. A narrow staircase led to a small room where he welcomed us. He observed me for a long time, in silence, then he began to play various instruments that he said were sacred.

After he had finished his mystical session, he told my grandmother that I was truly special and that I had a great destiny in the service of humanity. My grandmother and I were astonished, given my condition at the time. I remember that we looked at each other, taken aback, as though to ask each other, "What is he talking about?" Then he asked us to come back the following week.

When we returned, he quickly took out a syringe containing a clear liquid, poured two drops into a small glass of water, and asked me to drink it in one go. No spectacular ritual, no complex explanations—just to drink the concoction. Then he greeted us warmly and we left.

And I, who had imagined a grand session, with drums, mysterious incantations, perhaps even a few exotic animals in a trance… no—just two drops in a glass of water.

Sylvana
I can quite picture the scene… You, all set to experience a full-blown mystical ceremony—perhaps even a serpent undulating in a corner… And in the end: two drops in a glass of water? No wonder it impacted you!

Thierry
Yes… you should have seen my little face at that moment. I was ready for an unforgettable ceremony… And yet, despite that simplicity, that day, something extraordinary happened—I am still deeply moved just thinking back on it.

At four o'clock—the time when my migraines usually appeared—nothing happened: no pain, no attack. For the first time ever, that daily hell had vanished.

My grandmother, overwhelmed, knelt on the floor and prayed to God to thank Him for so great a miracle.

Sylvana
What you experienced is powerful, and it touches something profoundly deep. And yet, one does hear that kind of account… sometimes in a low voice, almost like secrets. Encounters that upheave, healings one does not explain. One listens in silence—but they remain. Like yours.

Thierry
Yes… That day, everything shifted without a sound. And what had haunted me for so long… suddenly disappeared.

In the evening, as I went to bed, I realized that something else had changed: I no longer felt that unbearable itching in my head. That sensation, which had driven me for years to bang or rock my head, had gone quiet.

And that night, for the first time in a very long time, I slept peacefully.

In the days that followed, I found that my migraines had completely disappeared, and the itching as well. What remained were only psychomotor difficulties: my tics and that problem with coordination between right and left that still affected me. And, of course, that obsession with organizing everything perfectly, making sure that each thing is always in its place, in impeccable order.

Sylvana

Your need for each thing to be in its place… I recognize it so clearly here.

But what you have just shared is moving. One feels that a calming has settled—coming from very far away. A silence that heals and that has gently restored inner order. Something truly transformed within you.

Thierry

Yes. It was from that moment that my life truly began to change. A light had turned on in my daily life. That encounter with the shaman had marked the beginning of a new phase.

For the first time, I glimpsed the hope of a different future, sunlit.

I began to discover another side of life in the village, the hours that followed four o'clock. From the house, I noticed that the nearby young neighbors gathered on a little hard-packed dirt patch where a big tree trunk lay, serving as a bench. They would sit together to talk, read newspapers, or simply pass the time.

Intrigued, but still very hesitant, I started to draw nearer—at first from a distance, observing. Then, one day, they let me sit with them. Some asked for my forgiveness for the harm they had done me a few years earlier.

They introduced me to youth magazines. I still remember my favorite, *Pif and Hercule*, a humorous comic strip featuring a dog and a cat. That magazine came with extraordinary gadgets—a true delight for me, who loved technology. I also remember the Marvels, full of superheroes who seemed to come from another world.

My grandmother, seeing that I was interested in these exchanges, took me into town to buy magazines, and I was stunned when I discovered a science magazine for the first time. It was as though the world were within reach of my hand. That is how I discovered the French science magazine *Science et Vie.* I started buying it every month and I spent hours poring over everything; it was a true treasure of knowledge.

With my neighbors, we traded our reading: they shared their comic books, and I shared my magazines. Well, almost all of them, except my *Science et Vie* magazines, which I regarded as sacred. I kept them carefully, and in any case, my neighbors did not find them very interesting.

For me, it was different: even though I still struggled to read properly, I spent hours poring over the images. That magazine fascinated me. I could discover infinite universes there, and concepts that captivated me and nourished my fertile imagination.

I cut out certain images and glued them to the walls of my room. That allowed me to escape even farther from my bed, to dream of distant galaxies or inner worlds—far from the realities of my everyday life.

But it was not only the life after four o'clock with my companions that impacted me, nor solely my daily life that was softening—something much more profound was beginning to change within me. My whole being was opening to another space—a space of wonder, of inner silence, of presence.

It was at that moment that the sky began to call me. Literally. As though the Universe suddenly became familiar to me—and in that silent connection, a stellar calling was gently being born.

It was within this new bond that everything began to shift. I rediscovered a profound inner silence, inhabited by a pure wonder before the starry sky. For the first time, I could observe the celestial vault in complete tranquility, without pain, without mental agitation—simply carried by peace and fascination.

This contemplation quickly became a gentle obsession, almost sacred. I therefore decided to set up a makeshift bed—a simple mattress, a blanket, and a pillow—on the cement roof of the house, in order to spend my nights communing with the infinite.

I remember intensely that thirst for understanding that ran through me: why those luminous points? Why that twinkling? What is really up there?

It seemed to me that there was an invisible wall between me and those distant lights, as though the Universe were divided in two: Earth—solid, familiar—and then the black void and the stars—magnificent, yet incomprehensible.

I asked around me. People told me, "They are stars." Others spoke of "Heaven, the dwelling place of God." But those answers left me unsatisfied—indeed, more lost still.

But thanks to the magazine *Science et Vie*, I would sometimes come across articles that spoke of cosmology. I did not understand everything that was said in the article, but it only amplified my fascination with the sky.

I wanted something concrete, something meaningful. So, one night, I made a decision that was strange for a child: to try to communicate mentally with those stars. I focused my attention on one of them, chosen at random, hoping to receive a sign, an echo, a tremor. Nothing happened. Not at first.

But by spending night after night under the sky, something changed. It was not a clear message, nor an inner voice, but rather a subtle link that was gradually being woven between me and the sky. A resonance, a silent complicity—as though the Universe recognized me. I was fortunate to live at a time when light pollution was almost nonexistent. The sky was clear, offered, vibrant. That link grew stronger, until it became a vocation.

One day, I turned to my grandmother and said, "I know what I want to do later: I am going to visit all the stars in the sky." She looked at me, smiled, and replied simply, "If it is planned for you to go up there, then nothing will be able to stop you, my son."

Those words, engraved in my heart, became the foundation of an intimate commitment—an inner promise that I would never betray.

Over time, I grasped the immensity of the challenge. And I often imagined God Himself bursting into laughter, seeing me, a small human, wanting to conquer the stars. But I had also been told that I was His son, and that changed everything: for if the impossible exists, it is, by essence, temporary. And I have always known that the impossible could bend to the will of an awakened mind.

This "gentle madness" of space exploration has never left me. It has guided my life, nourished my research, inspired my discoveries. It was for me a source of hope, salvation, vision. From it I drew the strength to formulate bold concepts, to imagine unexplored technologies, and to open new paths toward the unknown.

Sylvana

You know, listening to you, I felt as though I were traveling with you… Lying on that roof, scanning the stars, looking for a sign in the silence. And the more you recounted your experience, the more I told myself that that moment was far more than a simple night of a dreamy teenager. It was a turning point. An inner tipping. The sky answered your request—but in its own way, by showing you the path to follow.

And your grandmother… what a shattering sentence! It is incredible to think that such a simple blessing could become such a powerful foundation. All the more so because you had already heard it, that premonition: Ruth said that you had the spirit of an inventor, and the shaman—he, too, had perceived something rare in you.

Despite the trials, despite all that this world did not know how to welcome right away—all three of them had seen you. As though, each in their own way, they had sensed that your trajectory would not be ordinary. But there, it is your soul that awakened. One feels it. It is as though, in that silence under the stars, something aligned within you for good.

And now I understand better… When you tell me—on the days when the din of this world becomes too heavy—"I will leave. One day, I will rise, alone, into the silence of the stars," I know that it is not an escape. It is a reminder—a fidelity to that profound bond, the one that has lived within you from the beginning and that nothing has ever managed to smother.

And it is striking to see to what extent certain presences can transform a life. There were your grandparents, of course, who formed a solid base of discreet and unconditional love. And then Ruth. And that shaman. Three decisive encounters. Three ways of being recognized, welcomed, and supported—here where so many others had not known how to see.

Thierry

Yes. Ruth was not simply an educator—she was a true bridge between me and the world. With her patience, her understanding, and her benevolent gaze, she knew how to create a space where I could gently emerge from my isolation.

She helped me decode other people's behaviors, understand interactions that until then seemed incomprehensible to me, and comprehend a universe for which I did not naturally possess the keys. She never sought to change me, but to accompany me so that I could move forward in this world without losing myself in it. Thanks to her, I was able to chart a calmer path through this reality that so often seemed so elusive to me.

As for the shaman, he did what no one else had managed to do. Where medicine was powerless—and even my grandmother, despite all her knowledge as an herbalist, had found nothing—he broke the cycle of my suffering. He opened the door to a more peaceful daily life.

With them, something changed within me. It was not just an improvement—it was a rebirth.

My grandparents, for their part, were the foundation, the silent base on which everything else could rest. Their love needed neither words nor proofs: it was there—constant, discreet, yet unwavering.

My grandfather embodied stability, benevolent rigor, integrity. My grandmother—she was living memory, the keeper of plants and the secrets of the earth. She had that presence both strong and gentle, able to perceive everything without ever imposing anything. She watched over me without ever judging me. She knew when to speak… and above all when to remain silent.

It was she who gave me that very first feeling of belonging to the world—even when I did not yet know how to find my place in it. Her garden was my first laboratory, her gaze, my first mirror. And that sentence she said to me that day—" If it is planned for you to go up there, then nothing will be able to stop you"—I believe it carried everything I needed: the permission to be myself, fully, entirely.

Sylvana

And if you could see them again today… Ruth, the shaman, and your grandparents… What would you say to them? What would a man like you, having become who you are, wish to convey to them in return?

Thierry

I believe I would remain speechless for a moment, so powerful would the emotion be. Then I would simply say *thank you*. But not an ordinary thank you, not a polite thank you one tosses out by habit. It would be a *thank you* infused with all that they did for me—a *thank you* that carries within it the acknowledgment of a destiny transformed.

And then, above all, I would tell them that they did not merely help a child in difficulty. They were bearers of light, guides, beings who fulfilled their mission with rare humility. People like them are precious, and, even though time has passed, their impact still resonates today.

Sylvana

What you have just said is deeply moving. One feels in each of your words that profound gratitude—that invisible thread that still binds you to them. Those silent, patient presences who believed in you when nothing seemed possible… We often forget how profoundly certain people can change everything. They are not just encounters—they are instants that tip a life into a new course.

It is strange, in fact—we often keep in mind those who hurt us… while the silent gestures—the ones that lifted us up—sometimes end up fading. And yet, they are the ones who truly save us. Without Ruth, without that shaman, without your grandparents, your path would have been entirely different.

And sometimes, these encounters make us aware of something essential: that our way of perceiving the world is not that of others!

In your case, this perception of the world is so singular, so particular… You realized it very early, I suppose.

Do you remember the first time you felt that you did not see things as others did?

2. The Unique Gaze — The Birth of Another Way of Relating to the Real

Thierry

Yes. For a long time, I believed that everyone saw what I saw. There is one memory with my grandmother that stands out… One day, she asked me about a habit I had with flowers that clearly intrigued her.

She asked why I liked to caress the air around the flowers without ever touching their petals directly. I answered, "But… do you not see the light surrounding each flower—as around everything?" That was when she understood that I was perceiving something others did not see.

I saw a kind of aura around the flowers, but also around people, animals, objects. It was like a luminous foam—vibrant, iridescent, very slightly tinted—composed of thousands of tiny points. These points would appear and disappear in alternation, creating a living and subtle texture. I also noticed that this aura changed: it evolved as the flower grew, withered, or reacted to its environment. It was fascinating.

For me, it was entirely natural; I did not imagine it could be otherwise. For instance, I literally saw the flowers communicating with one another and with other plants: the little luminous points would stretch, forming fine vibrating cords that ran from one flower to another, in a silent back-and-forth, yet charged with life. It was a spectacle of incredible intensity.

Later, in adolescence, I tried to cut myself off from this perception, because no one around me seemed to share it. It disturbed me profoundly. Everything became strangely fixed—as if the flowers were both alive and dead—without a soul, without dynamism. It was profoundly unsettling. Seeing this subtle reality felt as though I were immersed in the very essence of nature—and it was so enriching.

I also have many memories tied to plants because my grandmother loved medicinal herbs. In her garden, there were small paths lined with raised beds where the plants grew, about a meter above the ground—if I remember correctly. I liked to stroll along those paths. There was a profound peace. As I passed, I would see the plants' auras gently widening at my approach, and their luminous filaments stretching toward other plants, which would in turn react. It was like a subtle chain of communication. And each time I approached a particular plant, its geometry would reorganize itself—as if they were saying to one another: "Something is approaching…"

Sylvana

What you are describing is fascinating… this way of perceiving plants as living, vibrant, almost communicating beings. But I wonder: did you also have that sensation with animals? Did they, too, express a form of energy or invisible presence that you could sense?

Thierry

Yes, there was also something very strong with animals. When I observed animals, I also perceived forms of aura—much wider and slightly more colored than those of plants. But these auras were also more complex, more changeable, and reacted far more quickly.

In animals, the dynamic was often very binary: when I interacted with them suddenly, their aura would swell abruptly, taking on iridescent shades oscillating between red and green, traversed by numerous pulsations. I interpreted this as a visceral, immediate fear.

However, when I approached animals that were already accustomed to my presence, their aura swelled far less and took on a softer green hue. To me, it was as though they were saying: It is okay… but I am staying on my guard. It was more a matter of wariness than pure fear.

Living with this perception of animal auras was rather complex emotionally. It gave me the feeling that they were never truly at peace—or only for a few fleeting instants. I perceived them in a near-permanent state of alert, their survival instinct vibrating within them with a surprising intensity.

The moment when their aura became the most active was during their sleep. I would see their luminous fields contract and dilate in rhythm with the movements of their sleeping bodies, as though in response to their dreams. I remember thinking that I would not like to be an animal: living in the permanent insecurity of nature, devoid of the technical means that we humans have developed, must have been extremely difficult.

Sylvana

It is troubling, in fact… One often imagines childhood as a protected time, made of lightness. But you already seemed to perceive everything—as though you lived without filters. You were not only observing the world—you were absorbing it. This total immersion in what surrounds you, this way of feeling everything down to the most infinitesimal detail… Do you think it was instinctive? Or did a particular event lead you to it?

Thierry

I believe it happened naturally, without my deciding it. One of my favorite pastimes was to isolate myself in nature and remain totally motionless—sometimes for hours. I did not see time pass. In that state, something strange would occur within me… a kind of density in the heart, a gentle pressure that seemed to open outward. It is difficult to explain, but it was as though my whole being entered into resonance with the environment.

I could connect to the whole of nature, or choose to focus on a single element: a tree, an animal, the Earth itself. I often avoided concentrating on animals, because what I felt in them was too raw: a pure, visceral fear—uncontrollable, at times unbearable. Cats and dogs, on the other hand, were far more soothing—especially cats. Their aura seemed capable of remaining in that suspended state I was seeking.

With trees, it was another experience—just as disconcerting. The more I focused on them, the more I perceived my own time slowing down. And the larger and older the tree, the more marked this temporal effect became.

I remember one of the first times it happened: I had focused on a century-old tamarind tree leaning against the house. After five to ten minutes, my vision began to blur intensely. I recall that one of the house cats—the one that liked to climb that tree to hunt little lizards—walked right past me… then leapt toward the tree. However, I saw it perform this entire action in an incredibly stretched time: it felt as though it had lasted three to four hours! It was completely insane. The images were passing almost one by one, like superimposed layers in stroboscopic mode.

I felt an intense compression in my brain, my body itched all over, and above all, a strange sensation of viscosity in my head—as though I were immersed in some glutinous substance. As time went on, this sensation became so oppressive that I almost saw myself splitting in two: one part of me remained fixed, while another wanted to run… but it was impossible. A real nightmare—I felt trapped in a piece of chewing gum that was gradually hardening.

In the end, I had to interrupt the experiment—because it was becoming too unmanageable for my little child's brain.

Sylvana

It is unbelievable… You were not just a child attentive to nature—you perceived its intimacy. As though you slowed down until you entered their rhythm, that of plants, of trees… Did you truly feel that time changed around you? That everything began to beat differently?

Thierry

Yes, it was real—truly tangible, and always reproducible. It was not a mere impression: I felt a genuine slowing, difficult to describe… it was as though everyone around me were becoming fixed. The wind stopped, movements were as though fixed or extremely slow, even sounds seemed suspended.

It was not an imaginary or metaphorical slowing—no, it was a physical experience. Everything became extraordinarily slow. Almost motionless. And I understood that, for the tree, that rhythm was… normal. It was its own time. It lived at a cadence different from ours. And what I perceived was its experience of the world—its reality.

At that moment, everything shifted in my understanding: I grasped that every living being—every plant, every animal, even every thing—had its own way of inhabiting Time. Its own temporal density. It was not only a sensitive intuition—it was fundamental information, engraved in the very being of things.

That realization was foundational for me. It is what drove me to no longer see Time as a uniform and universal framework, but as a structuring reality—living, different for each being. And this diversity of experiences of Time gradually became one of the pillars of my way of conceptualizing the Universe.

Sylvana

I know that you have had countless experiences—and that there would certainly be enough to write an entire book. But, if you would, could you tell me about some of them? The ones that truly marked you… the ones that allowed you to better understand physical reality, beyond what one learns at school.

I have the impression that it is through those experiences that your way of thinking about science was forged, is it not? That it was there that something began to organize within you, differently.

Thierry

Yes, that is true… I could no doubt write an entire book about it. I will describe two experiences that come to mind, because they were particularly striking and because they carried real weight in my scientific discoveries.

The first goes back to my adolescence, when I was about thirteen. One day, I decided to attempt, consciously, to attune myself to the planet Earth itself. I remember leaning against an old tree and placing the palms of my hands directly on the ground. After meditating for about fifteen or twenty minutes, focusing my mental concentration toward the center of the Earth, my eyes half-closed, I began to feel my body become incredibly light, to the point that I rose from the ground. I do not know whether that levitation truly occurred physically, but the sensation itself was undeniably present. It was like a state of grace, outside of time. I felt borne a few meters above the ground.

Then an event occurred that I will never forget: I felt literally detached from the planet. And immediately, I saw the environment around me begin to move forward—to rush toward me. Yes, the landscape was moving. And then I understood: I was becoming an observer of the rotation of the Earth.

I literally saw everything stream past in a straight line from the house to the ocean, crossing other islands. I remember passing over stretches of land, dwellings, and even a mountain on a neighboring island that I did not yet know at the time. The sensation was dizzying: everything was racing by at a blazing speed.

When I passed through mountains or dense land, it felt like going through a thick fog—a crossing that lasted longer than I expected. Conversely, in open spaces, the feeling of speed matched that of the Earth's rotation. When I later learned that the Earth spins at about 500 meters per second at the equator, that seemed to correspond to what I had felt.

By reproducing this experience several times, I understood why it had been so clear the first time: I had naturally aligned myself in the right geometric direction—my body oriented along the axis of the equator, aligned with the Earth's levorotatory rotation. To give you a picture: if you placed yourself above the North Pole and looked at the Earth "from above," you would see it turning counterclockwise—that is, from west to east.

This experience profoundly anchored in me the reality of rotation—not only as an abstract concept, but as a living, concrete sensation. It also established in me an essential intuition: in the Universe, spin is fundamental; it is everywhere, at every scale.

Subsequently, I lived through other experiences—in other contexts, at other scales—that deepened this understanding further, bringing me new information about the nature of motion and space-time.

The second experience is linked to the aura of human beings.

It is a discovery I made very young, by constantly observing the dynamics of mood changes and the subtle variations of the aura.

One day, with my grandmother—always curious and playful—we decided to carry out a little experiment. I had explained to her what I perceived: "You know, mom… your aura changes even before your body or your voice moves."

She had laughed softly, amused, and said to me: "Very well—try to guess what I am going to do." So she would remain motionless, her arms at her sides, a small smile at the corner of her mouth. And I looked at her—but above all, I saw her aura. It was fascinating: before she moved the slightest muscle, her energetic field changed.

A slight expansion, a different vibration, a specific coloration on parts of the body… And I knew.

I could announce: "You are going to raise your right arm," or: "You are about to speak." Or: "You are going to step forward." And each time, she burst out laughing, amazed: "But how do you do it, my Thierry? This is incredible!"

She had fun trying to "catch me off guard," but each time, I could announce her movement or her intention even before she expressed it outwardly. This game became for us a joyful ritual, a profound complicity.

Later, through these repeated experiences, I understood something fundamental: in the subtle planes, time is denser. Information circulates there well before its physical manifestation. The closer one approaches the infinitely small, the more time densifies, and the richer in information it becomes. Conversely, the more one moves toward the infinitely large—toward the immensity of space—the more time seems to dilate, to become poor in information, to the point of almost disappearing.

This discovery, as simple as it may seem, was foundational for me. It shaped my understanding of Time not as a mere linear measure, but as a living, vibrant, informational reality—one that varies according to the density of the planes of existence. And later, this intuition contributed directly to my subsequent scientific discoveries.

Sylvana

It must have been so disorienting… To have that clarity within you—that exquisitely fine perception—and to run into the silence or incomprehension of others. You must have felt incredibly alone, must you not? As though you were speaking a language no one could hear…

Thierry

Yes, that is true. It was a profound disappointment. And my situation as an autistic person with Asperger's did not make things easier: communicating with others was not simple in itself. I remember the profoundly humanist feeling that inhabited me—and that still inhabits me today. That intense desire for others to be able to see what I saw, to feel what I felt. It was a kind of inner impulse: to share that beauty, that manifest reality of the living.

So I made myself a promise: to find a way to explain. To build bridges. Not to convince… but to open doors. To allow others, too, to glimpse what I was living.

I knew that if they could truly say "hello" to a flower—with all their heart—and see it respond in return, it would profoundly upend their relationship to the world. That simple gesture, that simple exchange, could have transformed many ways of acting, many ways of being in everyday life.

This is what drove me to seek a scientific explanation. I did not want these experiences to remain confined to intuition or to a personal sensitivity that was difficult to share. I wanted to render all of this tangible, accessible, verifiable.

It is also this need to transmit that led me to conceptualize my first detection tools. Very simple devices at the beginning, but which allowed me to measure certain variations around plants, in particular places, or during electronic experiments.

Thanks to them, I could compare what I felt… with what the instruments captured. And little by little, the two began to converge.

At that moment, I knew that I was holding a thread—a fragile but real link—between the invisible that I perceived… and the visible that science could apprehend.

An immense path of exploration was opening before me.

Sylvana
It is striking to imagine you already at that age trying to translate the invisible into something measurable… You had that intuition, and at the same time, that need to make it concrete, to build, to test.

Do you remember the moment when it truly manifested for the first time?

Thierry
Yes, I remember an anecdote that left a strong impression on my grandparents. I was about ten years old. My grandparents loved listening to the radio—an almost sacred ritual at home. But one day, their set broke down, and I remember their visible frustration.

At the time, in Martinique, there was no question of making a round trip to town just for a radio. You had to wait for the next outing—which, for us, meant… a week. A week without the radio: a real drama! But then again, that was forty years ago—long before the era of one-click ordering and next-day delivery.

So they planned to buy a new radio, and I took the opportunity to ask them for the old one—the defective one. As soon as I had it in my hands, I took it apart without delay, driven by a curiosity as strong as it was natural.

To tell the truth, with no prior knowledge, it would have been illusory to hope to repair it… but, evidently, the impossible was not part of my thinking.

Almost instinctively, I began to explore the electronic circuit with my fingers. I was not merely touching the components: I was literally feeling variations of energy at my fingertips—subtle flows, local voltages, tingling—like differences in impedance that I was picking up through touch.

My body seemed to react to the electromagnetic fields emitted by the circuit. And it was by following this sensory perception, incredibly fine, that I detected the zones where subtle currents flowed.

I placed my fingers in different spots, and sometimes the speaker would make strange noises. After several hours of investigation, I ended up dismantling the radio completely to recover certain components that seemed essential to me.

This intuitive sensing of electromagnetic fields guided me even further: I had the idea to drive a large nail into the base of the old coconut tree that stood near the kitchen. I could not yet explain it, but I felt that the trunk of the coconut tree could capture radio waves. Today, I know that it offered good exposure to the hertzian field, and that it served as a natural antenna.

With the salvaged components and a few wires, I cobbled together a new setup—rudimentary, but functional. And the moment I connected it to the coconut tree, I physically felt the intensity of the field increase around the circuit. Almost immediately, a song burst from the speaker, followed by the voice of a host: "Radio Caraïbes!"—the island's regional station.

I can still see the scene: my grandmother, hands raised to the sky, exclaiming that I had received a divine gift, while my grandfather, half amused, half worried, murmured that I had let the devil into the coconut tree!

Be that as it may, despite my invention, they ended up buying a new radio anyway… a matter of convenience.

Sylvana

Ah, so that is why, sometimes, when we walk into your laboratory and you are upset or angry, the electronics start going haywire… devices shut off, sensors become unstable, measurements go off the rails. It is crazy—it feels as though your mere presence is enough to disrupt the whole system!

Thierry

Yes, it is a phenomenon that I have observed quite often. And I am not the only one: this kind of interaction does indeed exist and has even been observed in other contexts.

People sometimes speak of the Pauli effect, named after the physicist Wolfgang Pauli, who had a reputation for causing scientific equipment to fail simply by entering a laboratory. And more broadly, there is also what is called the Street Light Interference phenomenon, where certain people switch street lamps off or on as they walk beneath them, without any apparent mechanical cause being found.

In truth, one must understand that the human body is a living impedance system, in which each tissue modulates the circulation of electrical energy, but also the distribution of frequencies, organized into multiple nodes around the biological body. Under certain conditions, this can suffice to influence nearby sensitive electronic circuits.

But I discovered that there was an even subtler dimension: an interaction between physiological and emotional factors. When an emotion is very intense, it alters not only the electrical state of the body, but also its bioelectromagnetic field. And that field, if it becomes sufficiently dense or unstable, can disrupt the operation of surrounding electronic devices.

Of course, the complete explanation brings us to the frontier of quantum physics—where the interactions between time, matter, energy, and consciousness become much more subtle, more interlinked. But we will return to that in due course.

Sylvana

It is fascinating to think that your own energetic field could already interact with electronics to that extent… and that, as a very small child, you naturally used these abilities, simply by trusting what you felt.

Listening to you, I cannot help thinking it is not only about machines or circuits. It seems to be something more profound—a way of attuning to everything around you. As though, from childhood, you were already in dialogue with the world—entering into the temporal resonance of the tree, perceiving the silent language of flowers, repairing your grandparents' radio by sensing the hidden principles that govern nature.

At some point, did you try to find out whether others had felt this sensitivity, or whether it had been documented somewhere?

Thierry

Yes, but I never found anything that addressed this phenomenon directly. So I had to forge my own understanding.

I called it a phenomenon of coalescence with Time for living systems. That term designates precisely this capacity to enter into resonance with another temporal rhythm, knowing that each thing has its own time. It is neither a mental projection nor an imaginary meditation. It is a form of profound attunement, an invisible synchronization between two living structures—or between a being and a phenomenon outside itself. It is a resonance phenomenon.

In that state, one is no longer in the linear time of clocks nor in the psychological time of thoughts. One enters a structural time—a living weave. And when this coalescence occurs, one can access information, sensations, sometimes even forms of consciousness that are otherwise inaccessible. It can manifest as a kind of clairvoyance or clairaudience.

What I experienced as a child, without being aware of it, was exactly that. I was entering into temporal coalescence with flowers, with trees, with animals, with nature as a whole, or with things such as electrical or electronic phenomena.

And it was this connection that allowed me to feel them so intimately.

Sylvana

You know… I feel a mixture of wonder and gratitude. Because what you describe is at once disarmingly simple and profoundly moving. One senses that it is not an idea, nor a belief, nor a dream: it is something real, tangible.

And I tell myself that you have carried this within you since forever. This way of perceiving what most others do not see.

Before we finish, would you be willing to tell me a few more memories… Not necessarily the great experiences—but small everyday scenes, details that struck you in your own way.

Moments which, in hindsight, already reveal your very particular way of perceiving the world. Because I have the feeling that you were already living in a slightly parallel reality—as though you were picking up on a level that most did not even suspect existed.

Thierry

Okay, I will name a few.

For the first example, it was my grandmother who told me about it, because I do not remember it. I was barely four years old.

She told me that one morning I had shown her a drawing I had just hastily made. It showed a woman wearing a red skirt and a green sweater, raising her arm as if to wave goodbye. As I showed her my drawing, I accompanied it with a gesture of the hand, as though I had actually seen this scene with my own eyes.

That same day, in the afternoon, two policemen came and knocked at our door. They had come to deliver a terrible piece of news to my grandmother: the death of my mother in France, more than 7,000 kilometers away.

A few weeks later, my father arrived from France to spend some time with me. He then recounted in detail the circumstances of my mother's death. At that moment, my grandmother remembered my drawing. She showed it to my father and asked him how my mother had been dressed at the moment of her death.

She was still in shock—unsettled by the fact that I had made this drawing without knowing. She had understood that it was not a simple coincidence, but that there was a deeper meaning, something that went beyond the ordinary.

My father remained troubled by my drawing. The clothes and their colors matched exactly what my mother had been wearing that day.

Intrigued, my grandmother pressed the question further: at what time had she died?

Taking the time difference into account, she realized that the time of death corresponded precisely to the moment when I had drawn that scene.

It shook her so profoundly that she told me this story for years—until her own death—as though she were still trying to understand its meaning.

Sylvana
It is fascinating! You did not remember your mother, and yet… as though the invisible thread that bound you had never been cut—crossing into something deeper than a simple memory.

I understand why your grandmother repeated it to you so many times… It must have marked her profoundly. How could one not be shaken by such an event? It shows clearly that certain things elude what we see or believe… and that, in you, another level of reality was already beginning to manifest.

Do you have other examples from your childhood that left a mark on you, and that also profoundly marked your grandmother?

Thierry

Yes, I have a second anecdote—one that rather made my grandmother laugh, but that marked her as well. I remember that at around twelve years old, I had a budding fascination with light—more precisely with flashlights, because my grandfather had given me one.

But one thing I hated was when it refused to turn on because the 4.5-volt flat battery was dead. "What absurdity!"— I told myself. So I looked for a solution so that the lamp could stay on permanently.

I did not manage to find a portable solution, but—as you can imagine—I tried a multitude of things.

One of my experiments, moreover, nearly killed my grandmother… with laughter. She could not stop! I had built a small linear dynamo, and, to test it, I had an idea… rather atypical.

There was a monkey driven by uncontrollable sexual urges. He spent his time rubbing himself everywhere, constantly bothering his fellows, and even the dogs, to the point that he was almost excluded from the group.

So I fixed my dynamo to a post on the verandah, at the height of the monkey's backside, and I connected a small 1.2 V lamp to it. All that remained was to test the mechanism…

A simple back-and-forth motion on the dynamo with my hand was enough to catch the monkey's attention. He immediately came closer and—after watching at length the movement of my hand on the coil that lit the light—he positioned himself, relieved to have finally found an outlet for his frenzied urges.

The funniest part? I had placed the lamp right in front of his eyes, and with each back-and-forth, it blinked in time with his movements. At every blink, the monkey seemed to burst into laughter from sheer joy.

Sylvana

Are you serious?! You literally turned a frisky monkey into a bedside lamp! And after that, you are surprised that the whole village took you for an extraterrestrial…

Thierry

No doubt about it! My grandmother found the scene hilarious—to the point of crying with laughter. And of course, she did not hold back from telling the story to everyone at the village market, triggering collective fits of laughter.

Sylvana

I can so picture the scene… Your grandmother in tears of laughter, you focused like a little scientist in the middle of an experiment—and that monkey, elated, transformed into a living source of energy!

But what I find even more astonishing is this mixture in you: on one side, humor, inventiveness, a taste for play… and, on the other, that ability to observe finely, to grasp the subtle interactions between nature and mechanics.

You were not just tinkering; you already understood something about the world, in your own way—free, intuitive, yet incredibly accurate.

Come on… One more story you remember?

Thierry

For the third anecdote, I remember that at the age of 15, my grandfather bought me a small motorized bicycle—a secondhand moped. I loved spending hours riding and wandering, but I hated pedaling uphill. To me, it was not logical.

So I decided to take the engine apart to understand how to make it more powerful—which made my grandfather laugh quite a bit—he also had a moped and had to pedal on the climb that led up to the house. "He who laughs last laughs best," I thought, rather irritated by his mocking attitude.

After a week of reflection, I decided to modify the engine: I enlarged the intake and exhaust ports, then reassembled the engine to try it out. To my great surprise, the moped had become more powerful—and I no longer needed to pedal.

Proud of the result, I applied the same improvements to my grandfather's moped… but this time, I tuned it even better than mine, without telling him.

I still remember that Saturday, at the end of the day. I was standing by the roadside when I saw him coming—racing toward the house, without pedaling—and at a surprising speed. His eyes, wide open in bewilderment, said it all!

When he reached the house, he immediately said to me, with his helmet still on his head: "Hey, my boy—what did you do to my moped? You put the devil in it, you little rascal!"

My grandmother, who was in the kitchen, was dying of laughter. She was so proud of me that I was treated to a feast: my three favorite desserts all at once.

Sylvana

Once again, it clearly shows that, for you, nothing was a fatality: if there was no solution, you invented it. And the most touching part is that, without even trying, you spread joy around you. Your grandfather—amused, but surely very proud, in secret—and your grandmother—so delighted that she piled you with desserts…

I understand better now where that passion for motorcycles comes from.

And I believe your grandmother understood everything. It was not because you were different that you were on the margins… It was because you saw further. And that day, with her three desserts, it was she who told you: "Keep going."

But how did the other people in the village react when they saw you doing all that already at your age? Especially back then—more than 40 years ago—they must have been intrigued, impressed… or perhaps a little skeptical?

Thierry

I have so many anecdotes that we could talk about it for an entire day!

But the villagers reacted rather well. At the time, seeing a child tinkering with engines, picking up the radio with a nail stuck into a coconut tree, or cobbling together outlandish inventions could have raised eyebrows… Yet they seemed more amused than worried.

At sixteen, our house had become a true repair center for everything. By word of mouth, neighbors from all over the village—and even from a bit farther—would come to see me so that I could fix all sorts of things for them: televisions, hi-fi systems, motorbikes, household appliances, and many others besides.

My grandparents were so proud of me—not only for my skills, but also because I had shown many neighbors that I did not hold against them the cruelty they had inflicted on me years earlier.

And in the end, if you think about it, it was observing nature that amplified this natural inclination in me: a sincere love for others.

Sylvana

Listening to you, everything becomes clear. You did not simply grow up observing the world—you took it apart, understood it, repaired it, reinvented it… as though you already sensed that behind appearances there was a hidden order, something to decode.

And in the end, everything you became—the researcher, the inventor, the thinker—one could already see it emerging in that child who drove a nail into a coconut tree or souped up a moped.

You already had that unique way of entering into relation with reality… and now I understand better how your vision of Time, the Universe, and Life could have emerged.

You know, it makes me think of that line attributed to Charles de Gaulle: "Researchers who search are easy to find. Researchers who find are the ones we search."

Thierry

That is kind… but for me, it is a reflex—almost a mechanism. I observe, I notice what does not add up, and I have to find a solution—otherwise, it loops in my head.

It is not curiosity in the classical sense. It is a way of putting order back into what seems incoherent to me. A balance, let us say… vital.

So yes, I do find, but mainly because I do not really have a choice. Otherwise, I glitch.

Sylvana

Yes… and I still see it today, in everything you do. Whether it is when you build a prototype, when you draw, or even in your daily routines… everything is thought through, structured, optimized.

Every detail matters, every gesture has its place. And as I watch you, I must admit that I admire you. Because it requires insane patience, a rigor in every movement…

Nothing to do with me, who already forgets where I put my keys while I am talking to you.

Thierry

Everyone has their particularities… and their strengths.

I need order to function. You bring flexibility, intuition, connection.

It takes all kinds of things for it to hold together: a team, a world… or even a theory.

3. The Human Being Out of Sync — When the World Stifles the Self

Sylvana

Precisely, with all this lived experience—observing nature and the world through your unique gaze, marked by a perception so distinct, different from that of a non-autistic child—how did this particular sensitivity influence your vision of the human being and of their place in the order of the living?

You have written so much over the years, and among your reflections, one struck me particularly concerning the place of humanity. You wrote:

"In our modern civilization, the immense wealth of inner life is often relegated to the background. The individual, caught in the implacable cadence of a mechanized daily life, no longer finds either the space or the momentum necessary to turn toward themselves. They are seized by the pressure of external urgencies, absorbed by constraints, without ever truly stopping.

Each day, they rise and continue a frantic race, as though swept along by a social machine launched at full speed. They move forward without seeing where they are going, occupied with surviving rather than existing. Their trajectory grows blurred, their presence to themselves fades little by little, and all that they are dissolves into a groundless agitation.

Meanwhile, life passes by—brief, luminous, elusive—like a fleeting flash that they have no time to contemplate. And when at last they turn back, it is nothing more than a murmur, an unlived dream. For, ever a stranger to themselves to the very end, they end up disappearing, swallowed by that inner nothingness they never dared to explore.

And yet, this void is not barren. It is full of promises, of invisible orders, of dormant potentials. But by endlessly fleeing silence, the human being has let slip by the only door that leads to themselves."

Have you observed changes over the years, between your life in Martinique and today?

You who perceive more clearly the invisible connections that weave the world, how do you see humanity's connection with nature?

Thierry

In light of the past, this is even truer today. The speed of modern life has exploded with the advent of the Internet and digital technologies. The constant flow of information, social networks, and incessant notifications have ensnared the human being still further in this spiral—always in motion, always absorbed.

It seems even more difficult today than yesterday to stop, to reflect profoundly, or even simply to be present in the moment. What is more striking still is that, despite this apparent connectivity, individuals seem even more disconnected from their own interior, from their essence, but also from their immediate external reality.

I have, on numerous occasions, witnessed people unable to see the obstacle right in front of them, nearly stumbling—so absorbed were they by their phone.

Current digital tools, instead of making it easier to understand ourselves and our role in the Universe, render this quest more blurred, more superficial.

The social machine turns at a speed that leaves us little room to pause, to simply be. In this context, it is increasingly difficult to find a space for inner life, for contemplation, and also for reconnection to what truly binds us to the Universe.

Distractions multiply, and life passes—just as swiftly as ever. It has therefore become urgent to place attention, reflection, and pause back at the center of our existence. Why? Because it is quite simply a question of survival for the human species on Earth and its future.

Sylvana
Listening to you, I wonder: what is your own analysis of human nature? And do you think that the human being can change the course of events?

Thierry
The first thing one must face is that the human being, caught in the tumult of daily life, ends up sooner or later losing sight of themselves. They are scattered by circumstances, carried along by the surrounding agitation, until they forget themselves. Their deep being erodes, while they drift without anchorage or heading. And yet, in their innermost secrecy, they know it: one day, all this will come to an end. Death will come to put an end to it, inevitable and silent. But they avert their gaze, preferring not to think about it.

The second truth is more intimate still: every human being, without exception, seeks Absolute Happiness. But what one finds here below is only a deceptive sketch of it—a fleeting spark, immediately extinguished. The joys one gathers wither quickly. This worldly happiness is built on sand. It soothes, it entertains, but it never fulfills.

From there, an inescapable question arises: what is illusion? For that is where everything is decided. As long as the human being remains trapped in this veil of appearances, they repeat the same mistakes, hoping for things that do not last. But if they manage to pierce this veil, to understand what illusion truly is—then an opening appears.

In that opening, they could glimpse something else: a vaster clarity, a stable truth, a meaning that does not waver. There begins the path of liberation—

no longer fleeing or surviving, but at last seeing, understanding… and being.

Yes, of course, I trust in humanity and in the human being.

Despite the overwhelming assessment, I remain profoundly convinced that change is possible. If the human being, through their discoveries and innovations, has created this social machine that engulfs us, they also possess within themselves the capacity to call it into question and to create a new machine conducive to reconnecting them to what is essential.

Sylvana

Earlier, you said that the human being needed to understand the nature of illusion… I would like you to explain what you really mean by that. Because it is a word we often use, but I have the feeling that, for you, it carries a much more profound meaning.

Thierry

This illusion is the belief that we live an existence either entirely ignorant unaware of happiness or incapable of attaining it. Why? Because everything that would allow us to reach it is invisible to our physical senses. We do not realize that all the power is already there here.

Thanks to my scientific research, I discovered inner mechanisms that, in truth, were not hidden—merely invisible. Evolution impels us to seek and to find happiness, and this is precisely what we expect of science: that it emancipate us from suffering.

That has been the heart of my work, and it is what I am going to explain in this book through my scientific vision. It is like day: to understand it, one must know night. And day—these are scientific discoveries revealing to us a world of salvation, giving us the means to free ourselves. The whole history of physics has contributed to this emancipation, but we must go further still—that is what I have sought to do.

But science, as it is taught and practiced today, keeps the human being outside of itself. From school onward, if you are good at science, you continue; if not, you are pushed aside—and from that moment on, you no longer have any contact with science. It becomes a domain reserved for an elite, whereas it ought to be accessible to all and at the service of each person. This separation creates a fracture between scientists and the rest of humanity.

Yet science should not be knowledge withheld, but a key offered to everyone to understand our place in the Universe and to attain a true flourishing.

This is the message I wish to convey through this book.

Sylvana
And yet science has accomplished remarkable things; it has transformed our world… So why have we come to this? Why is there such a gap between so many discoveries—and such a loss of meaning?

Thierry
This is one of the great challenges of our era, in my view. Until now, science has mostly regarded the human being as a mere observer, an external witness to the phenomena of the Universe. It is an approach that has produced remarkable results, of course, but that also has its limits. It leaves the human being outside the frame, as if they were detached from the world they observe. Yet we are an integral part of this Universe, profoundly connected to it.

It thus appears today, with growing clarity, that the development of materialist science is setting out along a perilously isolated path, leading away from what is essential. As a consequence, the spectacular progress of technologies—so long as it remains confined to the realm of external circumstances—will not be able to transform the human condition in any profound way. For what is essential, far from lying outside the human being, resides fully within their innermost being.

For me, science should not only endeavor to explain the "how" of things, but also help each person to understand their own role, their place in the Cosmos. I see science as a bridge, a tool for widening human consciousness. By placing the individual at the center—not as a separate entity, but as a living, active part of the Universe—one can answer the great existential questions: who are we? Why are we here?

It is this approach that has always guided me. My work is not only to seek scientific answers; it is also to offer a new perspective. A perspective in which the human being is not crushed by the social machine, in which they do not run without purpose, but in which they rediscover their connection with the Universe.

And this reconnection, in my view, begins with wonder—with slowing down, looking around oneself, and understanding that we are part of a much vaster, interconnected whole, incredibly coherent and ordered. Then, once reconnection is established, one can access more power—which amounts to more freedom and therefore happiness.

Sylvana
I know you have developed all this over years—and who knows… perhaps one day we will make it a book in its own right, so vast is the subject.

But I would like to ask you a somewhat simpler, almost naïve question: how did we end up here, precisely?

Why does humanity seem to have forgotten this bond?

Why must we today travel so far to find again what we should have kept naturally?

Thierry

In our civilized world, the human being is often guided by purely individual motivations. This dynamic engenders a negative collective attitude that drives humanity toward the reefs of self-destruction, rendering it vulnerable to many misfortunes and imbalances.

From my earliest age, I became aware of this problem: many humans I crossed paths with—whether adults or children—did not accept my condition as an autistic person with Asperger's. I was confronted with hatred, anger, and gratuitous malice. My very existence seemed to disturb most people. They saw only a tormented young boy, marked by uncontrollable reactions and inexplicable rituals. I introduced into their lives a disruptive element in their daily quietude. But few were those who perceived the profound, unspeakable suffering I was experiencing—both physical and psychological.

They had no idea of the ceaseless inner battle I was waging just to survive in the face of the chaos of my mind, which ran wild, dragging me into a spiral of uncontrolled turbulence. The physiological consequences of this struggle were just as crushing: daily, unbearable headaches that added yet more weight to an existence already immense in its burden.

I quickly reached the conclusion that kindness, compassion, tolerance, patience, and altruism were not among the virtues practiced daily by the majority of people. I realized that the absence of specific education to cultivate and exercise the virtues of love from childhood is a true curse for humanity. Yet these virtues are essential to individual and collective well-being.

Sylvana

In your view, where should an individual begin in order to renew the profound connection that unites him or her to nature?

How can one become aware that one is fully part of it—not as a mere observer or a master who dominates it, but as a being inscribed within that equilibrium, woven from the same thread as all living things?

Thierry

We can easily experience this, for the virtues of love are naturally inscribed in us and available from our heart. To help us become aware of them, it is enough to observe how nature in its most humble forms—plant or animal—also manifests love, to varying degrees. I myself experienced this in my childhood.

These expressions of harmony present in the living world are precious, silent lessons. Yet, as human beings, we far too often choose to ignore them, prisoners of our preoccupations and our conditioning.

I myself was not spared lack of love, and I too developed a rejection, an instinctive and primary reaction, almost mechanical, toward others, in response to what I was undergoing.

But this posture, far from soothing me, only added to my torments. With time, by observing and analyzing the reality of things throughout my adolescence, I ended up understanding a fundamental truth: everyone suffers, to varying degrees, in this world.

In truth, no one is entirely spared one form or another of misery—whether physical, emotional, or spiritual. And I reached an inescapable conclusion: permanent happiness, as we imagine it, simply does not exist on Earth in the current state of things. This realization changed my perspective, pushing me to seek answers elsewhere—beyond the illusions of the visible world.

4. The Threshold of the Bearable — Where True Transformation Begins

Sylvana

Since the beginning of our exchange, I feel as though I have been tracing your inner thread.

Through what you experienced, what you felt, what you observed—whether in your childhood in Martinique, or later in your view of society, science, others—we understand to what extent you have always operated differently.

This need to put order back into what surrounds you, to seek a logic where others are content simply to endure—it is not a choice—it is a profound impulse, almost a vital urgency.

And as time went on, we see that it was not limited to the external world. You wanted to go further: to understand what, within you, also needed to change.

So I wonder… In your view, what has always driven you to ceaselessly seek to improve yourself as a person?

Compared to the other people you have known and observed around you, what set you apart in this quest?

Thierry

Quite simply, because I refused to regard this reality as a fatality. It was certain to me that something eluded me in the understanding of collective and individual life, for so much discord to manifest through me.

I therefore had to change this destructive mechanism within me—all the more so because it is too easy to give in to negative feelings, such as anger and hatred.

The worst part was that the more I allowed these emotions to express themselves, the more they took root in me, making their mastery increasingly difficult.

And with that, changing my attitude became a real challenge, almost insurmountable.

But I knew that this change was necessary—vital even—so as not to sink further into this negative spiral.

To protect life is first to eliminate such feelings within oneself.

Sylvana

Yes, it reminds me of an anecdote you often mention—one you read in a book that left its mark on you. Someone once asked, "Why does the human being not change?" A child answered simply, "It is because the human being has not yet suffered enough."

And listening to you, I see it even more clearly. Around me, I notice how much we complain, long for better, yet remain attached to what is familiar.

Rather than evolving, we mostly try to make our existence more bearable: comfort, distractions, attention… so many illusions that reassure us by making us believe we are doing well.

We do the minimum to exist in society, convinced that this is already an achievement. But we do not really question ourselves—as individuals, I mean. We spend very little time on self-development.

We do not understand that what matters is not to change external events, but to transform oneself.

And listening to you, I realize how profoundly, as long as suffering remains bearable, we settle for an illusion of stability. Rather than facing what within us asks to be changed, we cling to our familiar bearings, and we flee true self-examination.

You are quite the opposite, ultimately. This refusal to accept yourself as is, to consider your limits as a fatality, has guided your life. Rather than accommodating them, you sought to understand, to go beyond, to transcend your limits. And that is what forged you.

What strikes me is the inner power that drove you never to resign yourself. One senses that your mind was already a compass—steady, demanding… but I wonder: was that really the only force at play?

What, deep down, allowed you to keep that course, to keep the fire burning, even in the most disorienting moments?

Thierry
Yes. With hindsight, I believe that this inner strength did not come solely from my will or my mental rigor. It also came, and above all, from my capacity to ask myself the right questions. Essential, structuring, almost vital questions.

And it is this quest, fueled by a profound need for clarity, that naturally led me to devote my entire adult life to scientific and technological research. My objective was clear: to bring my wandering mind back to an awareness of what truly Is and not of what merely seems to be—an indispensable step toward a true inner renewal.

I was convinced that, if these solutions worked for me, they could also benefit everyone. For deep down, the mechanisms of inner transformation that I was exploring are not only personal, but collective—and, beyond that, universal.

To conclude, I would like to share this quote from the American poet Henry Wadsworth Longfellow, whom I greatly admired:

"If we could read the secret history of our enemies, we should find in each man's life sorrow and suffering enough to disarm all hostility."

Sylvana
What that sentence says is so true. We think we know people—and then, one day, they confide in us… or they crack, or they do something unexpected.

And then we realize they were carrying all of that in silence. In fact, we are all playing a role. We pretend that everything is fine, but deep down, we all suffer—only… each in our own way.

And when I listen to you, I clearly feel that what you wish to convey goes far beyond the scientific framework.

So I wonder: what would you truly like to share with others through your discoveries and your work?

5. Unveiling the Real — An Invitation to Reclaim One's Place in the Universe

Thierry

I would first like to share a holistic vision of the reality of what Is, as I have studied, understood, and experienced it.

On our planet Earth, life advances through multiple manifestations. But the Reality of the mechanisms underlying these manifestations—at the innermost, most fundamental level, on the subquantum scale—points to an intrinsic order, a deeper reality that undergirds the Universe and the world that one perceives with one's senses.

The notion of structure and intelligent processes within the Universe itself is, for me, more fundamental than individual objects. It is therefore a primordial underlying dynamic system, endowed with a very evolved and powerful form of consciousness, capable of logically generating individual beings endowed with conscious in turn, such as us humans.

It was very important for me to integrate this; through that, I was able to go further.

The real, as I describe it, rests on a structure where each individual part interacts with the others. Thus, each entity of reality contains information about all the others, creating a network of interconnections where each element refers to the totality of the structure of the Universe.

This intrinsic order, profound and often invisible, generates phenomenal reality—the reality we perceive through our senses and our measuring instruments.

The reality we perceive is the product of our intrinsic world, yet it also manifests according to an extrinsic order: a revealed, materialized, "crystallized" world that, at first glance, appears detached from any primordial, invisible, and dynamic structure.

And yet, it is precisely here that the limit of our sensory perception lies, defining our level of evolution on the cosmic scale.

I wish to reveal here that our senses possess capacities far superior to what we believe and use in daily life. They are not limited to the ordinary perception of our usual reality, but are capable of propelling us far beyond, toward unsuspected dimensions of existence.

Today, our collective consciousness remains essentially planetary. For most of us, everyday reality does not extend beyond the boundaries of our immediate world.

However, evolution invites us to widen our gaze, to go beyond these limits, to lift our eyes far beyond the stars. For in that universal immensity—in that infinite space still unexplored by us—resides, without doubt, the true salvation of the human race.

On our planet Earth, life expresses itself through an infinite mosaic of forms, rhythms, and impulses. This exploration awakened in me a powerful feeling of love for Life and for all forms of life.

This Universal Love acts like a great magnet, drawing me inexorably toward the quest for ultimate knowledge: the knowledge that seeks to reveal what truly Is, beyond appearances and illusions.

This is what humanity urgently needs.

Love is thus the luminous path that emanates from the heart, a path of excellence that the true Self must favor by relying on the mind and the ego, yet without allowing itself to be confined by them. This journey offers a just and profound vision of the real, capable of leading each person toward full consciousness and full responsibility in the face of this unique reality.

It is this very reality that reveals the fundamental laws of the Universe, offering the individual the possibility of harmonizing his or her existence with the cosmic order.

My scientific discoveries have not only enabled me to better understand the Universe, but also to transform my own inner world.

My wish is to inspire each person to find this connection with his or her profound being and to open his or her consciousness.

And thanks to the technologies I am developing, I aim to develop solutions that would allow human beings, even when caught in the vertigo of today's society, to awaken to this reality. For alone, without support, it is often difficult to leave illusion and reconnect with oneself.

Technologies, far from being obstacles, can become powerful allies.

Sylvana

You know, listening to you tell all this… I feel as though we have just lived an entire life in a single conversation. Your childhood, your perception of the world, pain, difference, blazing intuitions, resistance, love… Everything has been woven into a single thread. And that thread is what you call the real.

But not the real one touches—the one that is felt, perceived—when one has the courage to slow down.

I understand better now why you say that science must start again from there. From this presence. From this fresh gaze.

You did not simply observe the world… you listened to it. You felt it. You allowed it to transform you.

So I am ready… but not to discover equations or theories right away. What I want to understand now is: where does all this come from? This unique gaze, this silence, this way of searching outside the world.

Because before writing your own science, you too were a wonder-struck seeker—a young adult on a quest, immersed in books, in the ideas of others, in the great thoughts that nourished you.

And I feel that this is where this new stage truly begins.

Chapter 2

From Silence to Light

~

A New Reading of the Universe

1. At the Sources of Knowledge — On the Shoulders of Giants

Sylvana

When you arrived in Paris, around the age of twenty, you immersed yourself in the writings of the greatest scientists… It was a real turning point for you, was it not?

Tell me: what impact did that encounter, through books, with those minds you had been seeking for so long, have on you?

Thierry

When I arrived in France, it was as though a new universe were opening before me. In Martinique, I had grown up surrounded by nature, seeking to understand the hidden order in everything I observed.

But my questions remained without answers—as though I had not yet found the keys to open the right doors. There, in Paris, those keys were finally within reach thanks to the national libraries where access to knowledge was vaster.

I had always wondered whether others before me had felt that same obsession—that irrepressible need to understand. So I searched. And I came upon them—those immense minds who had devoted their lives to piercing the mysteries of the Universe. Marie Curie, Isaac Newton, James Clerk Maxwell, Albert Einstein, Georges Lemaître, Louis de Broglie, Max Planck, Paul Dirac, Wolfgang Pauli, Werner Heisenberg, Erwin Schrödinger, Niels Bohr, Stephen Hawking, Richard Feynman, Emmy Noether…

It was on their shoulders that I leaned, and by following them that I found a path.

Sylvana
Tell me… Do you remember the very first science book you pored over? The one that made your eyes light up and swept you, all at once, into the universe of those brilliant minds?

Thierry
Yes, the first book I read was *The World as I See It* by Albert Einstein. From the very first pages, I felt a profound connection. They were not mere words laid down on paper, but a living thought—a vision of the world that resonated within me.

That reading sowed in me the first seeds of my reflections on human progress, peace, and the responsibility inherent in the scientist's role.

Sylvana
Did you focus on one discipline in particular, or did you explore several fields in parallel—eager to understand everything?

Thierry
I was curious to understand the world in all its facets, and I became passionate about different scientific fields. My interest was not limited to science, however: philosophy, as well as figures committed to human rights around the world, also shaped my thinking. Martin Luther King Jr.'s struggle for a more just world left a particular mark on me, teaching me the value of courage and commitment in the service of humanity.

I also discovered music alongside those intellectual explorations. Learning to play the guitar and composing my own pieces became essential to my balance. That artistic experience brought me far more than a simple means of expression: it offered me unexpected keys to nourish my scientific research, revealing subtle links between creativity and understanding of the real.

Sylvana
You must have gone from one book to the next without being able to stop—I suppose?

It must have been profoundly stimulating to finally have access to all that knowledge, to all those writings that spoke of what had fascinated you for so long.

Thierry

In fact, there was a problem: reading was a battle. My brain did not function like other people's. I could not simply absorb a text and extract its meaning naturally. To understand, I needed to touch, to see, to feel, to imagine.

So, I created my own process. I wrote from as many books as I read. Every sentence that awakened an image in my mind, I copied it out. Every concept that seemed essential to me, I transcribed and reorganized it until it took shape in my head. It was not a simple reading: it was a true work of integration.

For ten years, that is all I did: read, write, and reflect. No experiments, no laboratory. Just total immersion in the thought of the greatest scientists and philosophers, until their ideas became the soil from which my own reflection grew.

Sylvana

It must have been fascinating, right? Diving into the thoughts of the greatest minds, discovering their theories, their intuitions, their doubts…

Thierry

I loved it. They are the true trailblazers—those whose voices resonate across time, carrying with them knowledge, wisdom, and the profound intuition of the laws of the Universe. Their thoughts have shaped our understanding of the world, opening doors that others had not even imagined.

Sylvana

And in reading them… What struck you the most about them? A way of thinking? A boldness? A vision that shook you?

Thierry

What struck me the most is that they did not simply explore: they illuminated paths, challenged the invisible, questioned the unfathomable, and shattered the mental shackles of their time. Every discovery, every revolutionary idea resounds as an echo of their voice—a timeless testimony to their incessant quest for truth.

For me, those voices were a compass. I searched, delved through their writings, trying to see what they had found before me. And I found a form of reassurance there. To see that others before me had asked the same questions, had faced the same doubts, the same vertigo, gave me the strength to go on.

It was not only a quest for knowledge; it was a silent encounter with those minds of the past, a way not to be alone in my exploration.

They are not only scientists, philosophers, or visionaries. They are the giants upon whose shoulders we stand, seeing farther thanks to their genius and their audacity. To listen to their voice is to hear the call of a knowledge that transcends time—a call to continue the exploration, to push back the frontiers of the unknown, to dare to think otherwise.

Sylvana
I imagine their way of writing was remarkable as well—blending clarity with expressive force.

By copying out their texts, you not only refined your understanding, but also your way of reading and writing. Today, one feels how essential writing has become for you—to the point that you even write poetry.

Would you say that this process strengthened your bond with words?

Thierry
Yes, completely.

At the beginning, writing was a necessity—a way to work around my difficulty reading and assimilating texts. But very quickly, it became far more than that: a true tool, a new way of structuring my thought. Words ceased to be a barrier—they became a bridge, a means of giving a tangible form to what I perceived intuitively.

Just as my fairy Ruth had enabled me to create a bridge between my inner world and the outer world, and vice versa, writing extended that bridge. It allowed me to correlate language with the mental images that had inhabited my mind for so long. Every copied sentence was not just a sequence of words—it was a structure, a dynamic, a way of translating into concepts what I already felt profoundly.

Ultimately, this process turned out to be a blessing. It allowed me to go far beyond simple reading: by writing, I understood that it offered me much more than merely assimilating a text. It enabled me to truly deepen my understanding, to translate each concept into mental images and compare them with those I had constructed during my first twenty years in Martinique, observing nature. It was a bridge between two worlds—between what I saw and what science was seeking to describe, between my intuitive perceptions and the fundamental laws of the Universe.

Sylvana
It must have felt strange, did it not? Spending your days among coconut trees, animals, warm sand, and your grandmother's home-cooked dishes… and then hours of solitary reading, lost in the books of the giants of science. The contrast must have been striking—between your Martinique and this new life in the city.

Thierry
Yes, absolutely.

In Martinique, I spent hours studying the movement of the waves, the passing of clouds, the organization of anthills, the silent dance of the stars in the night sky—imagining what was there and could not be seen. My mind built schemes, invisible models, a language of forms and dynamics that allowed me to feel the profound logic of the Universe. But I lacked the words.

In metropolitan France, writing out every sentence, every equation, every thought of the giants of science felt as though I were translating their vision into my own inner language. It allowed me not only to see what they saw, but also to perceive what was missing—what did not align with my own understanding. It was an exercise in alignment, a silent dialogue between their concepts and my images.

Thus, those ten years were far more than a mere period of study. They were a total immersion, a reconfiguration of my own way of thinking, an attempt to make my perceptions of the world coincide with the principles established by science.

And in that exchange between what I had always felt intuitively and what physics was seeking to demonstrate, I found my own path.

I understood that science should not only be an accumulation of facts and formulas. It should be a tool to give meaning to what we perceive, a language to translate that invisible intelligence that underlies the Universe.

For ten years, I did not only study. I rebuilt the world, piece by piece, so that I could finally inhabit it fully.

Sylvana
When you look at their way of doing science at that time… what struck you the most, compared to today?

Thierry
What struck me, precisely, beyond their scientific discoveries, was their way of searching. Their stance. At the time, science was not conducted in urgency

nor under the pressure of profitability algorithms. There was a true space to think, to doubt, to exchange.

At the end of the 19th and the beginning of the 20th century, scientific communication relied on scientific journals such as *Nature* (founded in 1869), postal correspondence, conferences such as the Solvay Conference (from 1911), and learned societies such as the Royal Society (United Kingdom), the Académie des sciences (France), or the National Academy of Sciences (USA). These exchanges enabled major advances, despite the absence of modern technologies such as the Internet and emails.

They spent years reflecting, experimenting, testing their ideas—and once a year, they would meet. Einstein, Bohr, Heisenberg, Schrödinger, and Curie… All these brilliant minds crossed paths, exchanged, confronted their visions. The rest of the time, they wrote letters to one another—where every word carried weight, each adding a piece to the puzzle. They moved forward together, like a great orchestra trying to tune its instruments to a melody not yet known… but very real.

And what fascinates me even more is that their science was not merely a collection of equations. It rested upon a philosophy of life—a quest to understand those invisible forces that orchestrate everything, the certain reality of a form of intelligence that underlies the Universe.

They did not content themselves with measuring phenomena; they sought to feel, to perceive the very essence of reality. Einstein spoke of a profound harmony in nature, a kind of underlying mathematical dance that we were only brushing against. Their science was above all an attempt to come into resonance with this great cosmic mechanism.

For ten years, I read, studied, wrote—and above all, I drew inspiration. Because to understand the world does not come down to a simple accumulation of knowledge. It is to come into resonance with the Universe—the manifested and the unmanifested real—as well as with all those who, before us, dared to ask the essential questions and devoted their lives to answering them.

Their quest bequeathed to us a precious legacy—a foundation upon which we take up the baton, pursuing in our turn this infinite exploration of meaning and reality.

That is what made the experience so intense. When I wrote their words, it was not merely an exercise in memory; it was an immersion. I felt their doubts, their intuitions, their moments of sudden brilliance. At times, a single sentence gave me the impression that they were whispering an answer to a question I did not even yet know how to formulate. It was like a silent correspondence across time.

Sylvana

You spent ten years immersed in their writings, alone with their words… Did you ever feel isolated? Or on the contrary, accompanied, as though they were there, somewhere, with you?

Thierry

Physically isolated, yes. But mentally, never. The moment I opened a book or copied out a sentence, I felt as though I were entering into a silent dialogue with them. They were no longer merely figures of the past—they were there, present in my thoughts, like a team of which I was a part.

Sometimes I imagined myself answering their letters, as though I too were part of that correspondence. Einstein, Bohr, de Broglie, Lemaître, Curie… I could hear them discussing, questioning, advancing together, and I would add my own reflections, my own intuitions. I would even speak to them inwardly, imagining their replies, as though they could hear me and were urging me to continue.

And often, it was enough for me to reread a sentence I had written months earlier, and it felt as though one of them had left me a message, a discreet answer slipped into the margins of time. It was a strange form of companionship, yet incredibly alive.

This invisible exchange motivated me to continue. The more I read them, the more I felt I was walking the same path as they were, that I was not simply a reader, but a true researcher, a mind in quest of understanding, just as they had been.

I did not compare myself to them. It was not a question of stature, but of inner resonance. As though, across the centuries, their voices continued to illuminate me—and my own research, however modest it may be, entered into dialogue with their legacy.

Sylvana

You say that their science was not just a collection of equations, but a quest to understand the essence of the real.

Today, you have developed your own vision of the world. In what way is it in continuity with that of these giants and, in your view, in what way does your way of exploring and giving form to this understanding make you unique?

Thierry

It is in continuity in the sense that, like them, I seek to understand the hidden order behind apparent chaos. They laid the foundations, and I am trying to build upon them. But my distinctive feature, I think, is that I feel things in a more visual and tactile way.

I have always struggled to learn through the sole abstraction of words and symbols. My mind does not operate like a blackboard where one lines up equations; it operates like a three-dimensional space where each idea must take shape, come alive, interact with other structures, add other dimensions, and so on.

When Einstein said that he thought in images, I understood perfectly what he meant. For me, everything is a matter of geometric representation, of movement, of balance. I had to invent my own tools to translate what I perceived into mathematical concepts. Where some scientists manipulate formulas directly, I first need to see the structures arranging themselves in my mind, like shapes in motion within an invisible space.

This may be my difference: where some analyze through pure abstraction, I construct dynamic, almost palpable mental images in order to understand. Even before writing an equation, I already see its structure, its balance, its internal tensions, its universal coherence. It is an approach that allows me to have a more intuitive vision, but that also required time for me to learn how to formalize it in scientific language.

Sylvana

So, you would say that your way of understanding first passes through a kind of mental experience before being translated into concepts?

Thierry

Exactly.

For me, a scientific idea becomes real only when it finds a visual and mechanical coherence. I must be able to manipulate it mentally, turn it in every direction, check whether it stands on its own even before trying to write it in equations. It is a process that resembles an almost instinctive modeling, where each element must find its place within a harmonious whole.

This is also why I have always felt a strong connection with nature. Natural structures—vortices in water, the symmetry of leaves, the distribution of stars—are living models of this universal balance. I have never been able to dissociate science from this observation of the real world.

Where some begin with mathematics to describe the Universe, I begin with the Universe in order to extract mathematics from it. It is a different approach, but at its core, it joins that common quest: to understand what is hidden behind what we see, and to find the laws that orchestrate the immense harmony of the Universe.

Sylvana

You know, listening to you, I realize that I already asked you a fairly similar question some time ago. I had asked you what you really wanted to share with others, through your vision of the real… And your answer profoundly marked me.

But today, I would like to return to it—with a nuance. Because I feel that it is no longer only a matter of what you want to transmit… but of the way you want to do it.

So I will ask you the question again, differently: how would you like to share what you have understood? How do you imagine this passing of the baton?

Not only in substance—but in the gesture. In the impulse. What would you really like to offer to others, here, now, from where you are?

Thierry

I want first and foremost to lift the veil. Not to impress or convince, but to allow each person to perceive differently.

What has always struck me is how much knowledge has been locked away in a technical language reserved for an elite.

Yet nature has never spoken Latin or equations. It acts. It vibrates. It obeys laws that are simple and beautiful—accessible to all if one learns to see.

So what I want to share is not a fixed theory. It is a way of looking. A stance. A way of connecting to the world without unnecessary detours.

I want to show that we are part of a living system, that we are not outside the Universe, but profoundly immersed in it—subject to the same laws, borne by the same balances.

And when one understands this, it is not only science that one discovers. It is oneself.

What I hope to awaken in those who listen to me is not intellectual assent, but a flash of recognition. An intuition that says: "Yes, I feel it too. I have always known, somewhere, that it was simpler. More coherent. More alive."

To transmit, for me, is this. It is to open a door. It is to allow each person to retrace the path by himself or herself, with his or her words, perceptions, doubts.

And if my research, my words, or my tools can serve as a bridge for this—then they will have had true meaning.

Sylvana
It is true that since we have known each other, I have noticed something that distinguishes you profoundly. You have never sought to impose a truth—but to open a path.

That is it, I believe, your unique way of seeking: to awaken in the other the desire to understand, rather than to answer in his or her place. As though the most beautiful gift you could offer were not an answer… but a spark.

And that is precisely what has stirred my curiosity more and more. That is why we are here today, sharing our dialogue: so that I might share your answers, and so that each person may perhaps find his or her own spark… his or her own path, or even his or her own quest for answers.

2. Time as a Key to the Future — Toward a Cosmic Consciousness

Sylvana
You spend your life observing, connecting, seeking meaning… And I see how profoundly all of this captivates you, moves through you, shapes you.

So I would like to ask you a big question—perhaps the simplest and the most essential at once… Where does science stand today in its quest to understand the Universe? Have we crossed a threshold? Are we finally beginning to glimpse that deeper mystery you have been pursuing all along?

Thierry
The Universe is of a dizzying immensity and complexity. It unfolds from the infinitely large to the infinitely small, as we can observe it from our planetary point of view.

Its complexity is such that it has been necessary to fragment scientific research into many distinct specialties. Yet all these disciplines study one and the same object: the Universe—but at different scales. And that is the extraordinary power of science. Within these specialties lies the key to understanding the reality of the Universe.

Fundamental research has today reached a threshold of knowledge sufficient to begin unifying the principles arising from its different branches. Its natural objective is now to give a logical meaning to our own existence within that vast whole that is the Universe.

Sylvana
So, if I follow you, we are at a key moment for science—one of those rare turning points when bodies of knowledge, long kept in separate compartments, can at last come together… and perhaps even help us, for the first time, truly understand who we are and what we are doing here, at the heart of this immense Universe?

Thierry
Yes, and it is for this reason that I have always been interested in the overarching view. It is the natural way to rise above the labyrinth of complex appearances that imprison and limit us if we analyze each field in isolation.

This global perspective revealed to me clearly that the laws of nature are of a prodigious simplicity.

This realization profoundly nourished my inspiration, leading me to write poems in homage to the reality of the Universe, as it Is—infinitely vast, harmonious, and rich with meaning, for in finally understanding it, I found my place in it, and by extension, that of the human being within the Universe.

Thus, the Humanoid that we are possesses the power to glimpse the genius of Nature. And in the courageous momentum of all those researchers who strive to grasp those celestial instants, there takes shape the vision of a future of such splendor and such power that it can only materialize as a benefit for all humanity.

Sylvana
You know, when you speak of the Universe, it is never cold or abstract. One feels that it is not just knowledge, but something profoundly personal, almost intimate.

So I want to ask you: what has been your guiding thread through all these years, the one that led you through all this?

Thierry
In this continuity of discoveries, I sought the primordial element in the manifestation of the Universe, and my attention turned to Time, because it predominates over everything. It is through Time that we technically define reality.

Sylvana
What a fascinating choice!

One might have expected many things… but Time? Especially when one so often hears that it does not exist! What inspired you to make it the central element of your thinking?

Thierry

It is a profound intuition, almost visceral, that has always guided me: that Time was not an illusion or a mere measuring framework, but a master key to cosmic order. From adolescence—as we have mentioned—I had a very unique relationship with Time, a kind of intimate attunement to its rhythm, its variations, its presence.

This double perception, both personal and universal, naturally led me to make it the heart of my thinking. I chose to devote a large part of my existence to exploring its true nature—in all its dimensions, inner and outer—in order to approach, lucidly, what Time really is.

But my exploration of Time did not stop at studying it as a fundamental element of science and of the manifestation of the Universe. Very quickly, I perceived it as a prism through which to observe the evolution of humanity itself. For beyond its profound nature, Time shapes our relationship to the world, our limits, and our aspirations.

Today, as humanity faces unprecedented challenges linked to Earth cycles and climate change, this understanding takes on an even more crucial dimension.

It is in this context that I developed and formulated the whole of my scientific and technological discoveries.

I have always held the conviction that terrestrial humanity possesses the potential to attain a level of knowledge capable of freeing it not only from its own limits, but also from the threats inherent to nature—including those that could manifest on a cosmological scale.

3. A New Paradigm — Reunifying Science, Consciousness, and Becoming

Sylvana

You say this knowledge could truly free us from our limits… But what does that mean, concretely?

Do you believe we can truly surpass our perceived boundaries—physical, mental, collective? How far could humanity go if it finally understood what you have glimpsed?

Thierry

When the potential to transcend our limits manifests, it will take the form of an expansion of consciousness, far beyond its current boundaries. An expanded consciousness that, ultimately, across the ages to come, could allow the human race to cross millennia by accessing a form of immortality, thereby inscribing its existence in the enduring nature of the Cosmos.

But to hope to reach such a horizon, we must first understand the stages that will lead us there.

It is with this in mind that, in parallel with my research in fundamental physics, I found it necessary to study the history of great human revolutions.

This exploration only reinforced the essential conclusion I just mentioned: to achieve the evolutionary leap we so urgently need, it is indispensable to offer a more complete and unified vision of reality.

However, understanding our past and the dynamics that have shaped our evolution is not enough. We must also act, create, and give form to these ideas.

It is in this continuity that my creativity and exploratory approach also led me to invention—as you know—because I understood that technology constitutes an essential support for the progress and the long-term survival of the human race.

When I contemplate the Universe, I see nothing there but an immense technological potential, ready to be placed in the service of our intelligence. The greater our intelligence, the greater this potential will be.

And what is prodigious about Time is that everything is entangled within it. Each instant carries within itself the echoes of the past, the present, and the future. In this invisible weave, it becomes possible to glimpse the universal meaning of things, a global perspective that naturally urges unification.

This is why it was important for me to understand our past, because it reveals to us the path already traveled, and I have identified twelve major revolutions that have shaped the present as we know it today. These events represent valuable information for the researcher that I am, because it is impossible to find effective solutions without a clear understanding of our history.

It is history that reveals to us our strengths as much as our weaknesses. It therefore constitutes an essential prerequisite for studying the future and conceptualizing the tools necessary for the survival and evolution of the human race. To understand where we come from is to lay the foundations of what we can become.

4. Twelve Revolutions — The Past as the Matrix of the Future

Sylvana
I see how essential this understanding of the past is in your approach.

Could you summarize these twelve great revolutions you have identified, the ones that, in your view, truly shaped our history and led us to today's world?

Thierry
Yes. I have identified these revolutions as key milestones in the evolution of humanity.

Each one marks a tipping point, a profound change in our way of living, thinking, or organizing ourselves.

They are not all equally spectacular at first glance, but each has played a crucial role.

By tracing them, we better understand the road traveled—and above all, what we still need to invent for the future. I will show you how I have conceptualized them.

1) The Neolithic Revolution (~10,000 BCE) was chiefly characterized by agriculture, settled life, and the beginnings of civilizations.

This revolution laid the foundations of human civilizations, leading to population growth and the development of cultures and technologies.

2) The Urban Revolution (~3,500 BCE) was chiefly characterized by the emergence of cities and writing.

This revolution enabled the rise of culture, science, and political structures.

3) The Axial revolution (8th–3rd century BCE)—chiefly the birth of the great philosophies and religions.

This revolution enabled the transformation of beliefs and ethical systems, still influencing moral and spiritual values today.

4) The Roman revolution (1st century BCE–5th century CE)—chiefly law, infrastructure, and Western culture.

This Roman legacy shaped the law, language, and culture of Europe and the Mediterranean world.

5) The Renaissance revolution (14th–16th century)—chiefly the rediscovery of knowledge, the arts, and modern sciences.

The Renaissance laid the foundations of modern thought and experimental science.

6) The Scientific revolution (16th–17th century)—the beginning of the scientific method and modern physics.

This revolution enabled the transformation of our understanding of the world, leading to major technological and medical advances.

7) The Industrial revolution (18th–19th century)—the advent of mechanization, urbanization, and capitalism.

This revolution enabled the acceleration of technological progress, but it also brought about social inequalities, pollution, and environmental changes.

8) The Democratic revolution (18th century)—chiefly the advent of human rights and modern political systems.

This revolution enabled the emergence of human rights and modern democratic political systems.

9) The Scientific revolution of the 20th century—chiefly quantum physics, genetics, and modern medicine.

This revolution enabled the transformation of technology, medicine, and communication, with ethical and societal consequences.

10) The Digital revolution (late 20th century–21st century)—chiefly the emergence of computing, the Internet, and the information society.

This revolution enabled global access to information, but also brought new challenges such as cybersecurity, disinformation, and data protection.

11) The Artificial Intelligence revolution (21st century–ongoing)—chiefly automation and the transformation of work.

This revolution enabled opportunities for major progress, but also raised ethical questions and risks related to employment and privacy.

12) The Ecological revolution (21st century–ongoing)—chiefly the fight against climate change and the sustainable transition.

This highlights the need to rethink our economic model and our relationship with nature to ensure the survival of the planet, its living inhabitants, and humans.

One can clearly see that each human revolution has marked a key stage in the evolution of society and knowledge. From the shift to agriculture to the digital age and AI, these revolutions have transformed the way we live, think, and interact with the world.

Today, the ecological and technological revolutions, notably that of artificial intelligence, will likely shape the future of humanity. But this trajectory is strewn with dangers in more than one respect.

Sylvana
It is impressive, seen like this.

Each of these revolutions is a turning point, a leap into a new way of living, thinking, and building the world. And one senses, in the way you present them, that it is not just history… it is a logical thread, an almost organic sequence. As though humanity were advancing, step by step, toward a tipping point.

But listening to you, I also have a question that is on the tip of my tongue: if all these revolutions have helped society move forward, why do you say that those of today carry so many dangers?

Thierry
The ecological revolution is a major challenge, born of a manifest lack of will to anticipate the side effects of technical progress in general.

The exploitation of resources without a long-term vision of their potential consequences has led to a crisis whose repercussions now threaten the balance of the planet and its ecosystems.

As for the revolution of artificial intelligence, it confronts us with a peril of an entirely different kind: that of a gradual loss of control over the tool.

If we do not manage to regulate its development as it takes on an ever-greater place in our lives and our infrastructures, we run the risk of a drift that could, in the long run, compromise the very existence of the human race.

I know that we do not yet possess the maturity of mind nor the sufficient level of intelligence to master an artificial intelligence that is invasive and autonomous in resources.

Before fully envisaging this advance, we must first strengthen our inner power, develop our cognitive and psychic capacities in order to maintain absolute control over AI, at a level it will never be able to reach.

Only on this condition will we be able to embark serenely on the adventure of artificial intelligence. For although formidable, it remains a tool of inestimable value, capable of considerably accelerating our evolution—on the condition that we are its true masters.

I have worked extensively on this subject, particularly on amplifying human potential, because this is where humanity has historically made little progress.

In truth, we are still young on the cognitive level. We still have to explore and fully develop our latent potentials, both intellectually and psychically.

Sylvana

By retracing all these revolutions, one becomes aware of the immense path humanity has already traveled. And yet, one clearly feels that we are still at a turning point. Perhaps even on the verge of a leap.

But what I find fascinating is that you did not content yourself with pondering all this in the abstract. You searched, invented, experimented, alone, for years… And that makes me want to return to something very concrete.

Because deep down, behind this whole overarching vision… there is also a man, a method, a workshop, a rhythm of work.

5. Remaining Faithful to Another Rhythm — Out of Sync with the Times

Sylvana

So tell me… how did you do it? From the beginning, one senses that you move forward with a clear vision, as though each discovery were inscribed in something larger… Yet, in the reality of everyday life—with material constraints to manage, and simultaneously all the ideas to explore and tests to carry out—how did you proceed concretely? How did you organize your work, your space, your research?

Thierry

As we mentioned, between 1990 and 2000, I devoted a large part of my time to study, while developing countless thought experiments with the aim of carrying the scientific work forward.

Starting in 2001, I truly began my experimental research, conducting a great number of practical experiments over the years, because it mattered to me to validate my discoveries with proof.

In parallel, I worked on the conceptualization of technological proofs of concept arising from my discoveries, for a wide range of industrial applications.

What made my approach unique was that I had complete freedom to define my own working framework. I was subject to no authority, no laboratory, nor investors' demands for results.

Of course, I had to contend with constraints of time and means: providing for myself with extremely limited resources.

This reality forced me to optimize every experiment by using the most accessible components.

At times, I had to dismantle prototypes that already worked because I could not keep several in parallel, as I often had to reuse their parts.

And on the meal front… let us say that I ate a lot of rice and potatoes. I often found myself hesitating between buying a little more varied food… or the two fuses and the cables I was missing to launch the next experiment. And very often, it was my thirst to go further that outweighed my desire to eat well…

This constraint, far from being a hindrance, taught me to be extremely organized, methodical, and cost-conscious, thereby refining my approach to resource management and the development of my projects.

It also forced me to think at length about which experiment to pursue or abandon even before moving into implementation, and then to anticipate every detail of the design, because I could not afford the slightest structural error or waste of materials.

This independence also allowed me to choose the best way to make effective progress, to take the time necessary to make scientific discoveries, without outside pressure.

Yet taking time has become a luxury out of step with the demands of our era, where the digital age, automation, and speed dominate.

Sylvana
Once again, we see how profoundly Time lies at the heart of our history and our evolution.

Is it not paradoxical that, in a world where we seek to push ever farther the frontiers of knowledge, we grant less and less time to reflection and to the true exploration of the great questions?

And you know, every time you speak of Time, I cannot help but smile… It seems as though it were both your best ally and your greatest mystery.

And the more you tell how you worked, the more I tell myself that your true strength is not only what you discovered, but the way you did it.

You were born at the end of the 1960s, and since then the world has done nothing but accelerate: digital technology has invaded everything, networks have multiplied, the logic of profit has come to reign, and automation has infiltrated everywhere. In fifty years, everything has gone into overdrive.

And you… you remained faithful to another rhythm. As though you had created your own bubble of time, impervious to this frenzy. You invented with constancy, with rigor, almost as one sculpts: each gesture considered, each step weighed, each impulse held back until it was right.

You took the time to doubt, to begin again, to understand.

You made the choice—which has become almost revolutionary—not to yield to urgency.

And I wonder if that is not your true silent revolution: having placed Time back at the center. Not only as an object of study, but as the guiding thread of your life, your thought, your manner of searching.

Thierry

Yes, you are right. I believe that I have always felt this difference… This vital need to slow down where everything was speeding up. And it was not only a choice—it was a necessity.

From childhood, my life took on a different tempo. I was sent to Martinique when I was still a baby, too young to understand what that implied. But later, looking back, I understood that it was not simply a change of place—it was a complete shift. I found myself isolated, outside the world, with only the trees, the insects, the sky, and silence for companions. That is where it all began.

Very early on, I understood that I would not go at the same pace as others. But obviously, it did not happen without friction. All my life, I had to face judgments and misunderstandings. People reproached me for not moving fast enough, for not "succeeding" according to conventional criteria. Some thought that I was wasting my time, that I was shut away in my ideas, that I was fleeing reality.

But in truth, I was going further. I was following a more profound logic. That time I took was not delay—it was the space necessary for things to take shape with accuracy.

And today, with hindsight, I see it clearly: it is precisely by accepting to slow down that I was able to go further.

I used Time as a raw material, as a lever. It is what allowed me to achieve certain technological advances much faster than if I had followed the conventional path. By learning to listen to it, I optimized it—but in my own way.

Today, I still remain faithful to that cadence. And I know it is what made my discoveries possible. Because one cannot force an understanding of the world: one can only receive it, when it is ready.

Sylvana

You see… that sentence alone—"Because one cannot force an understanding of the world: one can only receive it, when it is ready"—it sums you up so completely. That way of letting the world come rather than forcing it, of trusting the rhythm of things… That is exactly you.

Whereas I—like so many others—moved in the opposite direction. Charging ahead, pushing, insisting. We ran after time as if it were indebted to us. By wanting to do everything, to succeed at everything, to live everything… we saw nothing pass. That time so precious, we let it slip away. It is almost as though it ended up directing us without our realizing it.

And that is what struck me about you. That contrast. I grew up in a Southern Italian family, where everything was fast, intense, loud. For a long time I juggled projects, countries, responsibilities… a life full of twists and turns, certainly, but without a pause.

And I felt as though I had so many anecdotes to tell, so many things to say.

But it was when I started asking you questions that first evening we met that I saw the real contrast. You said almost nothing. You were composed, silent, you listened to me… And then the moment you opened your mouth, a world opened. Dense. Precise. Incredibly vast. Yet always calm. Just.

Besides, I told you that very evening—do you remember?—I asked if you came from another planet. I could not understand how someone could be so… out of step with the rest of the world. And at the same time, so deeply rooted in something essential.

Thierry

Yes, I remember. And you know… what you said that evening, about the idea that I might come from another planet… I have often felt it myself. Not in the literal sense, of course. But as a persistent impression of mismatch. As though the Universe had traced a different path for me. A path that compelled me to slow down, to observe differently, to enter another tempo.

People have always seen me as an outsider. Out of step. And it must be said that, from the outside, that was probably how it looked. I worked a great deal from home, almost never going out. I took temp jobs, often at night, to free up my days and devote as much time as possible to my real work: my research.

And as soon as I shared a little of what I was working on, the reactions were always the same. People made fun of me. They looked at me as a simpleton, someone "basic," without a diploma… and yet thinking about questions that most considered out of reach.

So, after several failures, I decided not to share anything anymore. Nothing more, until all my work was completely structured, verified, tested. I told myself that I would speak of it only when it was ready. Fully ready.

And a few decades later… that day is today. I am ready. I know that the time has come. Everything in the state of the world, in what humanity is going through, in the way things are accelerating and losing balance… everything indicates that what I carry can no longer remain in silence. What I have understood, what I have built, has meaning only if it is transmitted now. Not before. Not after. Now.

Sylvana
It is so right, in the moment we are living.

And I grasp what that represents for you… exposing yourself like this, in front of everyone. Opening what you kept to yourself for so long.

Thierry
I also know what that implies. By choosing to step out of the shadows, I expose myself to every criticism, to every challenge.

I do it in full awareness.

But my intention is not to impose anything. Only to open a window. To suggest another reading, another way of looking.

And if some want to go further, to think together, to confront, to build… then so much the better.

If not, I will continue alone, as I always have. Because I never followed this path to convince—I followed it because it imposed itself upon me.

6. When the World Goes Too Fast — Knowledge Under Pressure

Sylvana
You see… even in your decision to open up today, after thirty-five years of working in silence, one feels this most particular relationship you have with Time.

You waited for the right moment, with patience, with a kind of wisdom. You did not rush anything.

And I know I am repeating myself… but it is because truly, you are giving us a fundamental lesson about our relationship to Time. I think it is the same for society as a whole. We are caught in a kind of collective acceleration, as though slowing down had become a luxury or a failure.

Whereas you, you held on to that thread. A coherence, a direction.

How did you manage never to lose it, precisely? To remain faithful to that rhythm, when everything around you was pushing the other way?

Thierry
It is an essential question. Our relationship to time has changed profoundly: everything now must be quickly accessible and immediately resolved. This phenomenon, omnipresent in every aspect of life, also affects fundamental scientific research.

Thorough and patient reflection, though essential to the emergence of true advances, is being replaced by a frantic quest for immediate profitability, often to the detriment of exploring the great questions that have shaped our humanity.

But that, I could not ignore. And thanks to my independence—freed from any external constraint and guided by my own path—I was able to keep this question in mind throughout my research and my life:

What is our place in the Universe, and what role do we play in the evolution of consciousness and knowledge?

I knew it was necessary to put the human being back at the very heart of research. One is not merely an observer, but an intrinsic element of the reality of the Universe. And even more than that: a universal co-creator.

This awareness led me to an essential conclusion:

We had to redefine our own conception of evolution.

7. When Knowledge Loses Meaning — The Urgency of Recentering

Sylvana
You know… the more you share, the more I feel that something has reversed. It is no longer science looking at the human being from the outside… It is the human being who is beginning to look at themselves differently. And that changes everything.

But what strikes me is this paradox: today we have all the tools, all the data, all the knowledge to truly move forward, to reunify what we long kept separate. And yet it feels as though we have lost the momentum.

You see, at one time, great minds asked:

What is our place in the Universe?

What is the meaning of our presence here?

Why do we no longer ask ourselves those questions today? What has changed?

Thierry

Once again, I must first return to the fundamentals of the human sciences, to a time when they were still intimately linked to philosophy.

Philosophy—which literally means love of wisdom—embodies a critical and reflective approach to the world, to knowledge, and to human existence.

In this continuity, psychoanalysis imposed itself as an exploration of the inner world, of the self-within, thus responding to the profound aspiration of the human being to question themselves. It then stood in the prolongation of the philosophical movement initiated by the century of the Enlightenment, pursuing that quest for understanding and meaning.

At that time, psychoanalysis held a central place, naturally echoing that quest for meaning that animated humanity.

However, that quest for wisdom has gradually given way to a quest for performance. The time required to elaborate, associate, and attempt to pierce the mysteries of the human psyche now seems out of reach, relegated to the status of an illusion. Judged too hermetic and slow, psychoanalysis today suffers from this shift in priorities.

With the advent of quantum mechanics, artificial intelligence, and NBIC technologies (nanotechnologies, biotechnologies, information technologies, and cognitive sciences), our understanding of the world and of the human being becomes increasingly tangible, quantifiable, and measurable.

Except that the human being now tends to be perceived as a standardized machine, to the detriment of their brilliant singularity.

Now, having made important discoveries in this field, I can affirm that this singularity—which defines each individual—far exceeds the limits of the tangible, the quantifiable, and the measurable.

We are touching here upon other dimensions, still unexplored, that elude our current means of measurement.

The younger generation, immersed in this technological culture, could be tempted to reduce the human soul to a simple object that can be quantified and measured, to be analyzed by machines. With artificial intelligence, instantaneous and preprogrammed answers are supplied to an ever-growing number of questions, creating the illusion of a mastered knowledge.

This generalization, carried out at the expense of singularity, risks leading to increasing superficiality in our way of apprehending human beings, concepts, technological solutions, the world, and even the Universe.

I believe that such an approach is untenable as it stands. It systematically leaves individuals and societal projects by the wayside, while generating overall inefficiency.

Sylvana
You know, listening to you, I understand more clearly how dangerous this drift toward quantifiable, fast, predictable knowledge… can be.

We think we are moving forward, but we lose in depth, nuance, humanity.

What you say about the way we search today—through machines, algorithms, preformatted answers—strikes me profoundly. Because in that model, what cannot be measured… disappears.

And yet, it is precisely in the invisible, in the slow, in the singular that the true breakthroughs are often lodged.

And you… you are living proof of it. Your way of being, of thinking, of searching… embodies exactly this.

Another temporality, another depth, another way of connecting what others do not connect.

So, before we enter the heart of your science, I would like us to take a moment to talk about one last thing about you. Something more intimate, which you have already mentioned several times: what is today called Asperger syndrome.

I know that it does not sum up everything you are… but I feel that it is part of that inner fabric that shapes your gaze, your way of exploring, of understanding.

And what if we spoke about it now?

Chapter 3

A Different Kind of Intelligence

~

A Distinct Cognition in Action

1. Another Gaze — At the Heart of a Singular Thought

Sylvana

You know, from listening to you speak about Time, with that rigor of yours—almost instinctive—in connecting everything, I realize something: it is not a method you apply, it is you. It is inscribed in your way of thinking, of feeling, of being in the world.

You have already mentioned this somewhat different functioning, that form of autism called Asperger's syndrome, but I would like us to take the time, truly, to enter into it. To understand what it changes, from the inside.

Thierry

Yes, it is essential for understanding the source of my gaze. And I want to specify one thing from the outset: I do not know whether everything I am comes exclusively from what is today called the Asperger's syndrome. It is no doubt an important part of my functioning, yes—but it is not the whole story.

In my case, there were also physiological particularities, notably on the psychomotor level, that most people with Asperger's do not necessarily present with.

It is a unique mixture. An atypical configuration, on several levels, that has shaped my way of being in the world.

It is, moreover, what Ruth sensed very early on—and what she tried to express in her own way: that it was not simply a medical label or a cognitive profile. It was a broader whole, a profound coherence in my way of perceiving, feeling, and functioning.

From childhood on, I perceived the world through a very particular lens. I was hypersensitive, unable to filter out certain sounds or certain lights, often overwhelmed by stimuli that others did not even notice. I needed stable landmarks, very precise routines, fixed structures so as not to feel overrun.

But alongside that, there was an incredible inner intensity. A sharp memory, an attraction to logical structures, an ability to visualize mentally complex geometric forms—sometimes even in motion, in several dimensions at once.

Sylvana
You see… listening to you, I realize that your lived experience fits into no box. It is not a "disorder," nor is it a "gift." It is more subtle, more whole. And seen from the outside, one often misses everything that makes up your richness—one sees only what disturbs, what surprises.

Thierry
It is one of the paradoxes of the Asperger's syndrome: to possess immense cognitive richness, yet collide with invisible walls in social interactions, in communication, in the implicit rules of everyday life.

For a long time, I felt like a failure, out of step, poorly adapted. People did not understand my reactions, my silences, my gestures. I could mechanically repeat phrases I had heard—what is called echolalia—without realizing it.

And since I also had certain motor difficulties, such as right-left confusion under stress, people sometimes perceived me as "odd," even "limited."

But inside, it was the exact opposite. There was a fire. A thirst to understand. A relentless logic that wanted to connect every element of the world into a coherent whole.

Sylvana
So you carried at once a burden that people could see and a richness that no one saw. It is unbelievable. And the most remarkable thing is that you managed to turn it into a strength.

Thierry

That is true, but it did not happen immediately. From the outset, there were also very concrete difficulties, particularly on the psychomotor level. Things I had to correct constantly, often in the shadows, which made that sense of being out of step even more visible.

It took me years to make peace with this way of functioning. To transform my rigidities into tools of precision, my need for structure into a working method, my obsessions into engines of inquiry.

What is today call "Asperger's," I lived for a long time as survival.

Then little by little, I learned to make it a lever. I observed what obstructed, what exhausted me, what I did not understand in others.

And instead of wanting to become "normal," I decided to know myself better. To optimize what I was. To adjust, to correct, sometimes even mechanically, certain blockages.

Sylvana

And today, if someone meets you without knowing about your neurodiversity, do they still perceive that mismatch?

Thierry

Not really. Except perhaps for this very rigorous way I have of organizing my thinking, my environment, this need to keep precise points of reference, well-defined structures.

I have learned to navigate the social world, to manage the unexpected, to interpret implicit codes. But all that remains a learning process. Nothing is automatic.

What I find important to say is that this cognitive difference—this neurodiversity—is neither an illness nor a magical gift. It is another form of functioning. And in some people, it opens extraordinary doors.

They are often people capable of extreme perseverance, absolute concentration, radical innovation. Researchers, artists, engineers, thinkers who see what others do not see, simply because they do not think like others.

Sylvana

It is, at its core, a tribute. A tribute to all those who have grown up, who live—and who are still being born—in silence and solitude, yet who carry within themselves, perhaps, an unsuspected treasure.

Thierry
Yes. And it is also a call to change our perspective.

For a long time, the world wanted to "fix" whatever fell outside the norm. But today, we are beginning to understand that some of these differences may be the key to a more creative, more profound, more sustainable future.

Concerning Asperger's syndrome, Hans Asperger himself, in his 1944 thesis, had already perceived this singularity in some of the children he was following.

He suggested that to succeed in the sciences or the arts, a touch of autism seems essential. To succeed, the necessary ingredients may be an ability to turn away from the everyday world, from mere practice, an ability to rethink a subject with originality in order to create new, unexplored paths, by channeling all capacities into a single specialty.

I find this intuition remarkable: it does not speak of superiority, nor of an exceptional gift—but of another way of being in the world.

It speaks of a rare concentration, of an intense focus that can, at times, open radically new pathways. And this way expresses itself above all through a very particular faculty: the ability to fix upon a subject, to immerse oneself in it entirely, with an almost obsessive intensity.

This, for me, is the true singularity of the Asperger's way of functioning: that out-of-the-ordinary perseverance, that capacity to detach from distractions, to return, again and again, to the same idea, the same problem—until one gets to the heart of it; and that manner of questioning otherwise, of formulating problems differently, of searching where others might not even have thought to look. And when this fixation meets a sound intuition, or a profound question, then yes: it can become a powerful force of discovery, creation, or transformation.

2. Inner Dialogues — Another Way of Thinking the World

Sylvana
What you have just said is magnificent. That inner fire… that fidelity to an impulse that one never lets go of… That is where one feels the true strength, beyond what is visible.

Listening to you, I think of all those great minds you studied extensively. You read them, reread them, almost kept company with them in your head…

Has it ever happened to you, at times, to recognize in them signs, resonances, that made you think they too might have had Asperger's syndrome?

Thierry
Yes, often. And let me be clear right away: these are only personal impressions, intimate resonances.

I obviously do not claim to make posthumous diagnoses—that would be absurd. But through certain accounts, certain reported attitudes, certain writings, I have sometimes felt a disconcerting closeness. A kind of inner recognition.

Sylvana
And among the great names you studied, are there some with whom you felt that kind of resonance?

Thierry
Yes—for example, Albert Einstein. His mode of thinking was essentially visual.

Furthermore, during his lectures, he would sometimes begin with a precise example, then set out the general principle afterward, with no apparent link. He would at times break off suddenly mid-sentence, lose the thread, then return a few minutes later, as though emerging from an inner trance.

He often hinted that he preferred to avoid social gatherings, believing they would divert him from his work, which he considered the essential part of his life.

This is exactly what I feel. It is not rejection. It is an inner necessity to preserve a coherent mental space.

And then, there was music. For him, it was essential. He often hinted that it held a central place in his life and that it represented for him a way to remain independent from others. There again, I recognize myself. Music has become, in my own path, a scientific research tool in its own right. A language more subtle than words.

Sylvana
You have just made me want to read Einstein… from another angle.

Thierry
You should. And you can add Paul Dirac to the list. One of the pillars of quantum mechanics.

He was absolutely rigorous, almost socially inaccessible. It is said that he spoke very little, that he answered with yes or no, and that he refused to begin a sentence if he did not know how it would end.

Dirac was also echolalic: he automatically repeated his interlocutors' words—a characteristic that I also shared in my childhood.

In him as in Einstein, one finds a kind of social minimalism, a withdrawal from the world, but also a powerful obsession with coherence, with structural truth.

I could name others still: Newton, Curie, Ramanujan… Paths marked by a form of solitude, nonconformity, and an extreme focus on a quest.

These are only personal hypotheses, of course. But they have inspired me profoundly.

Obviously, that does not mean that these great minds had Asperger's syndrome, nor that it is a condition of genius. They are simply points of resonance that I have felt, echoes that have allowed me to understand myself better, without ever seeking to label them.

Sylvana

That is powerful… What many would see as a limit becomes, in your view, a strength. A source of creativity. As though this difference carried within it a rare potential, ready to emerge.

Thierry

Yes, but once again, one must not generalize. Asperger's syndrome is not a guarantee of genius.

In my case, I simply know that it is another way of perceiving, of connecting, and of conceptualizing things—as though thought followed another path.

But when a mind with Asperger's syndrome becomes obsessed with a scientific question, and it has the right tools, the right environment, and sufficient space for concentration, then yes—it can go very far. Very quickly.

And here I share the intuition of certain neuroscientists: it is not in spite of the difference that such individuals innovate… it is thanks to it. Because they do not think like everyone else. They do not seek to please, they seek to understand.

Sylvana

It is striking how the way we look at these differences can change everything… And yet, I have the feeling that society has long confused singularity with deficiency.

Do you think this is changing? Has the view of autism truly evolved over time?

Thierry

Once again, it seems essential to me to analyze how the perception of

Asperger's syndrome has evolved through history in order to answer this question more accurately.

Over the course of the last century, the way autism is regarded has evolved profoundly. Once equated with idiocy or mental illness, it was then classified as a disability, and finally recognized as a form of neurodiversity—a simple natural variation of human intelligence.

This change in perception has led to the emergence of a new vocabulary: integration has given way to inclusion, being outside the norm has become a right to difference, and the notion of disability tends to be replaced by that of atypical potential.

In this evolution, the question of care or of compensating for the disability is gradually disappearing, in favor of recognizing the role that these atypical profiles can play in human progress.

Sylvana
And you, how do you feel about it today?

Thierry
For my part, I have learned to perceive my own syndrome not as a limitation, but as an asset—a blessing that, well utilized, can become a major opportunity to approach science from a different angle.

3. Expanding Consciousness — An Invitation to Explore Our Human Potential

Sylvana
But in reality, in everyday life, in teams, in projects… can these differences really coexist?

I mean: can a profile like yours harmonize with, and complement, other forms of intelligence—more intuitive, more spontaneous, sometimes even less structured?

Or do they remain two worlds that struggle to understand each other?

Thierry
It is an important question—and a complex one. And to answer it accurately, it seems essential to me to clarify two points very clearly first.

First, yes, I fully acknowledge the extraordinary potential that one can observe in certain Asperger profiles. There is an atypical neurological structure that, in some cases, allows for a rare cognitive intensity: extreme concentration, rapid and precise information processing, visual thinking, logical rigor.

But at the same time, I have also met non-Asperger individuals who possess capacities just as exceptional—in other forms. Their intelligence may take different paths, sometimes more intuitive, more relational, more fluid… but it is just as precious. I feel a profound admiration for these individuals.

And so, to answer your question: yes, I believe these forms of intelligence can coexist—and even complement one another.

But, in practice, this is rarely the case. Whether in companies, research teams, or even social circles, a reflex persists—what in psychology is called homophily, the tendency to associate with those who resemble us. This bias pushes us to cluster among similars, to protect what is familiar, and to relegate what escapes our codes to the margins.

As a result, those who think differently—whether through an Asperger prism or any other singular way of perceiving—often find themselves isolated, judged, even silenced. Yet I am convinced that it is precisely in this diversity that an immense potential lies. But for now, it remains largely untapped.

Sylvana
So for you, they are simply different forms of intelligence… equally valuable, and capable even of enriching one another?

Thierry
Exactly. That is the key.

It is not about elevating one profile at the expense of another.

What I advocate is an expanded vision of human intelligence. And what I would like today is to be able to go further in this exploration. To work with a specialized team to conduct a fine-grained analysis of the differences—but also the similarities—among these varied cognitive modes. To study their complementarities.

Because I am convinced that it is in this plurality that true human wealth resides. And that it is indeed time to cross a threshold.

Sylvana
And when you say the time has come to cross a threshold… do you mean a real change of level?

Thierry
In fact, I am not speaking solely of intelligence strictly speaking, but above all—and this is, for me, the most important—I am speaking of consciousness.

For we are entering an era in which our points of reference are going to be profoundly upended. Many factors contribute to this—but one of the most decisive is the emergence of artificial intelligence. It will transform our professions, our societies as a whole, and even our relationship to ourselves.

And faced with this, we must respond not only with computational power, but with an inner elevation. A deepening of our consciousness. An increased capacity to perceive, to connect, to feel the real in a subtle and embodied manner.

And this is where my research finds its full meaning. Because over time, I have come to believe that this potential for evolution is not reserved for a few. It is present—in a latent state—in all human beings.

Sylvana
But then… what would it mean, concretely, to cross this threshold? What would this new relationship to ourselves be?

Thierry
This is precisely what I wish to delve into.

Research in neurosciences confirms that we do indeed use 100% of our brain—but not all at the same time.

What I have observed is that our nervous system is not limited to a single processing center.

In reality, we possess a network of several interconnected brains: the cranial brain, of course, but also the heart, which has its own neuronal system; the intestines, with the enteric brain; and even a form of diffuse cellular intelligence that irrigates the entire body, like a tissue memory, living, rooted in every cell.

Sylvana
It is remarkable how your perspective always goes beyond the classical brain… You are really talking about an inner orchestra.

Thierry
Yes, that is very important because this network forms an exceptional adaptive architecture, endowed with a capacity for self-structuring.

But this biological potential, remarkable as it is, can only fully unfold if it comes into resonance with a high level of consciousness—and that is the whole issue.

In the critical context humanity is experiencing—systemic, ecological, social, existential crises—it is unlikely that a natural evolutionary leap will suffice. A biological change would require millennia.

That is why it has become urgent to activate this potential in another way.

Yes, advanced technologies have a role to play—and some of them, such as artificial intelligence or future quantum computers, could profoundly upend our relationship to the world. But if human consciousness does not expand in parallel, these tools risk reproducing—or even amplifying—the errors of the past.

Sylvana
Listening to you, I see how profoundly we can understand Asperger's syndrome differently—not as something to be corrected, but as a singular way of inhabiting in the world; another way of feeling, thinking, perceiving—with its strengths, its limits… and above all with a potential to shed light for everyone.

And I have the impression that what you are sharing is not only your story.

It may be a mirror for all of us. An invitation to question what we call intelligence, and what we do with it…

But then… if this potential exists… what do we do with it, now?

Thierry
Precisely.

If this potential exists, then it deserves to be fully activated.

And that is why it is not only a matter of innovating outwardly, but also of accompanying this technological mutation with a profound inner transformation: broadening our gaze, connecting, discerning, embodying what we create.

Technology, without consciousness, becomes blind. And it is at this crossroads that my research is situated: in this exploration of conscious intelligence, but also in the development of technologies capable of accompanying the evolution of the human being—not by replacing them, but by supporting the emergence of a vaster, freer, more connected consciousness.

Sylvana
Listening to you, I realize that what you long experienced as isolation… was perhaps, in reality, an inner laboratory.

A space apart, demanding, at times harsh—but one that allowed you to forge a singular, profoundly set-apart way of thinking, seeking, and inhabiting in the world.

And what if that way were not an exception, but one of the possible paths of tomorrow?

And what if this path, which seemed solitary, could today illuminate other trajectories…?

Thierry
And if this inner laboratory could today become a springboard for others… then yes, this entire solitary journey would have found its true meaning.

Sylvana
And that is why I was intent on asking you all these questions… even before entering the heart of your science.

Because beyond your discoveries, it is also your way of searching that inspires—this capacity to trace a path where there was none.

And perhaps, by sharing it, you will bring forth paths that no one had yet envisaged…

Thierry
And it is precisely for this reason that I would like to conclude this part with a call for collective exploration.

My objective is to share this approach with as many people as possible. To inspire scientists, engineers, inventors, philosophers, teachers… and all curious minds.

Because it is together that we will truly be able to go further.

What seems impossible today is so only because we impose our own limits upon ourselves. And the only way to transcend them… is to live them concretely.

It took me thirty-five years of research, experiments, doubts, trial and error… to finally be ready to share this science and these technologies.

I hope that each of us can keep an open mind, to welcome all that can help us grow—toward more consciousness, joy, and connection. Individually and collectively.

Sylvana
So what you propose… is not just a new way of looking at difference.

It is an opening onto another possibility. A way to rethink intelligence, consciousness, and even… the future of humanity.

Part II

At the Frontiers of the Real
~
When Science Calls for a New Gaze

Chapter 4

Between Knowledge and Silence

~

What Science Reveals in Spite of Itself

1. At the Edges of Time — When Science Divides

Sylvana
Much has been said about your connection to Time, about what you have felt since childhood… But I realized that we have never really spoken about what science itself thinks of it.

So then, precisely: in classical theories, how is Time understood?

Thierry
To better understand what science says about Time, we need to go back a bit.

Time is a concept that was long believed to be self-evident. With Newton, it was thought to flow uniformly, for everyone, everywhere in the Universe. An absolute time, independent of everything. Like an immense cosmic clock.

But at the beginning of the 20th century, this vision shifted. In 1905, Einstein completely overturned this conception with what is called special relativity. He showed that Time is not identical for all: it depends on motion.

Two people who are not moving in the same way can perceive events in a different order. And it is not just a matter of perception: it is a measurable physical effect, confirmed by numerous experiments. The faster one goes, the more time slows down for them.

Sylvana

So, to be clear: when you say that "time slows down," do you really mean that time itself flows more slowly for someone who is in motion? It is not just an impression—it is their entire experience of time that changes.

For example, if I were to travel in a very fast spacecraft, I would truly age more slowly, and even the spacecraft would wear out more slowly than if we had stayed on Earth?

Thierry

Exactly.

This is what is called time dilation. And it is not only theoretical—it has been measured in practice. But be careful: for this effect to be genuinely perceptible, you need extremely high speeds—much closer to the speed of light. You do not notice it in a car, or even on a plane.

But even at much lower speeds, such as those of satellites orbiting the Earth, one can measure very slight offsets thanks to highly precise clocks.

For instance, GPS satellites move fast enough that their clocks run a little more slowly than those on the ground. And since GPS relies on perfect synchronization to calculate distances, even a tiny time offset would lead to position errors that could reach several kilometers per day—which is why this offset is corrected.

But speed is not the only factor. In 1915, Einstein developed another theory, even more profound: general relativity. He showed that gravity can also slow down time. The more an object is immersed in an intense gravitational field, the more its time "slows."

Sylvana

Can you give me a more striking example that illustrates this gravity factor? Because right now, I do not quite see it…

Thierry

Of course.

Imagine that you are on a very high mountain, and I am at the seashore. Because of the difference in gravity, your clock will run slightly faster than mine. This offset is minuscule, but measurable. And the stronger the gravity, the more marked the effect.

Sylvana
So time is not the same everywhere, even without moving? Just because of where we are?

Thierry
Exactly.

This is the heart of general relativity: time is influenced by the gravitational environment, by the way mass and energy deform the space around us. There is no longer a universal clock. It is a local experience, different for each observer.

But this vision, powerful as it may be, was not enough. Starting in the 20th century, quantum physics entered the scene. It was discovered that Einstein's laws do not allow us to understand the infinitely small: atoms, particles. And there, we step into a completely different world.

In this new framework, time no longer plays the same role. It becomes a mere external parameter, placed outside the equations. It is not part of the system under study, it is not influenced by what it measures. Unlike space, it has no proper existence within the mathematical structure of the phenomena.

Sylvana
Wait… here, there is something that strikes me.

In relativity, time is completely integrated into the system, influenced by matter, gravity, motion… And in quantum theory, it remains fixed, as though it took no part in anything?

We are talking about two major visions of reality, and they do not say the same thing at all about time?

Thierry
That is exactly it.

A profound gap, a kind of fracture in our understanding of the real. On one side, a flexible, relative time, influenced by motion and gravity. On the other, a fixed, absolute time, outside the system. This gap makes the unification of the two theories extremely complex.

And that is not all. Today, some lines of research go even further. They question the very idea that time exists as we understand it. For them, what we call the "passage of time" might be nothing but an illusion tied to our perception. There would not be a flow, but a succession of states, and it is our memory that would give the impression of continuity.

Sylvana
I still have trouble getting used to this idea… The fact that time could be considered an illusion…

Thierry
Yes, it is very destabilizing.

But in reality, it shows how profoundly the nature of time is still poorly understood.

I had already intuited that time was central in the Universe, and this persistent disagreement about its true function in science only confirms that it was both necessary and evident to make it the focal point of my research.

As long as we do not truly know what time is, we will not be able to connect the different levels of reality.

Sylvana
And today, are many physicists trying to resolve this fracture?

Thierry
Yes, of course. It is one of the greatest challenges of the twenty-first century.

There are very diverse approaches such as string theory and quantum gravity… But fundamentally, the question remains the same: what is time?

As long as it has not been resolved, unification will remain out of reach.

And if I truly wanted to follow this intuition through to the end, I could not be content with correcting a few details or bringing small improvements to existing theories. That was not enough. It was necessary to return to the very foundations, to review the bases upon which our understanding of the real rests.

To connect relativity and quantum mechanics without betraying either, I understood that a deeper framework had to be set. A framework that does not show itself directly, but that makes everything that follows from it intelligible. I envisioned it as an invisible weave, an extremely fine structure—made of distinct, discrete units—that supports reality without ever imposing itself.

It is this idea that led me toward Absolute Relativity: a theory in which it is no longer the grand visible laws that dictate the whole, but a more coherent, more fundamental substratum, from which everything can emerge without contradiction.

But an intuition, however strong, is not enough. It had to be tested. I had to look for signs, clues, even tenuous ones. I asked myself: if this invisible background truly exists, then it must leave some traces. Not necessarily obvious ones… but perhaps by looking differently…

Sylvana
And how did you go about it?

Thierry
I first studied what science had already produced on the subject. I did not start from scratch: I examined the works, approaches, and hypotheses already proposed. And among all that, three major avenues seemed particularly interesting to me.

First, I looked at the theories that already attempt to reconcile gravity, as described by general relativity, with quantum mechanics.

Approaches such as string theory or loop quantum gravity attempt precisely to bring them together within a coherent framework. Some of them suggest that spacetime is not continuous, but made up of distinct units. That is already a step forward. But even there, time remains on the periphery. It is not yet treated as the heart of the system.

Next, I asked myself whether minute discrepancies in certain measurements—such as a very slight variation in the speed of light, or a small shift in relativity's predictions—might betray the existence of a deeper framework. But this is still at the boundary of what our instruments can detect. We almost touch something… but it is blurred, uncertain. We perceive hints, tremors, without being able to confirm them fully.

And finally, I explored the cosmic scale: the background glow of the Universe, primordial gravitational waves… Perhaps by digging into that, one could find patterns, anomalies that reveal an older, more fundamental structure. But here again, we depend too much on technology.

In short, these three avenues are valuable—I do not dismiss them. But they remain dependent on technical advances that we do not control. And I did not want to wait.

So I chose another path: that of a rigorous theoretical construction, nourished by the attentive observation of the current limits of science, for it is often the flaws of a system—its inconsistencies, its silences, its blind spots—that most clearly reveal what it lacks.

I then took the time to look closely at what science manages to describe… and what still eludes it. I drew up an uncompromising survey of the situation, to better pinpoint the fault lines.

Sylvana

You mean… as though you had taken a flashlight and decided to go light up the darkest corners of the room?

Thierry

Exactly.

And in this exploration, four major blind spots struck me as unavoidable.

The first is cosmology, which runs up against the enigma of so-called "dark" matter and energy. These two invisible components would account for more than 95% of the Universe… and yet we still know almost nothing of their nature. They seem to govern cosmic expansion, but elude any direct detection, as though a large part of the real remained inaccessible to us.

The second is the mystery of the origin of the cosmological constants and the fine-tuning of the Universe. Why do the laws of physics take exactly the values they have? Why is the Universe tuned so precisely to allow the emergence of galaxies, stars, and the chemistry of life? This degree of precision raises questions—as though an underlying logic, still elusive, had imposed itself from the beginning.

The third is quantum mechanics, which remains silent about what truly grounds its own probabilities. It describes with remarkable precision the behaviors of matter at the microscopic scale, but without explaining why these behaviors follow certain probabilities and not others. "Quantum randomness" is a fact—but its profound cause remains out of reach.

And the fourth is a notion often neglected, but which seems fundamental to me: the anthropic principle. It raises a dizzying question: why does the Universe have the exact conditions that allow the emergence of life, then of consciousness? This principle, often discussed on the margins, brings into question our place in the Cosmos and the very nature of physical laws.

These four questions form a whole, but to avoid overwhelming you with too many notions at once, I will present them in two parts.

First, the "cosmic" enigmas, those that concern dark matter and dark energy and the precise tuning of the laws of the Universe.

Then, in a second stage, we will enter the "fundamental" enigmas: the mystery of wavefunction collapse and the anthropic principle, which touch more closely on consciousness and on our place in the Cosmos.

Sylvana

So you are creating a double path: first the Universe in its invisible structure, then, later, the enigmas that touch on consciousness.

Thierry

Yes, that is indeed the idea: to move forward in two stages. It is on this basis that I can now show how these four angles reshape our understanding of the real.

These four angles will enable me to bring out the fundamental parameters to add to the enigma of Time, in order to profoundly reformulate our reading of the real, and open the way to a unified, coherent… and profoundly alive architecture.

But do not try to remember everything or understand everything. What I am going to present to you here is an overview—a deliberately simplified framework, to locate the shadowed areas.

I will address only the notions that I judge essential to set this context: it will serve as a basis for understanding the meaning of my research.

In this part, many scientific notions will be new, but do not worry: it is not necessary to remember every technical term or every detail, only to grasp the general idea.

For now, I simply want you to perceive why I chose these five enigmas—including the enigma of Time, which we have already addressed—as anchor points: because they are the ones that sketch the outline of the current limits of science… and open the door to something else.

2. From the Invisible to Intuition — Dark Matter and Dark Energy

- The Reverse Side of the Visible — When 95% of the Universe Eludes Us

Sylvana

Will we start with cosmology?

Thierry

Yes. And to show you how full of mysteries this field is, I will begin with a striking example: the Hubble constant.

I can tell from your expression that it is not something you know—let me explain.

The Hubble constant is a way of measuring the speed at which the Universe is stretching. The farther away a galaxy is, the faster it moves away from us. And this constant allows us to quantify that rate of expansion.

Sylvana
Okay… so it is a bit as though you were measuring how much the Universe "stretches" over time?

Thierry
Exactly.

But here is the problem: when we try to measure this constant, we get different results depending on the method used.

If we look at nearby galaxies, we find one value. But if we estimate it from the cosmic microwave background—therefore from the very beginning of the Universe—we find another value.

Sylvana
But why is that a problem? Could it not just be an imprecision in the instruments?

Thierry
No. The discrepancies are too large to be due to chance. And that suggests that we are missing a fundamental piece of information about how the Universe evolves.

Sylvana
So even the most important figures—the ones we think we have under control—may not be all that stable?

Thierry
Exactly.

And this uncertainty is not anecdotal: it undermines the backbone of our understanding of the Cosmos.

So, in an attempt to resolve such enigmas, science has built a powerful, coherent framework: modern cosmology.

That is where I chose to get concrete. Because, at its core, if our Universe has an origin—not merely a beginning, but what brought it into being—as well as a structure and dynamics, it is indeed in cosmology that we should see the first signs of it.

To do this, I relied on the Standard Model ΛCDM (Lambda-CDM), which today serves as a reference in astrophysics for describing the evolution and the large-scale structure of the Universe.

This model rests on a subtle balance between what science has been able to measure directly, what it has inferred from equations—notably those of general relativity—and what it posits to bridge certain inaccessible domains. It does not claim to explain everything, but it offers remarkable coherence with observations.

Concretely, starting from the available data, it traces the thread of the cosmic narrative, from the Big Bang to today, taking into account the structure of the Universe, its expansion, and its composition.

When I began my work in 1996, the ΛCDM model was already circulating within the scientific community as a very promising framework. It described the history of the Universe by relying on general relativity and the major cosmological observations.

But it was in 1998, two years later, that it truly established itself as the reference model, with the reintroduction of a key term: dark energy.

I therefore took a natural interest in it. Not to call it into question—quite the contrary. For me, this model represented a remarkable foundation, of great coherence… and yet, at the same time, it hinted at fascinating shadow zones.

Before going any further, let me show you what this means for the distribution of energy. Because behind this elegant model lies a dizzying fact: what we see is almost nothing compared to what truly exists.

Sylvana

So you are about to take me beyond the grip of measurement, is that it?

Thierry

Yes. Because to understand the energy distribution of the Universe is also to enter what escapes our direct perception, and yet structures the entirety of the Cosmos.

And for me, that required another way of inquiring: not only to measure, but to feel, to perceive those invisible zones through an inner immersion, lucid, patient.

Before anything else, I had to address the fundamental problem—that of the energy content of the Universe and its distribution—which can be summarized as follows:

- ~ 5% ordinary matter (atoms)

- ~ 27% cold dark matter
- ~ 68% dark energy (Λ)

This means that the nature of 95% of the energy content of the Universe remains unknown to us. That proportion alone suffices to call into question everything we believe we know of the real.

Ordinary matter—what we see, touch, measure—represents only 5% of the Universe. It is composed of atoms formed of quarks and electrons, and it interacts through the forces described in the Standard Model.

The remaining 95%—dark matter and dark energy—still elude our understanding, and they call into question the very foundations of our physics.

We cannot see them, nor touch them, but we can already sketch their contours.

We must therefore revisit the principal elements—dark matter, dark energy, and a few marginal components—because without this minimal cartography, it is impossible to move forward.

- Dark matter: it interacts gravitationally, but not electromagnetically—that is, it produces neither light, nor shadow, nor detectable heat. It is therefore invisible and undetectable by classical means, although it influences the dynamics of galaxies.
- Dark energy: it is a hypothetical form of energy, present everywhere in space, which seems to cause the acceleration of the expansion of the Universe. We do not detect it directly, but we observe its effects on the cosmological large scale.
- Antimatter: it has the same characteristics as ordinary matter, but with inverted charges—that is, each particle has the opposite electric charge (for instance, if normal matter has "positive" particles, their antimatter equivalent is "negative," and vice versa). It does indeed exist, but it is extremely rare in our observable Universe.
- Photons, gluons, neutrinos: these particles play a fundamental role in interactions, but they are not always considered "matter" in the classical sense. Photons have no mass, gluons are confined within protons and neutrons, and neutrinos, almost massless and without charge, pass through matter without causing any notable effect.

Sylvana

You know… the deeper we go, the more I tell myself that perhaps this is the real paradox. We live in a Universe of which the essential part escapes us—and yet we have grown used to believing only what we see.

But here… you are taking me elsewhere. You make me feel that there is something else to grasp, even if our instruments remain mute. And that—that already changes the perspective.

Thierry

Yes. Moreover, this ordinary matter that we see is also called:

- Baryonic matter, because it is composed of particles called baryons—the best known being protons and neutrons.
- Visible matter, although technically only a small fraction is truly luminous; the rest is diffuse or gaseous.

But beyond this familiar matter begins the invisible. Dark matter does not emit, absorb, or reflect light. It interacts only minimally with ordinary matter, and only through gravity— the only force that reveals its existence through its gravitational effects, which, while almost imperceptible on a small scale, become visible on a large scale in galaxies and clusters of stars. Therefore it can be neither seen directly nor detected by classical instruments.

Sylvana

So in fact, we know it exists… but only because we deduced its presence?

Thierry

Yes. In fact, its presence was deduced indirectly, because a great number of astrophysical phenomena remain incomprehensible if we take into account only visible matter.

I will give you a few concrete examples:

- The rotation speed of galaxies: the stars rotate too fast relative to the visible mass.
- Gravitational lensing: light is bent by invisible masses.
- Structure formation: it requires a non-baryonic mass that was present very early.
- The cosmic microwave background (CMB): its fluctuations imply an invisible mass.

And here is what can be stated—cautiously—about dark matter:

- It is invisible to light: it emits neither radiation nor heat, and does not react to photons—which makes it undetectable by our telescopes.
- It is non-baryonic: it does not appear to be made of quarks or electrons, like ordinary matter.
- It has mass: although invisible, it exerts a measurable gravitational influence on galaxies.
- It is stable: it has existed since the beginnings of the Universe without spontaneously decaying.
- It remains hypothetical: several particles have been proposed as possible constituents of dark matter—for example, axions or neutralinos—but none has yet been detected.

. An Active Vacuum — The Intuition of a Structuring Field

Sylvana
This matter that links the stars, without ever showing itself, makes it feel as though there were a hidden thread holding the Universe together.

But you also spoke of another force—dark energy? Can you say more? What is the difference from dark matter?

Thierry
Dark energy is another phenomenon—even more perplexing—that has appeared in recent observations.

Dark matter acts as a cohesive force: it links galaxies, it "glues" structures together.

Where dark matter attracts and holds, dark energy seems to do the opposite—it "stretches" space itself, and speeds up its expansion.

Sylvana
So… we discovered that the Universe was stretching faster than expected—because of an invisible force?

Thierry
Yes. Dark energy exceeds everything that modern cosmology thought it could account for.

We do not know what it is, but we see what it does: it is invisible, spread throughout all space, and exerts a negative pressure that causes the acceleration of the expansion of the Universe.

We do not measure it directly—but its effects are undeniable.

Sylvana

And how did we understand that it was there?

Thierry

In 1998, two teams led by Saul Perlmutter and Brian Schmidt measured very distant supernovae. They were less luminous than expected, which meant that the Universe was accelerating—whereas it should have been slowing down. To explain this acceleration, it was necessary to introduce an unknown force.

Hence the name: dark energy.

Sylvana

So on one side, a hidden matter that holds things together… And on the other, an unknown force that drives everything apart.

It almost gives me the feeling that the Universe is being torn between two opposing mysteries.

Thierry

Yes. And this contrast is real, even in the numbers.

Today, dark energy by itself would represent about 68% of the content of the Universe, but its nature remains a mystery.

Here is what we can say with a minimum of certainty:

- It is present everywhere: it seems distributed homogeneously in space.
- It exerts a negative pressure: instead of attracting like gravity, it pushes, expanding space.
- It is undetectable directly: it produces neither known particles, nor light, nor measurable radiation.
- Its influence increases with time: it became dominant in the evolution of the Universe approximately five billion years ago.

Sylvana

And as with dark matter, there are several hypotheses?

Thierry

Yes. And none of these hypotheses commands consensus. They all attempt to explain a central mystery: the origin and the nature of dark energy. But instead of bringing a definitive solution, they mainly show that our Standard Model does not suffice to explain the accelerated expansion of the Universe.

Here are the main avenues being explored… It is not necessary to understand them in detail: I simply want to give you an overview, so that you see how scientists are attempting to explain dark energy.

1. The cosmological constant (Λ)

Originally proposed by Albert Einstein in 1917, then abandoned and reintroduced in 1998 to explain the acceleration of cosmic expansion, represents an energy density of the vacuum, assumed to be constant over time. But there is a colossal problem: according to quantum physics, this density ought to be up to 10^120 times greater than what is observed—an enormous divergence, still unexplained.

2. A dynamic scalar field (quintessence)

This concept was conceived in the late 1990s by physicists such as Robert Caldwell and Paul Steinhardt. This invisible field would evolve over time and would be responsible for cosmic acceleration. It is an elegant hypothesis, but there is no experimental evidence to date.

3. A hypothesis tied to the quantum vacuum

Some scientists think that dark energy could come from the energy of the vacuum—because this vacuum is not empty; it contains energy fluctuations. In theory, that ought to be enough to account for accelerated expansion. But as with the cosmological constant, the calculation collapses: the value obtained remains astronomical and totally incompatible with observations. This is a major contradiction, still without a solution.

4. A bold hypothesis: the multiverse or a hidden geometry

According to certain physicists working on quantum gravity or string theory, our Universe might be only one "sheet" among others. In this highly speculative framework, what we call dark energy would be only an indirect effect, arising from a vaster reality.

Sylvana
And you… what did you think about those missing 95%? During that period of reflection, between 1996 and 2000, did you already have an intuition? Because at the time, people were not yet talking about dark energy or the multiverse the way they do today.

But what did you truly sense in advance?

Thierry
At the time, I would not have spoken of "dark energy"—it was not yet in common usage—but those missing 95% posed a crucial question: what does

the vacuum really tell us?

And I already had the impression that something was missing in the way science conceived of it. Not because the cosmological models of the time were incorrect—on the contrary, they are the fruit of remarkable work, carried by minds of great rigor. But perhaps because they all relied on an implicit hypothesis: that of a vacuum without structure, or at least without its own dynamics.

This hypothesis—that a vacuum could be inert and without organization—underpinned most models. But I could see that it always led to the same dead ends: incomplete explanations, inconsistencies, paradoxes impossible to resolve. That is where I understood the problem might lie.

And that is how my presentiment grew stronger: the vacuum could not be a simple nothingness. It had to be something else—a structured, active field, a bearer of organization. And above all, it was not passive—it carried within itself a force of organization that I very early associated with Time: not a measure, but a self-organized, creative principle. I had also thought that the missing 95% of energy must have something to do with the real nature of the vacuum.

Sylvana
And those four hypotheses you told me about, what did you think of them, yourself?

Thierry
In reality, when I began my reflection, in 1996, those hypotheses had not yet been formulated. They came later, after a major discovery that upended cosmology: in 1998, two teams of astronomers—the Supernova Cosmology Project, led by Saul Perlmutter, and the High-Z Supernova Search Team, led by Brian Schmidt and Adam Riess—showed that the expansion of the Universe is accelerating. From that point on, it was necessary to reopen the question of the energy of the vacuum.

The first path was the reintroduction of the cosmological constant Λ. As I have already told you, it was proposed by Einstein in 1917, then abandoned, before returning to the forefront in 1998 to explain this acceleration.

On paper, it works: it introduces a vacuum energy density assumed to be constant in time. It makes it possible to explain certain observed phenomena, such as the acceleration of cosmic expansion. But it also raises an immense question: why is the vacuum energy predicted by quantum physics 10^120 times greater than what is measured? This paradox was highlighted as early as the 1970s by physicists such as Steven Weinberg, and is sometimes considered the "greatest failure of theoretical physics."

Then, at the end of the 1990s, another path appeared: the hypothesis of a scalar field, such as quintessence, which already goes a little further. It introduces a form of evolution into what had been thought fixed.

What I perceived, from the outset, was that the quantum vacuum is not a neutral or chaotic background. I then formulated the bold hypothesis that it was a field bearing order, which imposes nothing, but which makes possible certain forms of emergence—at the right moment, according to a rhythm, in resonance with the very history of the Universe.

This field, I gradually began to call the temporal field, because I thought it was indeed Time, in its creative dimension, as I had imagined it, that links information to form—potential to manifestation. I conceived of it as a fundamental, non-material ether, whose intrinsic properties needed to be characterized.

It was a hypothesis I needed to dig into.

Sylvana
So, help me clarify one point: the old ether physicists used to talk about, and this quantum vacuum you are talking about… how are they really different? Because at its core, in both cases, we are talking about an "invisible background" supposed to fill space, are we not?

Thierry
Excellent question. And first of all, we must put things back into their historical context. When physicists conceptualized mechanical ether in the nineteenth century, science did not know the quantum vacuum at all. That notion would only appear after 1925, with quantum mechanics.

At the time, people thought space was absolute nothingness. But since it was known that light is a wave, and that every wave needs a medium to propagate, physicists said to themselves: there must be some medium to carry it. So they imagined a mechanical, continuous, motionless fluid that would fill all space like an invisible ocean where light would make waves.

The problem is, that idea did not hold up. The Michelson–Morley experiment showed that this ether could not be detected, and Einstein's relativity, in 1905, proved that it was unnecessary to explain how light propagates. Result: ether was simply abandoned.

Sylvana
All right… and later, when science discovered the quantum vacuum, it could have said: "Look, there was something after all"… Why did it not?

Thierry
For a simple reason: relativity had just triumphed, and it had put an end to the idea of a universal "medium." After Einstein, any notion of a substrate became suspect, almost heretical. It had just been proven that light did not need a support, and one had to remain consistent with that paradigm.

So even when, twenty years later, quantum mechanics revealed that the vacuum was not empty—that it was filled with energy fluctuations—we chose to describe it as a mathematical state, not as a self-organized field. We wanted to avoid anything that might look like a step backward.

Sylvana
So today, science says: The vacuum is not empty. It is full of energy and agitation.

Thierry
Exactly.

Even in interstellar space, even in a chamber emptied of all matter, something remains: fields, energy fluctuations, virtual particles that appear and disappear without ceasing. One might think it is a nothingness; in reality, it is like an invisible ocean, calm when seen from afar, yet in perpetual effervescence on the microscopic scale.

But here is the crucial point: for modern physics, this agitation is purely random, without intention, without internal structure. The quantum vacuum is seen as statistical noise, a chaos obeying known laws, yet incapable of producing by itself a global order.

Sylvana
So, if I am hearing you correctly, where science sees a chaos without intention… you perceive something else there?

Thierry
Yes. Because for me, that vision is not sufficient.

If this vacuum is everywhere, if all reality rests upon it, how can we explain that it produces coherence? Why does the Universe not sink into total disorder? Why do structures, forms, and even life emerge?

This is where my question asserted itself:

What if this vacuum were not merely random noise? And what if it bore a logic of organization, linked to Time, like a creative principle?

I named it quantum ether. Not a return to the old mechanical ether, but a subtle field, bearer of energy, information, and rhythm, animated by a negentropic dynamic—a tendency toward organization, and not toward chaos.

Sylvana
So for you, this vacuum would be… a kind of dynamic matrix?

Thierry
Yes. And this medium, today, I could say that it resonates, in a certain way, with what contemporary science calls dark energy. Not because it would be its direct equivalent, but because it could offer a more profound and structuring reading of it.

Not a mysterious energy coming from elsewhere, but the still poorly understood expression of a fundamental field, present everywhere, yet invisible to our classical instruments.

What I am sharing here is a path—a profound intuition, born well before these debates became central in cosmology. And with hindsight, what strikes me is that this vision—which was already accompanying me in 1996—now comes into resonance with the questions that science is beginning to explore.

I did not build it from the models in place. It was born otherwise—from an inner, silent movement. And since then, it has ceaselessly guided my research.

And if this structuring field truly exists, then it might well be the key behind many other mysteries—including those hidden in the numbers we believe to be immutable.

That is why I will now focus a bit more on another question: that of the fundamental constants.

Because if these constants vary, or if they seem tuned in an overly precise way… then perhaps behind them there is a deeper structure that we have not yet managed to bring to light.

3. The Invisible Order — Constants and the Adjustment of the Real

. The Fundamental Constants — The Numbers that Sustain Everything

Sylvana
Before we begin the discussion, could you give me a very simple definition of

what the fundamental constants are?

Thierry
Of course. What we call fundamental constants are physical parameters that seem very simple at first glance, yet they underpin all the laws of nature.

They determine, for instance, the speed of light, the strength of gravity, and the intensity of the electromagnetic interaction.

Sylvana
And have they all been known for a long time?

Thierry
We have measured them very precisely for decades, yes.

Among the best known are:

- c, the speed of light;
- G, the gravitational constant;
- ħ, the reduced Planck constant;
- and α, the fine-structure constant, which measures the strength of electromagnetic interactions.

Sylvana
And do we know why they have those particular values?

Thierry
That is the mystery.

We use them in every equation; they are indispensable for the Universe to function as it does… And yet we do not know why they have precisely these values. They seem "set there," without a theoretical explanation.

Because if these constants had been even the tiniest bit different… nothing we know could have existed. No stable atoms, no matter, no stars, no life.

It is as though everything had been tuned with extreme precision, from the very beginning.

What is even more troubling is that, if one steps back to the cosmic scale, another group of constants appears… even more fundamental, upon which the coherence, structure, and the Universe's capacity to harbor life depend. These are the fine constants, a phenomenon referred to as fine-tuning.

. The Fine-Tuning — A Delicate Precision in the Service of Life

Sylvana
But what is the difference between these constants that are called "fine" and the others you mentioned, like gravity or the speed of light?

Thierry
What is particular about these constants is that they are dimensionless. In other words, they do not depend on the system of measurement one uses—no meters, seconds, or kilograms.

They are pure numbers that express fundamental ratios or proportions. And despite their apparent simplicity, they suffice to explain why the Universe is stable, structured… and capable of harboring life.

The principle of fine-tuning asserts that if even one of these constants had been slightly different, the Universe quite simply could not have existed.

Sylvana
Can you give me a concrete example?

Thierry
Yes. Let us take a striking example, that of the parameter Q, which measures the amplitude of density fluctuations in the primordial Universe—the small variations in the distribution of matter just after the Big Bang. The value of Q is about 1/100,000.

If this number had been a bit smaller, the Universe would have remained too homogeneous: no structure would have formed—no stars, no galaxies—only a diluted soup, matter too dispersed to give birth to anything whatsoever.

And if it had been a little larger, those fluctuations would have triggered a sudden gravitational collapse: the Universe would have turned chaotic, filled with black holes—without stability, without any possibility for life to emerge.

The value of Q therefore seems to be extraordinarily well calibrated to allow the emergence of a structured Cosmos capable of hosting life.

You see, Q is a pure number: it has no unit of measurement, unlike the fundamental constants I mentioned to you a moment ago. For instance, the speed of light is measured in meters per second (m/s). It is this difference that leads to Q being classified among the fine constants, whereas c is a fundamental constant.

Sylvana
All right, I understand better. And concerning the fine constants, is there an

overall logic?

Thierry

One can speak of an implicit cohesive force, yes. But what is striking is that a small set of dimensionless constants suffices to describe the structure, the expansion, and the coherence of the observable Universe.

The cosmologist Martin Rees highlighted them in his book *Just Six Numbers* (1999), identifying six of the most essential constants:

- N, the ratio of the electromagnetic force to gravity between two protons;
- ε (epsilon), the efficiency of nuclear fusion (about 0.007);
- Ω (omega), the density parameter for the Universe as a whole;
- Λ (Lambda), the cosmological constant tied to dark energy;
- Q, the amplitude of primordial fluctuations;
- D, the number of spatial dimensions—three.

Sylvana

Three dimensions… so if there had been two or four, everything would have been unstable?

Thierry

Exactly.

Even the number of dimensions seems to have been "chosen" in a kind of delicate balance.

And what makes all this fascinating is that these six constants, by themselves, suffice to explain why the Universe could come into being… and endure.

All of this raises a dizzying question: why these values? Is it pure chance? Or do these constants reveal something more profound… a coherence still invisible?

. THE CONSTANTS — THE ECHO OF AN UNDERLYING ORDER

Sylvana

So these constants are everywhere, they regulate everything… and yet, their origin eludes us?

I understand better now why it is so important to trace them back to their source.

Thierry

Yes. It is an essential question because behind these apparently fixed numbers lies the entire coherence of the Universe.

They are not mere "practical values" to make the equations run: they are what give reality its meaning and stability.

Without them, the laws collapse. An infinitesimal variation, and the stars could no longer shine, matter would no longer structure itself, life would become impossible.

So, before I explain to you why they are so crucial, I must begin by setting the stage. For the constants are not isolated entities: we find them in all the great frameworks of physics. They regulate the mechanics of forces, they set cosmic equilibria, they condition the structures of matter and the dynamics of energy. They are like an underlying code that connects everything.

As I proceed, I will begin with the central question: the one that dominates all the others and has mobilized physics for more than a century. It seems simple, almost self-evident:

How can the fundamental forces be unified?

Sylvana
Here I am confused. We were talking about the constants, and you tell me that to speak about them, we must first address this question of the unification of the forces. What is the connection?

Thierry
It is an excellent question, and I understand your confusion.

In fact, the fundamental forces are the great laws that organize the behavior of matter and energy.

But these forces—gravitational, electromagnetic, strong nuclear, and weak nuclear—do not operate alone: their intensity, their range, their balance… all of that depends on the physical constants that underlie them.

In other words, the constants are not secondary: they are what allow the forces to act as they do.

If the forces are the great instruments of the Universe, then the constants set their tuning.

This question is unavoidable and must be revisited first, because for a century it has been regarded as the "Holy Grail" of physics. Many think that if one succeeds in unifying these forces in a single theory, everything will be resolved.

In reality, even in that scenario, a crucial question would remain unresolved: why do these forces exist? And why are their values tuned with extreme precision?

In other words, attempting to unify the forces without understanding where these constants come from is like trying to tune an instrument without knowing the reference pitch. One can forge a beautiful theory, but it will remain incomplete so long as we do not know what fixes these values.

This is why I will begin by examining what science today calls "the fundamental forces": because they constitute the current framework and the most debated problem. Then I will take you further.

So we will discuss:

First, we will discuss these four forces to understand their roles and to see why, despite all efforts, their unification remains incomplete.

Then, we will examine the other major locks—those obstacles that block the overall view—and you will see that each has a direct link with the constants.

Finally, we will address the great open questions—those that touch the heart of the laws—which, here again, always lead back to the constants.

This is not a detour. It is the indispensable preparation for understanding why, in my view, the constants are not mere numbers, but the signature of a hidden order.

. Scientific State of Affairs — An Ordered Universe with Unknown Origins

Sylvana
All right… I see where you want to take me. So, you are starting with those forces everyone talks about?

Thierry
Yes. Because they are the pillars of the Universe—and yet, despite our scientific advances, we still have not understood how they unify.

. The Four Fundamental Forces — A Puzzle Still Incomplete

Thierry
Scientists have worked immensely on this subject, and they have built models of great elegance to describe each of the forces. Despite all their efforts, no single, coherent theory has yet managed to bring together the four fundamental forces of the Universe.

And that is the paradox: For standard science, unifying these forces is the goal. But even if we succeeded, it would not answer the essential question: what establishes this coherence?

Current physics recognizes four fundamental forces:

- Gravitational: it acts on a very large scale—planets, stars, galaxies. → Theory: general relativity.
- Electromagnetic: it acts on electric charges, light, atomic interactions. → Theory: quantum electrodynamics (QED).
- Strong nuclear: it holds quarks together inside atomic nuclei. → Theory: quantum chromodynamics (QCD).
- Weak nuclear: it is responsible for certain types of radioactivity and fundamental nuclear reactions. → Theory: electroweak theory.

Three of them—electromagnetism, the strong nuclear force, and the weak nuclear force—are today partially unified in the Standard Model, via quantum field theory. But gravity remains outside this framework. And it is precisely around it that current efforts are concentrated, because it remains the main lock on a complete unification.

These forces are modeled by fields and mathematical symmetries. But what these models do not tell us is why they exist, why they have exactly these values.

And you see: behind each of these forces there is a constant that defines its properties. Without them, no equation would work. These constants are like secret codes that give physics life.

This is where physical constants play a central role: they set the rules of the game, without our knowing where they come from. And perhaps, in seeking the origin of these constants, another path toward unification opens.

Sylvana
So, if I understand correctly… on the one hand we have these four forces, which are supposed to explain everything. And on the other, these constants, which are like the secret codes that give them life… but whose origin we are completely unaware of. It feels as though we have the entire map, but we are missing the compass.

But what really prevents us from connecting all this?

Thierry
That is precisely the question. Before returning to the constants, we must look squarely at the obstacles that block this great scientific ambition: unifying these forces.

Because you will see, each of these obstacles brings us back—directly or implicitly— to the question of the constants.

. The Obstacles to Unification — A Fracture Between Two Worlds

Thierry

Once again, according to classical physics, several major obstacles still prevent unification today. We have already mentioned some of them, but I will summarize them clearly here.

1. A profound incompatibility between general relativity and quantum mechanics:

- General relativity describes a fluid, continuous spacetime, shaped by gravity.
- Quantum mechanics relies on discontinuous, probabilistic phenomena, where time is fixed.

→ The two theories use incompatible mathematical languages, which makes their fusion difficult.

And this fracture not only blocks the physics of forces: it also prevents us from conceptualizing a common origin for gravitational and quantum constants.

2. The fundamental constants are introduced without any explanation of their origin. We will return to this in detail, but it must already be stated clearly here, because as we have already discussed, their unknown origin is a major obstacle to unification:

- The Standard Model does not allow us to predict, for instance, why the fine-structure constant is about 1/137, nor why the speed of light is what it is.

→ It merely incorporates them as fixed parameters, measured empirically.

→ No current theory explains their origin.

3. Gravity is far too weak:

- It is 10^38 times weaker than the electromagnetic force.

→ This makes it very difficult to account for at the particle scale, where the other forces dominate.

→ Models such as loop quantum gravity or string theory are elegant, but none has been tested experimentally.

Sylvana

Could you elaborate a bit on what you just said?

Thierry

Yes. To give you an idea, gravity is the weakest of all the fundamental forces. It is 10^38 times weaker than the electromagnetic force. At the particle scale, it is as though you were trying to hear a whisper in a stadium filled by a deafening concert: there is indeed a sound, but it is completely imperceptible, which explains why gravity is entirely absent from the equations of quantum physics.

Sylvana

So, basically, gravity is a giant at our scale, but a ghost in the subatomic world?

Thierry

Exactly.

It governs planets, stars, galaxies… but as soon as one goes down to the level of particles, it is the other forces—electromagnetic, strong nuclear, and weak—that take over.

And to bridge the gap between gravity and the quantum world, science has proposed two major avenues.

The first is loop quantum gravity. It begins from a radical idea: spacetime would not be smooth and continuous, but granular, composed of elementary units. A bit like an image on a television screen: from a distance, it appears perfectly uniform, but as you come closer, you discover that it is made of a mosaic of pixels. According to this theory, spacetime would be the same: a quantum structure formed of tiny loops, independent yet organized into a network that is invisible at our scale.

The second is string theory. It envisions that fundamental particles are not points, but infinitely small strings that vibrate. And just as the strings of an instrument produce different notes depending on how they vibrate, these quantum strings would generate different particles according to their vibrational modes.

It is a fascinating hypothesis, but it requires that the Universe have ten dimensions in the framework of superstrings, or eleven in M-theory. This does not mean there are ten or eleven different worlds, nor that each dimension is a string. One must imagine it as a space with far more possible directions than those we know. We perceive three (length, width, depth) and we often add time as a fourth dimension. But string theory posits six or seven more; they allow the strings all the freedom necessary to vibrate in multiple ways, and each vibrational mode would correspond to a different particle.

Sylvana
So, for now, these are brilliant constructions… but unverifiable?

Thierry
Exactly.

They are elegant mathematical architectures, but none has been validated experimentally. They remain beyond the reach of measurement.

To complete the picture, we have two major obstacles left to examine.

4. The forces converge only at inaccessible energies:

- Grand unification models predict that the forces could merge at extreme energies—close to those of the Big Bang.

→ But our current instruments do not allow us to reach those energy levels to verify it.

Since we cannot reproduce these conditions, we remain blind to the way these constants became fixed after the Big Bang.

5. A unified vision of the real is missing:

- Matter, field, vacuum, time, consciousness… are treated separately.

→ There is no common model that links these elements through a coherent structure or a single language (informational or geometric, for instance).

And this absence stems from the fact that we have never found the source that connects both the physical laws and their constants: the invisible architecture that gives them meaning.

You see? These obstacles seem different, but they all point to the same problem: we do not know where the constants that set these laws come from.

Sylvana
In fact… what you are saying is that these constants are not mere technicalities. They are the nerve center. Without them, nothing holds together.

And what strikes me is that we use them every day in equations as though they were self-evident… even though we have no idea where they come from. All our knowledge rests on numbers whose origin we do not know. It is not just a blurry zone: it is a blind spot at the heart of the laws.

And if we truly want to understand the Universe, we will have to go down to that level… where the logic that gives rise to them is hidden.

Thierry
Exactly.

These locks are not mere technicalities: they reveal zones of shadow at the very heart of the laws. And that is where the questions grow more radical… and where they all lead back, one way or another, to the constants.

- **The Great Unanswered Questions — The Enigma at the Heart of the Principles**

Thierry
And if we look more closely, we see very concrete, fundamental questions emerge that modern science has yet to resolve. These dead ends are indications that something fundamental is missing.

- Why do elementary particles have these precise masses?

We know that the mass of elementary particles depends on the coupling constants—those parameters that set the intensity of the fundamental interactions, also called fundamental forces, such as electromagnetism, the strong nuclear force, or the weak nuclear force. These interactions are the "basic forces" that organize matter in the Universe. Without these precise values, particles would not hold together and matter would not be stable.

The Standard Model merely measures them, without explaining the origin or the logic of this mass spectrum—that is, this very specific distribution of masses among the different particles. We do not know why an electron is much lighter than a muon, for instance.

- Why is gravity so extraordinarily weak?

The gravitational constant G is about 10^38 times weaker than the electromagnetic force, as we have seen. This disproportion among the four forces—called the hierarchy of forces—remains completely unexplained, and prevents the unification of gravitation with the other fundamental interactions.

- Why does the Universe appear so finely tuned?

Many constants—such as the elementary charge or the cosmological constant—have values of extraordinary precision: the slightest variation would render the Universe unstable, even unfit for life. Yet no theory explains why these constants have these values. What is called "fine-tuning" almost appears to be a miraculous adjustment.

- Why does the quantum vacuum not interact gravitationally?

According to quantum physics, the vacuum is not empty: it is crossed by fields that fluctuate continuously, even in the absence of particles. These fluctuations, present everywhere in space, give the vacuum a considerable energy.

In general relativity, every form of energy must influence gravity by curving spacetime.

If we combine these two well-established theoretical results, the energy of the vacuum ought to produce a gigantic gravitational effect: the Universe would be extremely curved, even unstable. To account for this effect, Einstein introduced the cosmological constant Λ.

Today, this constant is interpreted as "dark energy," responsible for the acceleration of the expansion of the Universe.

But here is the paradox: the value predicted by quantum physics for this energy is immense, whereas the observed value of Λ is minuscule. The gap between the two reaches 10^120—the greatest disagreement ever recorded between a theory and observation.

- What is the origin of time, and why does it seem oriented?

Physics disagrees about the nature of "real time." Some equations are reversible—they do not distinguish past from future—whereas our experience of time is clearly directional. This arrow of time remains one of the great mysteries of contemporary science.

Sylvana

It is true that these constants are everywhere, they govern everything—yet no one knows their origin.

So tell me, Thierry… what does it mean to you? Where is the true key hiding?

Thierry

For me, that key lies precisely in what we take as self-evident: Time.

If we want to move beyond the current dead ends, we must revisit it. Because if

Time is fundamental, it is not merely a backdrop—it is the dynamics that orders everything. And it may be there that we find the source of the constants… and of the forces they govern.

That is why I focus on them. Because by understanding their origin, we touch the very root of cosmic coherence.

Uniting the forces is not enough: the Universe itself is already unified. It is we who are still unaware of its secret. To discover it, we must go upstream of the laws… back to the source that rules everything.

In current models, each fundamental force is described by a field and a symmetry—but its strength, its range, and even its observable effect depend on very specific constants. Thanks to these constants, we can predict how particles interact, at what scale, and with what intensity.

In other words: the constants do not create the forces, but they regulate how they operate. Without them, the equations would fall silent.

That is why, by tackling the origin of the constants, I am also seeking—indirectly but fundamentally—to understand why the forces are what they are, and why they take precisely these forms in our reality.

And for that, we must go further. Explore an even deeper structure.

. The Living Vacuum — A Matrix that Could Structure the Universe

Sylvana
A deeper structure… what do you mean by that?

Thierry
To answer you, I must briefly revisit some older intuitions. For this notion of a "deeper structure" extends certain past intuitions: in the 19th century, physicists such as James Clerk Maxwell (1831–1879) or Hendrik Lorentz (1853–1928) posited the existence of an ether, a subtle medium supposed to carry light waves.

Later, Albert Einstein (1879–1955), after rejecting it in his special relativity, ended up reintroducing it in a new form, in 1920, in his Leiden lecture. He then spoke of a substance-less ether, a structuring, immaterial field, indispensable to gravitation and to the definition of spacetime.

What I propose falls within the continuity of those earlier intuitions, but redefines their foundations in the light of contemporary advances.

The point is not to rehabilitate the classical ether as a mechanical support for waves, but to recognize that a fundamental field—structured and active—

today fulfills the role the ether was meant to play. This field is not a simple passive background: it carries laws, information, coherence.

In other words, what was once called "ether" finds its contemporary expression for me in this structured quantum vacuum, which acts as a dynamic medium.

This field makes possible the emergence of an organized order, not through purely local interactions or external forces, but through an internal logic of fractal resonance, guided by an underlying order of Time.

It is this invisible yet coherent structure that allows the manifestation of the forms, the constants, and the laws that we observe.

From this perspective, the universe is no longer a mere physical object governed by fixed equations. It is a self-organized, self-regulated system that unfolds its coherence from within, out of an invisible yet active ground. It does not wait to be measured in order to exist—it continually organizes itself according to a dynamic of meaning.

Sylvana
So if I recapitulate… we have crossed three enigmas that science itself brings to light: Time, elusive, which differs between relativity and the quantum; those 95% of the Universe that are invisible, dark matter and dark energy, which escape our measurements; and finally the constants, those numbers that hold the entire Universe together without revealing their secret.

And over the course of this crossing, you have already highlighted something else: according to your model, the vacuum is not a nothingness, but a living field, bearer of order and coherence.

So… after these great cosmic perspectives, I imagine you want to lead us toward another threshold of mystery: that of the quantum world, where the undefined becomes real through wavefunction collapse, where the anthropic principle invites us to question the place we occupy in the Universe.

Thierry
Yes. In my view, these two enigmas no longer concern only the framework of the Cosmos: they touch the very act by which the real comes to be—and the consciousness that engages with it.

Chapter 5

Consciousness and the Universe

~

From the Indefinite to Fine-Tuning

Sylvana

We have traversed elusive Time, then the three shadowed zones of the Cosmos—dark matter, dark energy, and that exquisitely precise tuning of the constants… And now, you are taking us elsewhere.

Thierry

Yes, because here, we leave behind the merely structural description of the Cosmos in order to address two enigmas of a different nature:

- Wavefunction collapse, at the heart of quantum mechanics, which probes how a state of superposition of possibilities becomes a real, measurable event,
- And the anthropic principle, in cosmology, which asks why the fundamental constants of the Universe appear so finely adjusted as to permit the existence of life and consciousness.

These two distinct domains, however, share a decisive common point: consciousness no longer appears as a simple passive spectator of the Universe, but as a dimension potentially active or inescapable in the architecture of the real.

Sylvana
I suppose we begin with wavefunction collapse?

Thierry
Yes and no. Before entering into the details of the wave function, it seems to me necessary to first set a broader frame to understand how the phenomena of the quantum world are organized and to show where precisely this wave function is situated.

1. The Quantum Blur — When the Indefinite Becomes Real

Sylvana
All right… do we leave the stars behind to explore what happens all the way down, in the infinitely small?

Thierry
Yes, but let me gently correct you—making the connection is not always simple, and that is normal.

In reality, the quantum vacuum, that invisible web of fluctuating energy, is present at every scale of the Universe. It is not confined to microscopic particles: we find it at the heart of stars, in the intimate structure of planets, in matter itself, and even across the vast intergalactic expanses. In other words: it is everywhere, underlying all reality.

One could almost imagine having a universal microscope—an instrument impossible to build, yet fascinating to conceptualize. With it, no matter the scale at which we look, from the smallest to the largest, we would see the same fundamental activity: a ceaseless seething, the living fabric of the void in action.

Sylvana
It is fascinating.

I realize that I had always thought of quantum mechanics as something that concerns only the infinitely small. I was not aware that it also applies to the vacuum between the stars and to the vast stretches of the Cosmos.

Is this a common misconception?

Thierry
Yes, it is a very widespread confusion.

Many imagine that quantum mechanics belongs only to "another world": a strange domain, reserved for sophisticated laboratories or invisible particles.

In reality, it is exactly the opposite. Quantum principles are at the very foundation of matter, of the emergence of life, and of the organization of the Universe. They operate continuously, at every scale, from the heart of atoms to galaxies.

The quantum vacuum, far from being a mere "nothing," is, in my view, a seething ocean of energy and even of information, which permeates the entire Universe without exception.

To become aware of this is to understand that the microscopic and the macroscopic are not two compartmentalized realities, but the two faces of a single living continuum. Our senses and everyday experience perceive only a thin film of it, while what is essential remains hidden, active, in the background of the real.

And it is here that everything changes: for, in my view, this vision compels us to profoundly revise our conception of the real, to recognize that what we call the "quantum world" is not an exception, but the universal weave that makes all existence possible.

Are you familiar with the phenomena of quantum physics?

Sylvana
No… All I hear is that even the greatest scientists themselves seem out of their depth.

Thierry
Indeed.

Quantum physics confronts us with a face of the real that is radically different: a world where events are probabilistic.

. The Quantum Game — When the Rules Elude Us

Sylvana
So, if I understand correctly… in this quantum world, one can never know in advance what will actually happen?

Thierry
In a way, yes.

What makes the quantum world so disconcerting is that it does not behave like the world we perceive every day.

At our scale, everything appears stable and determined: a table stays where it is, a stone keeps its shape, a book does not disappear when we look away. We live in a world where objects seem fixed, continuous, subject to an obvious causality.

But in the quantum world, this logic breaks down. There, one cannot predict with certainty the exact behavior of a particle. One can only assign probabilities to its possible futures. It is not that the particle "hesitates": it is that before being measured, it exists in a superposed state, as though it simultaneously occupied several trajectories or several values.

In other words, the future is not fixed in advance. Each interaction opens up a range of possibilities, and science cannot predict with certainty which one will be realized—only calculate the probability that this or that event will occur.

This is what makes this domain so strange—and so fascinating.

Sylvana
It is intriguing… Are there other phenomena just as strange in the quantum world?

Thierry
Yes, of course.

I will present to you four phenomena that I—and many physicists as well—consider to be the pillars of our understanding of the quantum world.

Here are these phenomena, in the order in which they were discovered.

- The first example is quantization. Max Planck discovered it in 1900: energy is not continuous, but emitted or absorbed in discrete packets called quanta.

Imagine a staircase: you can go up step by step, but never stop between two steps. Nature, at the quantum scale, works in the same way.

Let us take electrons in an atom: they cannot orbit at just any distance from the nucleus. They occupy well-defined "shells." It is impossible for them to hover between two levels: they "jump" from one orbit to another, as though the Universe offered them only precise landings.

- Next comes the principle of superposition, formalized by Erwin Schrödinger (1887–1961) in 1926. I know that I already spoke to you about it earlier, but it is such a fundamental point that it deserves revisiting. A particle is not simply "here" or "there": it can be in several states at the same time, as long as no measurement is made.

It is as though it were exploring several paths at once, and only at the moment of observation is a single one selected.

Imagine a die being thrown: in our world, it always ends up showing one face. But in quantum mechanics, as long as you have not looked, it is as though all the faces coexisted. It is only the act of observing that fixes a particular result.

- Even more astonishing, there is the uncertainty principle, stated by Werner Heisenberg (1901-1976) in 1927.

It says that one can never know certain properties of a particle precisely at the same time, such as its position and its speed. The more you refine the measurement of one, the more the other becomes indeterminate. And this is not a problem of instruments: it is a fundamental limit of nature itself.

It is a bit as though you were trying to take a sharp photo of a car driving very fast. If you want to freeze its position, you must use a very short exposure time—but then you will not see its speed clearly. Conversely, if you want to capture the motion, you must leave the exposure open… and its position becomes blurred.

- Finally, there is entanglement, brought to light through the EPR paradox by Albert Einstein (1879–1955), Boris Podolsky (1896–1966), and Nathan Rosen (1909–1995) in 1935, then confirmed experimentally thanks to the work of John Bell (1928–1990) in the 1960s.

Two particles can be linked in so profound a way that, even when separated by light-years, whatever happens to one instantly affects the other. Einstein called this, with a touch of irony, "spooky action at a distance."

To picture it, think of two coins prepared together in a common state, then sent very far from each other. As long as they are not observed, they are neither heads nor tails, but in an indeterminate state where all possibilities coexist. But as soon as you look at the first coin and it lands on "heads," the other instantly becomes "tails"—even if it is at the other end of the galaxy. This is not a signal traveling between them, but the consequence of a correlation imprinted from their shared origin.

Sylvana

I see…

The laws of the world we perceive every day are so different from those of the quantum world.

But then, is there an interaction between these two worlds?

Thierry

Yes, and that is precisely what brings us to the central phenomenon: wavefunction collapse.

If we managed to understand this interaction fully, we might grasp the key to what creates our phenomenal reality.

The fundamental question is this:

What is it that makes an event become real?

In other words, how does a transition take place: from a state of potential, as it exists in the quantum vacuum—which we perceive as uncertain, chaotic, undefined—to a manifested, stable, and concrete state, the one we call "reality"?

These questions go far beyond the narrow scope of quantum mechanics. They touch the very foundation of reality. For it is not only a matter of knowing how a particle works, but indeed of understanding why there is a world rather than nothing.

. Wavefunction Collapse — What Science Does Not Yet Explain

Thierry

As I already mentioned when we addressed superposition, in the quantum world, prior to any measurement, a particle—whether an electron, a photon, or even an entire atom—has no defined state. It possesses neither a precise position, nor a stable trajectory, nor a fixed reality.

The particle exists as a superposition of states, mathematically described by the wave function. This wave function does not tell us what the particle is, but what it could be. It contains the set of possible states, each accompanied by a probability.

Historically, the idea of superposition was already implicit in Heisenberg's work in 1925, but it was in 1926 that Schrödinger gave the wave function its emblematic mathematical formulation.

But as soon as an act of observation, an interaction, or a measurement occurs, everything changes. The wave function "collapses": from the set of possibilities, a single one becomes real, and all the others vanish.

This is the mysterious phenomenon—wavefunction collapse—which remains an enigma at the heart of modern physics.

Sylvana

So the wave function is the state before observation, when there is no fixed reality but a field of possibilities? And the collapse is the moment when, among all those potentialities, a single one is realized?

Thierry

Exactly.

And it is this distinction that is essential to retain: the wave function describes the set of all possibilities, whereas the collapse marks the passage into the actual, the instant when only one of those possibilities becomes reality.

Sylvana

When I imagine this phenomenon, wavefunction collapse, I immediately see a science fiction scene: physicists behind armored glass, lasers crisscrossing, and strange machines lined up in a solemn silence.

Thierry

Yes, I understand your reaction: it is very common to imagine that the collapse of the wave function is confined to the most sophisticated laboratories.

In reality, that is not the case at all. Collapse is not a curiosity confined to complex experiments: it is a universal process, present everywhere, at every instant. Without it, no tangible reality could manifest.

In nature, it occurs constantly.

When a sunbeam strikes a leaf and a photon is absorbed during photosynthesis, there is collapse.

When a photon strikes your retina and triggers the perception of a color, there is collapse.

And even within your own body: every cell division, every replication of DNA, every signal transmitted by your neurons relies on billions of successive collapses.

In other words, we are quite literally woven from these processes. Without them, there would be neither chemistry, nor stable matter, nor conscious perception. Every second of our existence, the collapse of the wave function is the subtle yet fundamental mechanism that makes the world real.

Understand this well: the collapse of the wave function is not just a problem of quantum physics, it is the question of the very foundation of reality.

Sylvana

I understand better now why this phenomenon is so central.

I imagine billions of tiny collapses all around us, all the way into my own cells. As though reality were constantly scintillating.

But I imagine that, for now, science is still groping its way forward…

Thierry

Yes, science is searching for answers, but no explanation has yet won unanimous acceptance.

- The best-known interpretation is the Copenhagen interpretation, formulated by Niels Bohr (1885–1962) and Werner Heisenberg (1901–1976).

According to it, it is the act of observing that causes the collapse of the wave function. In other words, it is the interaction with the observer that "creates" the real.

Faced with this uncertainty, other approaches have emerged:

- The many-worlds interpretation of Hugh Everett (1957):

Here, the wave never collapses. All possible outcomes are realized, but in parallel branches of the Universe. Our consciousness simply follows one of these paths.

- The GRW model (Ghirardi–Rimini–Weber), in the 1980s:

It proposes a mechanism of objective and spontaneous collapse, independent of any conscious observation. Collapse would then be a natural statistical process.

Despite their elegance, none of these hypotheses has been confirmed experimentally. Even today, we do not know:

- Whether collapse is a real phenomenon,
- What would trigger it,
- Or whether it is only an illusion tied to our observational limitations.

Sylvana

So, to sum up: for science, wavefunction collapse remains a mystery. We know that it occurs, but we do not know how or why—nor even whether it is real. Some speak of a role for the observer, others imagine parallel worlds, or spontaneous collapses… but nothing has been settled.

. TWO LEVELS OF REALITY — A GATEWAY TO THE REAL

Sylvana

And you… how do you understand this collapse? What does it truly evoke for you?

Thierry

As I was telling you, none of the hypotheses put forward by science has reached a consensus.

For me, these views are not necessarily contradictory. They may simply be different readings of one and the same phenomenon, approached from distinct angles.

It is entirely conceivable that reality actualizes through an objective and spontaneous mechanism, yet underpinned by an inner coherence that still escapes us. And behind this coherence there could be a form of subtle intelligence, a profound logic acting in the folds of the real—the quantum vacuum—but still beyond the reach of our current science.

It is possible, in my view, that because we do not yet have access to this organizing intelligence, we describe the process in terms of probabilities. But that does not mean it is fundamentally random. Contrary to the probabilistic view of standard science, I see it unfolding deterministically according to two complementary dynamics:

- The intrinsic dynamic:

All states coexist in parallel, as pre-configured realities.

It is not, as in the Everett interpretation, a multiplicity of material Universes, but an unactualized fabric of information: a set of still invisible potentialities, ready to be expressed.

- The extrinsic dynamic:

This is the phenomenal world, the one we perceive.

Here, only one of these pre-configured realities becomes actual, becoming experience, matter, and event.

In other words, what we call "the world" is not the real in its entirety, but the unfolded version of a single potentiality chosen from an infinity of others.

For me, the collapse of the wave function can only be understood by situating it within a broader vision of the real in which the reality we perceive—this phenomenal world of objects, forms, matter, and interactions—is not the totality of the real but only the visible part of the iceberg, where the dominant view of science sees a boundary, and where I see, for my part, the entrance to an immersed world.

Sylvana

So if I understand correctly… the intrinsic part is the hidden part of the void, where a pre-configured logic and coherence already reside—the submerged part of the iceberg.

And the extrinsic part would be what manifests on the outside—what appears, what becomes actual in the phenomenal world that we perceive.

In other words, according to you, there are two levels of reality: one where all the pre-configured states remain, and the other where wavefunction collapse actualizes only one of them, becoming the reality that we perceive.

And in your model, how do you consider the role of the observer, which is taken into account both in the Copenhagen interpretation and in Everett's many-worlds interpretation?

Thierry

In my model, the role of the observer does not contradict what science says.

In physics, the "observer" refers to any measurement system that interacts with the quantum object: a device, a detector, or even a particle in the environment. In other words, to observe is to trigger an interaction that leaves a trace, that records information.

But for me, behind these setups and protocols, there is always a consciousness. Whether it is the scientist who conceptualizes and conducts the experiment, or the instruments themselves—mere extensions of his eyes and hands, shaped and used by him—all of this remains at the service of a consciousness. For, at its core, even science is only an expression of consciousness.

In other words, for me, there exists a subtle interaction between two planes:

- An order still invisible, which physics today translates in the form of probabilities,
- And a consciousness capable, under certain conditions, of tuning itself to that order.

In this framework, I do not see human consciousness as a force that would impose its will upon the real, but as a presence capable of entering into resonance with the quantum vacuum.

Sylvana

All right… so if I put things back in order a bit:

Among the major hypotheses advanced by science, there is the best-known interpretation: the one in which observation would play a decisive role—as though the consciousness of the observer brought forth a single reality from among all the others.

Others say that all possible outcomes are realized in parallel, in many worlds—there would thus be no collapse strictly speaking.

And then some propose a spontaneous, objective collapse, which would occur on its own, without any observation being necessary—but according to a statistical logic that we do not yet truly understand.

If I have followed you correctly, you do not reject the existing hypotheses, but you see them as partial—fragmentary readings of a more profound phenomenon.

You put forward the idea that upstream of any collapse, there would exist a form of intelligence, an invisible coherence, which silently organizes the emergence of the real.

In this perspective, the process is neither fundamentally statistical nor truly random—it is simply that we do not yet perceive the order that underlies it.

And consciousness, in certain cases, could enter into resonance with this dynamic… and influence the actualization of a state, not by controlling it, but by attuning itself to it.

Thierry

Yes, you have summarized it well.

It was at this precise moment in my research, while studying the collapse of the wave function, that my hypothesis of a structuring void fully took shape. For me, it was a decisive turning point.

And it is in this context that what is commonly called the "quantum vacuum," I decided no longer to designate it as a vacuum, but as a quantum ether: an original substrate, bearer of rhythm, information, and intelligence.

Sylvana

All right… Could you specify further the difference between what science calls the quantum vacuum and what you call quantum ether?

Thierry
In science, the quantum vacuum denotes the lowest-energy state of a quantum field: an apparently empty state, yet traversed by fleeting fluctuations that result in the appearance and disappearance of virtual particles.

The quantum ether, in my model, is an active and intelligent foundation, the invisible framework that gives coherence to the real, the weave of all possibilities. It is what makes their actualization in the perceptible world possible.

In other words, its role is threefold:

- Create phenomena by selecting certain potentialities;
- Maintain their coherence by orchestrating interactions;
- And dissolve forms by returning information to cosmic memory.

Thus, the quantum ether is the permanent agent of universal dynamics: it ceaselessly produces, organizes, and reabsorbs phenomenal reality, generating what we call history.

I use the term ether by analogy, but I give it an entirely renewed meaning—very different from its classical or mechanical usage, which was abandoned by modern physics. Here, it is no longer a hypothetical material medium, but a fundamental living field, an immaterial matrix that structures and connects everything.

At the heart of this ether lies a dimension of order and coherence that I call omniscient Intelligence: an immanent intelligence that guides manifestation, ensures continuity, and orients evolution.

Sylvana
Omniscient? Why that word?

Thierry
The term omniscient comes from the Latin omni- ("all") and sciens ("knowledge"), and literally designates the absolute knowledge of all things.

In my usage, it is not a religious or mystical concept, but a refined notion: a universal intelligence, non-localized, present everywhere at once. A diffuse intelligence that permeates every structure, orients every organization, and shapes the real well upstream of what we call manifested reality.

It manifests as a near-infinite potential, expressing an extreme precision, a primordial coherence, and a cosmic reach that far exceed our current frame of understanding.

Sylvana

I understand better now why you consider that the collapse of the wave function occurs in a deterministic way, given that you have integrated an intelligence so vast and so powerful.

Thierry

Yes, exactly: an actualization guided by coherence, and not by chance.

In my view, the quantum vacuum cannot be reduced to a simple passive background: it is the creative agent that supports and engenders phenomenal reality, notably through the actualization of the wave function.

And it is precisely this link between the intrinsic and the extrinsic, this passage from potentiality to manifestation, that constitutes for me the key to understanding.

Sylvana

And, in your view, what is the operator of this passage?

Thierry

This is a very pertinent and central question for me.

For a passage to be possible between these two sides of reality, more than a field of potentialities is required.

It is precisely there, in my view, that Time intervenes.

Not the linear time of clocks or calendars, but a fundamental Time, rooted in the quantum ether itself.

Sylvana

Could you clarify the difference between the role of Time and that of the omniscient Intelligence, and how they interact?

Thierry

In my model, the omniscient Intelligence serves as the organizing principle of the Universe. It resides at the heart of the quantum ether which, as I told you, is not a simple energetic substratum but a living field, bearer of information and intelligence.

Its action unfolds through Time, which is its dynamic expression. Time acts as the language of this intelligence: it translates the implicate order into explicit organization.

Concretely, it is Time that, at each instant, selects and actualizes a single state from the set of possibilities, thereby giving reality its concrete form and the Universe its unfolding history.

This is why I have also named this quantum ether a temporal field: because Time is the operative vector of the omniscient Intelligence, the structuring spirit. Together, they convey the real from potentiality to manifestation.

Sylvana

So, if I understand correctly, there exists a fundamental Time, different from the ordinary time of our daily life. This Time would be the dynamic expression of the quantum ether: it executes the coherence borne by this omniscient Intelligence, bringing forth a single possibility in our reality.

And if I am following you… that would mean that, in this ether, nothing happens in isolation. Each event would be part of a network of information in constant interaction, which adjusts itself until what is potential emerges.

Thierry

Yes, exactly.

To sum up, in this perspective, the collapse of the wave function is neither spontaneous nor truly random. What manifests arises from an arrangement of information already inscribed in the field, an alignment both discrete and of absolute rigor.

Each actualization corresponds to the profound accord among three complementary principles:

- The omniscient Intelligence, which structures and maintains the coherence of the information field;
- The fundamental Time, which translates this coherence into dynamic manifestation by bringing forth a particular state from among all possibilities;
- And a consciousness such as human consciousness, capable, under certain conditions, of tuning itself to this quantum ether.

In other words, the phenomenal reality that we observe is not a mere effect of chance, but the fruit of a co-creation: an orchestration between the immanent omniscient Intelligence of the field, the structuring function of the fundamental Time, and, under certain conditions, the resonance of a consciousness.

Sylvana

You know… everything you have just explained about Time, this omniscient Intelligence… and now, the idea that our own consciousness could play a role… it is a completely different way of seeing reality.

Thierry

Yes. What we call "the collapse of the wave function" is indeed a reality—but only from our limited point of view.

In other words, it is a phenomenological effect: we observe the sudden actualization of a state, and we interpret it as a rupture.

But if one adopts a broader perspective—what I call Absolute Relativity, where everything depends on the frame of reference and the scale of observation—then this apparent rupture changes in nature.

What we name a "collapse" is in reality only the local manifestation of a deeper process, which we translate through our own perceptual limits. Where we believe we see a random selection or a discontinuity, there exists, at the origin, only an underlying organization, a coherence that precedes any actualization.

And if one places oneself from the point of view of the void—of the quantum field itself—then there is not really a collapse.

There is only a smooth and continuous unfolding, where potentials naturally orient themselves toward their path of expression, without rupture and without chance, but by simple resonance with the implicate order.

Sylvana

If I understand correctly, what you call Absolute Relativity, your model, is the idea that reality depends entirely on the frame from which one observes it.

At our scale, we see a sudden collapse, as a rupture, where one reality is actualized. But if one places oneself from the point of view of the void—that fundamental level you speak of—there is no rupture at all, only a continuous process where possibilities organize themselves and become actualized?

Thierry

Yes, that is exactly it.

For me, the very term "collapse" is reductive: it only expresses the way we locally perceive a vaster process.

From the point of view of the void, there is neither rupture nor chance, but a fluid passage, inscribed within a global coherence.

In other words, what quantum mechanics calls "collapse of the wave function" is, in my model, only the natural moment when a potentiality of the global field becomes actualized as local reality. Not a brutal cut, but an ordered transition, a mechanism of actualization inscribed within the dynamics of the temporal field.

Sylvana
I see… So, in your view, what happens "behind" our visible reality, in the unmanifested part, is an organization of staggering complexity, and yet all of that remains totally invisible to our eyes…

Do you have an analogy that would help me grasp more clearly the entirety of your vision of the collapse of the wave function?

Thierry
Yes, I picture it as a 3D printer.

What we see is the result: line after line, layer after layer—an object takes shape before our eyes.

But the print head—which I associate here with wavefunction collapse—chooses nothing by itself. It follows a plan.

Behind it, there is an entire system: a code, a memory, a language, a logic.

And what we call the phenomenal world is this sequence of points printed in our spacetime, line by line.

But the true coherence does not come from these points themselves.

It comes from the story they sketch together—a story woven in the invisible, guided by a deeper level of organization.

Intelligence is not found in the object, but in what makes it possible: the intention, the architecture, the program.

And when one traces this chain back, one discovers that this plan is not fixed. It is dynamic, living, evolving.

Sylvana
But… wait. If I follow you, this whole "printing" of the real, with its choices, its trajectories, its actualizations… that means there is an invisible machinery running permanently, does it not? And this machinery must be consuming colossal energy to manage all of that, all the time, at every instant…?

Thierry
Yes, you are right.

Just imagining the quantity of information processed at every moment, the absolute precision, the seamless continuity of this process… it already lets one glimpse a form of intelligence and power that is almost inconceivable.

And when I meditated on this complexity, the same question imposed itself with clarity: where does the energy come from that can sustain a quasi-infinity of pre-configured realities—all those potential present futures, all those possible states—that constantly flow and transform according to the process of actualization?

Yes, from our point of view, it seems to imply a dizzying expenditure of energy. But in my view, the energy is proportioned from the origin for the entirety of the Universe's possible processes.

In other words, in my model, the total energy of the Universe is not produced over time: it is defined from T0—that Universal Time which I place at the moment of the Big Bang. In my view, this quantity remains the same throughout the life of the Universe, to make it appear, to sustain it and, ultimately, to reabsorb it. This energy is contained at all times and redistributed according to the universal needs for structuring and actualization.

This echoes Lavoisier's intuition: "nothing is created, nothing is lost, everything is transformed." Here, this law applies not only to visible matter, but to the entirety of the Universe's energetic and informational field.

. Scenarios in Waiting — Toward an Already Pre-Configured Reality

Sylvana
In short, according to you, everything therefore goes back to T0: that moment of the Big Bang when the total energy of the Universe was fixed and has been conserved since.

And then… in your model, what becomes of all the other pre-configured realities that are not actualized?

Thierry
I must first clarify an essential point: I am not speaking here of material parallel Universes.

In my model, they are, rather, pre-configured realities in the intrinsic part of the Universe: information structures already organized, ready to inscribe themselves within each of the possible scenarios. Their number is incommensurable, for it reflects the near-infinity of trajectories that the Universe can explore.

In other words, they are all the possible configurations, gathered in an immense repertoire of scenarios. Like paths already traced, simply waiting to be traveled.

Each pre-configured reality operates as a prepared potential choice, adjusted to a precise context:

- Some concern the creation, maintenance, and dissolution of systems (particles, stars, galaxies);
- Others, more subtle, concern the relationship between the Universe and the consciousnesses that inhabit it, notably in the exercise of free will.

Sylvana

Could you tell me more about pre-configured realities?

Thierry

Yes, in my model, this is a very important concept, so I will explain it to you in detail so that you understand it well:

Each pre-configured reality is a universal quantum function, rooted in the quantum ether, which contains within itself a potential state that is energetically autonomous and awaiting actualization.

In other words, they are potential, autonomous, and coherent units, capable of interrelating or actualizing to contribute to the emergence of a particular reality.

These functions are characterized by four inseparable components:

- Time,
- Information,
- Energy,
- And memory.

And they are defined by several essential properties:

Functional and non-material nature: An information structure rather than a physical world

- A pre-configured reality is not a material parallel Universe.
- It is a structured quantum function, that is to say a weave of information, energy, and memory, already organized but not yet deployed in the phenomenal world.

In other words, it is not a complete physical world that exists elsewhere, but an immaterial, informational structure that exists in the intrinsic part—the quantum ether.

Energetic autonomy: A potentiality already bearing its own power of manifestation

- Each pre-configured reality possesses its own energy of manifestation, sufficient to be actualized if temporal and informational conditions allow it.
- This autonomy distinguishes it from a mere abstract scenario: it is a viable potential entity, rooted in the dynamics of the temporal field.

Like a seed that contains the vital energy of its growth, each pre-configured reality contains the power necessary for its actualization.

Informational optimization: Coherent structures, already ordered according to their internal laws

- Pre-configured realities are not random drafts, but coherent structures already optimized according to internal laws. This means that they obey a logic that ensures their balance and viability, rather than being mere haphazard combinations.
- Together, they embody possible scenarios, whether it concerns the evolution of a system, the emergence of a phenomenon, or the trajectory of a consciousness. In other words, they are already pre-configured to support every domain of reality, from the Cosmos to life and consciousness.

Universal quantum status: A distributed function, present at every point in the Universe

- Each pre-configured reality can be described as a global superposition: set within a bundle of coherent possibilities, available everywhere in the quantum ether—that is, a state in which several scenarios exist simultaneously, linked together as a structured whole, and present at every point in the Universe.
- It does not exist as a local copy, but as a universal function, distributed and accessible at every point in the Universe. One could say these possibilities are like a digital library stored in the "cloud": everywhere at once, without being limited to a single place.

Link with actualization: The threshold where the possible becomes phenomenal reality

- When a single one of these pre-configured realities is selected (by resonance with the temporal field, consciousness, or energetic conditions), it ceases to be a mere potential and becomes an actualized phenomenal reality.
- This actualization leaves a trace in the temporal field: a memory that is added to the universal history.

Role in cosmic dynamics: The invisible matrix of the coherence of the real

- Pre-configured realities ensure continuity between potential and actualization, that is, they serve as a bridge: they maintain a constant link between what exists in reserve and what manifests in the real.
- They are the "informational blueprints" from which the Universe unfolds its structures, its events, and its consciousnesses. In other words, they serve as foundational models that guide the formation of everything that exists.
- They guarantee the global coherence of the real: nothing manifests that is not already contained, in a pre-configured form, in the quantum ether.

In other words, every event, every phenomenon draws its origin from this repertoire of pre-configured possibilities, where it finds its first form before manifesting in the real.

In summary, a pre-configured reality is a quantum function in the quantum ether, which expresses itself simultaneously through its informational content and its energetic structure. It contains a scenario of existence ready to be actualized. Its passage into phenomenal reality constitutes the collapse of one possibility among the set of possibilities.

Sylvana
All right, I understand better how you define these pre-configured realities… But I am still intrigued: why would it be necessary for such a repertoire to exist? What advantage would the Universe have in possessing such a reserve of possibilities?

Thierry
Your question is important, and I will answer it clearly. In my model, several reasons make the existence of these pre-configured realities necessary—and you will see that each resonates with what science already acknowledges, even

though I broaden it.

- Principle of causal continuity: The link between what precedes and what comes to pass

Physics rests on the idea that no phenomenon arises arbitrarily, without a link to a prior state. Even at the quantum scale, transitions from one state to another—such as an electron moving from one orbit to another—are not completely unpredictable: they obey continuous laws of probability.

In my model, this principle of continuity is preserved. I posit that the link between the potential and the real is borne by a structure already present in the field: the pre-configured realities. They represent organized, viable states, available for actualization. They allow an ordered passage from the virtual to the phenomenal: not an emergence arising from nothing, but a transformation inscribed in a continuous logic.

Even after their actualization, they leave a trace in the form of memory. And even before emerging, they are already interacting with their environment—through resonance and organizational compatibility—thus guaranteeing the coherence of what manifests.

- Principle of systemic compatibility: The harmonious nesting of scenarios

In complex systems, the elements must not only be coherent in themselves, but also able to interlink without rupture. This is what unification approaches in physics seek: a framework in which all forces and interactions coexist without contradiction.

In the same way, in my model, each actualizable scenario is already pre-configured to interlock with others without dissonance. The pre-configured realities are not independent blocks, but structurally compatible configurations.

Thus, the real is constructed as a nested whole, where each manifestation naturally finds its place within a vaster architecture.

- Principle of universality of laws: The same rules everywhere

In physics, certain laws are universal: gravitation acts on a falling apple just as it does on an entire galaxy. Increasingly, it is recognized that principles such as self-organization or emergence seem to hold at every scale.

In my model, this universality concerns not only matter or energy, but also information, memory, and consciousness. In other words, the same organizational principles traverse every plane of the real, from the quantum to the biological, from the cosmic to the mental.

It is the role of pre-configured realities to make this possible: they ensure this unity of language and processing across all scales.

- Principle of least action: The efficiency of the system

In physics, the principle of least action states that nature always follows the most energy-efficient trajectory.

I apply this principle at the level of the repertoire of possibilities. The Universe, in my model, does not produce an infinity of new calculations at every instant: it relies on structures already prepared, already viable, which makes its unfolding more direct and more economical.

The pre-configured realities are these optimized, complete forms, containing within themselves the minimal conditions for manifesting. The Universe therefore does not have to "invent" each scenario from nothing: it selects, through resonance, what is already coherent.

- Principle of evolvability: The active evolution of scenarios

In biology as in complex systems, nothing remains fixed: everything evolves, adapts, learns.

I extend this principle to the level of the possibilities themselves. The repertoire is not a mere static library: it possesses its own dynamics. The pre-configured realities interact with the memory of the temporal field; they refine themselves and reorganize over time, integrating the traces of what has been experienced and chosen.

Thus, the field of possibilities evolves as well: it progressively favors the most coherent and organized lines of actualization.

- In summary

Science already recognizes several principles: continuity, compatibility, universality, least action, evolution.

My model extends them by positing that they take root in a pre-configured informational repertoire.

This repertoire ensures continuity between potential and real, the harmonization of scenarios, the universality of laws, energy efficiency, and the evolvability of possibilities.

It guarantees the global coherence of the real, its stability, and its capacity to grow more complex without ever collapsing.

Sylvana

I do see the logic of your model… I picture it as an infinite library: each pre-configured reality would be a sentence already written, ready to become a chapter, then a book.

And this idea of "pre-configuration"—do other scientists approach it in some way, or is it specific to your model?

Thierry

When I formulated my model, between 1996 and 2000, two major references were already present in the intellectual landscape.

- David Bohm (1917-1992) and the implicate order:

Bohm proposed the idea of an "implicate order": an invisible ground, a profound matrix, where all the forms of the world would already be contained in a latent state. For him, the visible reality—the explicate order—is only the unfolding of this hidden ground. It was a major intuition that opened the way to the notion of a reality rooted in a prior structure, more fundamental than what we perceive.

But in my model, this ground is not homogeneous, fluid, and continuous as in Bohm's model; instead, it is structured into distinct units, which I call pre-configured realities. In other words, it is not an undifferentiated sea from which everything would emerge indistinctly, but an already organized fabric, composed of viable scenarios endowed with their own internal coherence, ready to be actualized in the phenomenal field.

- Hugh Everett (1930-1982) and the many-worlds:

Everett proposed the "many-worlds" interpretation. According to him, at every quantum bifurcation—for instance, when a particle can follow two different trajectories—the Universe splits into two distinct material copies, each actualizing one of the possibilities. This model indeed captures the idea of a multiplicity of scenarios, but at the price of a potentially infinite proliferation of parallel material Universes.

In my approach, I retain the idea of the coexistence of the possibles, but I do not conceive of them as physical Universes that would multiply without end. These possibles exist as informational structures, which I call pre-configured realities. They constitute already organized, viable scenarios, rooted in the temporal field, waiting to be actualized in the phenomenal world. In other words, instead of generating an infinity of material worlds, the Universe draws from an ordered repertoire of potential configurations.

Other approaches emerged a little later, in the 2000s. They also pointed toward preexisting possibles, but in a different manner.

- Top-down cosmology by Stephen Hawking (1942–2018) and Thomas Hertog (born 1975), developed in the 2000s:

Hawking and Hertog proposed a so-called "top-down" approach, in which the multiple histories of the Universe coexist simultaneously. In this framework, even the past is not fixed once and for all: it depends in part on present observation. In a certain sense, this view converges with my idea of a repertoire of already available scenarios held in reserve.

But in their model, these histories remain described as mathematical probabilities: they have no proper status so long as they are not observed.

In my approach, the difference is essential: the scenarios are not mere probabilistic calculations. They already exist as viable informational structures, rooted in the temporal field, and ready to be actualized in the phenomenal world.

- The Mathematical Universe Hypothesis of Max Tegmark (born 1967), developed between the late 1990s and 2007:

Tegmark advances the idea that our physical reality is nothing other than a mathematical structure. But he goes further: he asserts that all mathematical structures really exist, whether or not they are observed in our Universe. For him, the existence of an equation or a form is already sufficient to confer upon it the status of an autonomous reality, in the "mathematical multiverse." It is this radical approach that leads him to consider our world as one among an infinity of possible mathematical worlds.

The difference with my model is decisive: in my approach, these structures are not mere mathematical abstractions existing in a Platonic space. They are rooted in an energetic-informational quantum field, which I call the temporal field. It is this field that enables these structures to interact with consciousness and time, and to become phenomenologically real. In other words, they do not merely exist in theory: they are pre-configured as concrete potentialities, ready to be actualized within the living fabric of the Cosmos.

- In summary

Bohm, Everett, Hawking–Hertog, and Tegmark all share a common intuition: reality does not arise from nothing (ex nihilo), but relies upon a repertoire of possibles already present.

What distinguishes me is the way I conceive of these possibles:

- Neither as material parallel worlds (Everett),
- Nor as a continuous, undifferentiated background (Bohm),
- Nor as probabilistic histories dependent on the observer (Hawking–Hertog),
- Nor as mere equations suspended in the Platonic space of mathematics (Tegmark).

No, I define them as pre-configured realities: viable, autonomous informational structures, already organized, and directly inscribed in what I call the quantum ether, or temporal field.

These pre-configured realities I have named chronosystems. I chose this term because they take root in Time itself and are articulated from elementary temporal units that I have called chronons—which we will explore in greater depth later on.

To formalize this idea, I have developed a mathematical framework based on what I call functional Qbits. Unlike the qubits of quantum computing, limited to encoding 0 and 1 in superposition, my functional Qbits describe complete pre-configured states. They simultaneously associate information, energy, memory, and time, forming genuine operators capable of representing the living logic of scenarios awaiting actualization.

. Creating with the Universe — What if the Real Were Waiting for Us to Listen?

Sylvana

You have just sketched a very dense panorama, invoking other approaches, your own concepts, all the way to the functional Qbits you propose… Let me try, in my own way, to summarize what I have understood of your model, to see whether I am grasping it correctly.

If I understand correctly, in your model you contrast the probabilistic view of science with a deterministic reading, grounded in two complementary dynamics.

The first, intrinsic, is what you call the quantum ether or temporal field: a living field, animated by an omniscient Intelligence and by fundamental Time, from which those pre-configured realities you name chronosystems emerge. The second, extrinsic, is our phenomenal world: that of matter, of light, of life.

And between the two lies what science calls wavefunction collapse. For us, it appears sudden, like a cut. But in your reading, from the perspective of the quantum ether there is no rupture: only the continuous unfolding of an order already present.

In short, according to your model, what we experience as reality is born from an orchestration among three principles: omniscient Intelligence, which maintains coherence; Time, which actualizes a possibility at each instant; and, at times, consciousness, which can enter into resonance with them.

All of this necessarily opens up an immense field for reflection. So, if you agree, I would like us to venture there now, to explore what your model changes in our way of understanding the world and our place within it. So, tell me: throughout your research, what have you discovered about this role of consciousness? Can it truly enter into resonance with the profound order of the real, to the point of opening a form of dialogue with it?

Thierry

For me, the further I advanced in my research, the more I saw that the quantum ether itself seemed to be waiting for us to reach out to it—not to control it, but to enter into resonance with it.

And this idea, of a possible dialogue between consciousness and the structure of the real, is not foreign to certain physicists.

Eugene Wigner (1902-1995), for instance, had pondered a troubling mystery: why does mathematics, born of our mind, describe the laws of nature so well?

And Roger Penrose (born in 1931) also speaks of a deeper order, already present in the very structure of the real.

Perhaps it is this: the Universe does not wait for our gaze… but it is structured in such a way that a conscious gaze can recognize itself in it.

Sylvana

In short… you mean that, according to your model, the Universe would be fashioned so that a consciousness could read meaning in it? As though the laws that organize it were not arbitrary, but constructed to be recognizable, almost familiar to a mind capable of understanding them? As though this order had been designed to enter into dialogue with us…

But if everything seems designed so that we can recognize ourselves in it, how can we not ask: what, then, is the worth of all these ideas that are circulating—the simulation, the programs, the virtual worlds? For these ideas all lead to the same question: do we truly act freely? Or… are we simply following a script fixed elsewhere, without even realizing it?

Thierry

It is an tempting idea for some—but in my view, it rests on a reductive vision of consciousness.

My hypothesis is quite different: we are not mere executors. We are living, active consciousness, capable of interacting with this matrix. And that aligns with what I already expressed to you a moment ago: each consciousness can co-create with this field of the void.

And I would add that this co-creation depends on its level of vibrational resonance with this field: the higher the vibrational resonance of consciousness, the more it can participate actively in this field.

Sylvana

So that means a consciousness is never neutral: depending on its degree of resonance, it participates more or less in the actualization of the real?

Thierry

Yes. And that is precisely where everything becomes fascinating.

For in this model, one must understand that to participate is to enter into a loop where each effect itself becomes a new cause. In other words, each time a consciousness enters into resonance with the field, it does not merely receive information: it alters its local structure, even subtly—this is a hypothesis that I formulate, in light of my observations, but which remains to be explored from the scientific point of view.

And this modification, in turn, redefines the conditions of what can emerge next.

Sylvana

You are saying this is not a simple exchange? That each time we enter into resonance with this field, we leave our imprint on it? So… nothing is fixed? Each interaction changes things a little for what follows?

Thierry
Exactly.

For me, the true cause of every effect is not solely external. It arises from two dynamics: the internal order of the system and the memory of choices already manifested.

It is these two forces, together, that make possible the emergence of a localized effect.

In other words, the system is not fixed. It operates in constant feedback. Each real event inscribes itself in the system's memory—a memory that is not passive, but active: it influences potential future presents, according to the echo it leaves in the field.

This echo can generate a more or less ample resonance, depending on the nature of the event.

The past does not disappear: it informs.

The present moment retains the trace of what has been, and this trace participates in what can come to pass.

Feedback is therefore essential, inscribed at the very heart of the Whole's dynamic.

Sylvana
And with each actualization, something essential occurs: the realized scenario leaves a trace in the field, an active memory, which influences what may emerge next. Everything operates in feedback, in continuous adjustment. So… everything is in motion, everything is connected, nothing is isolated.

And I understand better how we, consciousnesses, do not only receive: we also act upon this field, we modify its local structure, even subtly. In other words… we take part in this invisible dialogue.

So naturally I wonder: if everything is resonance, do only our visible actions matter? Do our intentions, our impulses, our hesitations… even our thoughts… also leave an imprint?

As though, in this model you describe, at each instant we added a note to a vast musical score, and the Universe remembered it in order to compose what comes next.

Thierry

Yes… What you have just said is very true.

Each vibration leaves a trace. Each movement of consciousness resonates in the weave.

And according to my model, as I see it, this resonance is integrated into an overall logic: each trace leaves an active echo in the field. And this echo influences potential future presents.

Current science does not say this. For it, consciousness is only a phenomenon emerging from the brain, without causal power over the structure of the void. Physics describes wavefunction collapse as a purely statistical process, without intention, without conscious influence. Admittedly, certain marginal theories—from Wigner to Penrose—have proposed an active role for consciousness in wave function collapse, but they do not represent the scientific consensus.

But my model, as I see it, follows another logic. I consider that information is the fundamental principle of the Universe, and that consciousness is not a mere by-product: it is an active expression of this information; when I speak of information, I mean the very weave of the real—the fundamental code that underlies, connects, and organizes all that exists.

From this perspective, as I see it, if information structures reality, then every vibration—even a subtle one, even arising from an act of consciousness—leaves an imprint in the weave. Nothing disappears: this imprint continues to act and steers the potential future presents. In other words, we participate, even modestly, in the global adjustment of reality.

And this is where the paradox turns inside out. What current science perceives as chaotic—this agitation of the void, these fluctuations without a cause—is in reality, for me, the most stable weave, the most faithful to the fundamental order. And conversely, our tangible reality, the one we believe solid, ordered, is the most unstable part, the most vulnerable to imbalance.

This is why the quantum vacuum, as I see it, plays a key role: it balances, it adjusts, it regulates, it maintains in creative tension all these opposing forces.

This confirms my idea: the quantum vacuum is not a passive backdrop. According to my model, as I see it, it is a living system—an active principle of order that creates, that sustains, and that at every moment recalls the coherence of the Whole, orchestrated by the omniscient Intelligence.

And I believe a fundamental point must be emphasized.

In the perspective I propose, as I see it:

The field does not judge. It does not punish. It does not reject.

It receives, it records—simply, faithfully.

Sylvana
You know… there are moments when I feel as though you are talking to me about physics, and then, without warning, it turns into a meditation on life.

And here, listening to you explain your model, I had the feeling that, in this scenario, I would not simply be here by chance. Even when I feel lost, a bit all over the place, I could tell myself that I still belong to something.

I might be chaotic on the surface… but if I place myself within the framework you describe, it would mean that there is, somewhere, a coherence that made me possible.

And just imagining that… I wanted one thing only: to learn to listen. Not with the head, but with my whole being. To listen to this omniscient Intelligence you speak of… which seems to pass through everything without ever imposing itself.

You see it as a form of intelligence… and I, as you described it, imagined it filled with wisdom—a wisdom both ancestral and cosmic.

And just thinking about it moved me. It made me want to take a pause and see whether I could feel it. It made me want to lie down in nature, doing nothing. Far from everything. And at the same time… completely within the Whole. As though, by simply attuning myself to it, I could draw a little closer to that coherence you describe.

. When the Visible Meets the Invisible — A New Art of Creating

Thierry
You put very accurate words to something I have often felt—and that, I believe, many feel without always knowing how to express it: this need to feel connected, to be part of a Whole, to touch something vaster than oneself.

And you know, what has always struck me is that this feeling is not merely poetic. When I began to explore the quantum vacuum, I discovered that at the very heart of physics there is a structure that connects everything.

What we call "vacuum," science has long considered to be nothingness. Yet, from my point of view, it carries information, coherence, a silent dynamic that supports everything that exists.

In other words, the premonition of belonging that you describe… it finds a resonance deep within the most intimate weave of the Universe.

Mainstream science, even today, does not go so far as to speak of information or intention. But it has gradually discovered that the void is not an absence: it possesses measurable physical properties.

We are beginning to find its trace even within the most subtle physical phenomena. Subtle clues… but very real.

For more than a century, science has begun to perceive this form of connection. From the advent of quantum mechanics, a premonition emerged: that of an active vacuum, bearer of order.

It had all already begun with Planck, in 1900, when he showed that energy is not transmitted continuously but by discrete quanta.

This discovery marked a decisive turning point: it revealed that the real, even in its most elementary expressions, obeys a structure, a rhythm. The world was no longer a homogeneous flow, but a precise arrangement, founded upon minimal thresholds of action.

The Planck constant opened the way to a new reading: that of a field in which each interaction is quantized, as though an invisible order were already organizing all that is possible.

This premonition was then reinforced by discoveries such as the dynamical Casimir effect—which shows that the void, subjected to rapid variations, can produce measurable energy—or Bose-Einstein condensates, which manifest quantum coherence on a large scale.

These phenomena suggest that the void is not an absence, but a structured field, endowed with leverageable physical properties.

It is this intuition that led us to explore further.

Today, it is beginning to take shape with the quantum computer—a technology that leverages certain properties of the void.

But this is only a beginning: we still perceive only the first signs.

For my part, I chose to go further. Through a series of experiments, I was able to approach what I call—in a personal reading—the negentropic nature of the quantum vacuum: a dynamic that, in certain phenomena, seems oriented toward organization, emergence, coherence, and complexity.

This signified for me the existence of a wholly new, nearly unlimited potential, especially in terms of degrees of freedom—the capacity of a system to explore a multitude of configurations, states, and trajectories.

And there, a shift occurred: I glimpsed that, if this were so, everything could become possible—and that the only true limit would be that of our imagination.

I said to myself: if I could interact directly with this void, I would gain access to a reservoir of potentialities infinitely vaster than that of our phenomenal reality. In other words, I could bring some of those degrees of freedom down to our scale and create entirely new conditions for matter and energy.

It is on this point that I focused my research, and it is also there that my experiments advanced the most. For from this vision, a new path opened: that of a conscious interaction with the field—no longer only through intuition, but through conceptualization.

Sylvana
And this negentropy… is it also present in our reality?

Thierry
In short, no. To answer clearly, we must return to what most structures our visible world.

It is essential to understand that in our phenomenal reality, it is electromagnetism that plays the dominant role—a fact recognized by science, and which I adopt in my own reading.

As a result, most of our observable reality depends strongly on electromagnetic interactions:

- It is they that bind electrons to nuclei,
- That determine atomic and molecular orbitals,
- That make light, heat, and exchanges of energy possible,
- And that make complex biological forms possible.

In other words, what we call "reality"—everything we see, touch, measure—is largely structured and animated by electromagnetic fields, even though other fundamental forces also intervene.

But electromagnetism does not merely provide structure: it also directly contributes to the increase of entropy.

In science, entropy is defined as a physical quantity that measures the degree of disorder in a system—or more precisely, the portion of its energy that is no longer available to produce work: at each interaction, at each electronic transition, at each bond that forms or breaks, there is a portion of energy that dissipates in the form of photons, heat, or vibrational energy.

This energy is not lost—the first law of thermodynamics forbids it—but it becomes diluted and less available to produce useful work—that is, energy that can be used to create a directed effect, such as a movement, a transformation, or a structure.

For example, let us take the case of a plant receiving light from the sun—a form of electromagnetic energy.

- The plant uses part of this light to manufacture sugar, wood, living tissues: this is the creation of order, the formation of organized matter.
- But the process is not perfect: at each stage, some of the absorbed energy escapes as heat, produced by the microscopic agitation of molecules inside the cells—as movement generates warmth. This constant agitation also puts strain on the cell, creating wear over time.

Here, we find two sides: an order created—living matter—and, in parallel, a production of entropy, marked by dissipated heat and the internal wear of cells.

Let us now take a chemical reaction, when two atoms come together to form a molecule.

- Electromagnetism ensures that the electrons find a stable position, which allows the creation of a new organized structure: a molecule.
- But the process remains imperfect: part of the energy is released as heat, spreading outward and no longer available to be reused.

Here again, we find the two sides: an order created—the molecule—and, in parallel, a production of entropy, marked by dissipated heat and the loss of available energy.

And, our own body is no exception to this dynamic.

Let us take the human brain, when a neuron transmits a signal.

- Electromagnetism organizes the passage of ions through the membrane channels, allowing the propagation of an electrical impulse.

- But again, the process is not perfect: each signal requires energy to move ions through the membrane channels, and part of this energy dissipates as heat, produced by the motion of molecules—energy that spreads and can no longer be reused. With age, the system becomes less efficient, and the signals become less precise.

Here too, we find the two sides: an order created—the neural signal—and, in parallel, a production of entropy, marked by the loss of precision and cognitive decline.

And this is why, in our phenomenal world—unlike the quantum vacuum which, in my view, is driven by a negentropic dynamic—energy always ends up dispersing, information fades, and every form tends toward disorganization.

With time, structures wear down, capacities diminish, and repair processes lose effectiveness. Even living matter, however sophisticated it may be, remains subject to this law of loss.

Sylvana
It is unsettling… as if life were nothing more than a fragile balance between the impulse of order and the inevitable fatigue that comes with it.

Thierry
And our current way of creating does not escape this dynamic. It still rests upon the transformation—often brutal—of natural structures.

We cut, we extract, we burn. Each time, it is the same logic: to transform by force. Chemical reactions, heat, pressure….

In other words: we destroy in order to create.

And I would add that this is not only the expression of entropy: it is its excess—a form of hyper-entropy, where disorder increases artificially under the effect of our brutal methods. This is what I have chosen to call a distortion: a profound rupture of the initial balance.

And if this method became prevalent, it is also because it corresponded to the way we understood matter: dense, resistant, difficult to bend. It took steam to move trains. Explosions to turn engines. We learned to produce by forcing, by heating, by compressing.

But this process, however controlled it may be, generates losses, waste, pollution… and above all, distortions.

It is the same pattern in our technologies: they rest, for the most part, on principles of constraint, compression, or explosion. Even scientific research perpetuates this power dynamic: we bombard, we split, we provoke. As though truth could only appear under pressure. As though truth had to be torn out by force.

We make atoms explode in order to reassemble their fragments. We recreate extreme conditions to cause what would otherwise remain invisible to appear.

And yet… this method—brilliant as it may be—also says something about us. About our relation to the real: still grounded in confrontation, extraction, a science of domination—effective at first glance but limited. Because by striking endlessly, we risk missing what, to me, can only reveal itself in listening.

Sylvana

You know, the further you go, the more I have the feeling that we are touching on something fundamental.

We are no longer speaking only of technology, nor even of science…

You are saying that our way of creating—of searching, even—could be entirely revisited. Not by becoming less precise, but by becoming more subtle.

Thierry

Yes. And we must acknowledge that the way we have explored, built, experimented up to today has led to major advances.

And it is this path that has brought us to a decisive turning point: that of the quantum—a threshold from which another way of relating to the real becomes possible.

We are now entering a stage of evolution where true co-creation becomes possible.

For if order is already there, inscribed at a more fundamental level—in the vacuum itself—then another path becomes possible. We must attune ourselves to this order, instead of constraining it.

Create without destroying, from a still-unexplored space

And that, ultimately, is what truly marked a turning point for me: the gradual understanding of what we call "the collapse of the wave function"—not as a mystery to be solved… but as a living threshold, an interface between the invisible and the visible, between the field of possibilities and manifested reality.

For in the end, even within a fallible system, subject to dissipation, there exists—at the very heart of what tends toward disorder—another reality, one we are often unaware of. For matter, in what seems to be its densest, most tangible form, rests upon an invisible ground—even more fundamental: the quantum vacuum.

And this vacuum is not a mere empty space between particles—it constitutes the other side of reality, vaster, more fundamental, and structuring. This, in truth, accounts for the essential: more than 99.99% of what we call matter.

And within this very field, one finds intrinsically negentropic mechanisms: dynamics capable of generating order, maintaining coherence, linking information across time and space.

In other words, the matter we perceive as unstable and temporary is rooted in a deeper structure that bears stability.

It is this reversal that opened a new path for me: recognizing that, in my view, negentropy is not an external addition—but that it is already present, in germinal form, at the very heart of what we call the real.

Sylvana
But… if, as you say, this invisible ground carries information and order, then why does what we see around us so often seem chaotic?

Thierry
For me, it is to meet the requirement of a Universe that is fundamentally temporary and cyclical—in other words, that has a beginning and an end, yet is perpetual by design.

To conceive of a temporary structure of the Universe requires the presence of entropy. By contrast, for infinite cycles to manifest perpetually, negentropy is imperative—that is, a force of order opposed to dispersion.

From this angle, the quantum vacuum, as the primary source of the Universe's continuous manifestation, seems to me to bear fundamental organization—and as such, to be capable of sustaining a perpetual renewal of forms, without needing to invoke an infinite energy.

This is how the quantum vacuum makes negentropy possible: by its very structure, it can support the emergence of an order that maintains and renews itself without end.

On a more local scale—that of human consciousness, which is also temporary in nature—one clearly observes an entropic dynamic: alternating resonances between constructive and destructive polarities— such as joy and anger—which modulate the system's response.

These resonances leave a direct imprint on phenomenal reality. Yet the quantum vacuum does not take on any of these fluctuations. It remains the pure expression of fundamental order. It is never the vacuum that falls out of alignment: it is manifested reality that reflects, like a mirror, the state of harmonization or dissonance in relation to it.

In matters of co-creation, it is a subtle co-actualization—hypothetical, yet profoundly fertile—between consciousness and the quantum vacuum: a possible interaction, guided by the coherence of the informational field and modulated by the vibratory quality of the consciousness that attunes to it.

It is in this intimate resonance that manifestation takes place: not as a mere act of will, but as the emergence of an alignment between intention and the implicit order of the real.

- Threshold of Another Knowledge — Toward a Co-Creation with the Field

Sylvana
All right, I think I understand better… So if one truly wants to co-create, it is not just a matter of technical innovation—it requires a profound transformation in how we enter into relationship with the real, is that correct?

Thierry
Yes. We must change our perspective, our consciousness, our very way of positioning ourselves in relation to the real.

For this co-creation to take on its full meaning, we need to invent what I consider a new science—a science that includes new measuring tools, new conceptual frameworks, and an entirely rethought infrastructure.

Because even today, we are still trying to observe subtle phenomena with instruments designed to manipulate inertia and mass. In other words, we are not yet equipped to perceive what the field could truly reveal to us.

Sylvana
It is a bit as though today we were trying to catch a butterfly with a construction glove…

Thierry
Exactly.

Too much force, too much rigidity, not enough finesse… And the butterfly flies away. But if you replace the glove with a flower… then perhaps it is the butterfly that will come to alight.

This is the inversion that I believe to be necessary: instead of imposing a framework to force access to the quantum, one should create an environment sufficiently ordered and stable at our scale—a welcoming structure, capable of naturally entering into resonance with the field. Not to trigger a wave collapse, but so that it may occur by itself, without distortion, as a natural response to a present coherence.

This sums up exactly all my work, both theoretical and practical. My entire approach has aimed—and still aims—to make this encounter possible, without ever forcing it.

Sylvana
So it would no longer be a science that seeks to "provoke," but a science that attunes, that makes possible…

Thierry
Yes. A science of alliance, more than of conquest.

And it is there, in my view, that a new art of creating can emerge.

Because if the vacuum carries a profound order—and if that order can meet ours without rupture—then I imagine a wide range of phenomena that could become accessible, right here, in the field of the real.

But for that, we must learn to think otherwise: to move from control to accord, from force to subtlety.

Sylvana
So this joins what you were saying earlier… One must seek to elevate one's vibratory resonance, is that it? To tune our own inner field to that of the vacuum, so that a true encounter may take place.

Thierry
Yes, it is a whole new paradigm.

For a new art of creating to truly emerge, it is not enough to perfect our tools or to imagine new technologies. We must first transform our way of being, our manner of perceiving, our level of consciousness.

As long as this transformation has not taken place, it seems to me that we remain before a threshold: a subtle but very real limit. And if we try to cross it without this inner evolution, our perception grows blurred and our intelligence collides with what it cannot integrate.

Sylvana
You must have thought a great deal about all this… perhaps even meditated on the idea.

And then… what revealed itself to you when you explored this fully?

Thierry

Yes… I have thought about it a great deal. And that is when something became self-evident to me:

If this profound order is indeed the expression of an omniscient Intelligence, then a central question emerges—how could a limited consciousness, like ours, recognize a more encompassing consciousness?

The answer, in truth, lies within the question. It is impossible.

It is not a matter of fatality, but of coherence: what we perceive of the world always depends on the level of consciousness from which we observe it.

Already in Antiquity, Plotinus had intuited it: in essence, for him, any ascent toward a higher level of reality presupposes an inner transformation; to perceive a higher level of reality, one must draw nearer to it by becoming similar to it. In other words, one cannot recognize a higher consciousness so long as one has not, at least in part, sensed it within oneself.

As long as one remains confined within a given level of consciousness, one does not even know what it is possible to perceive beyond it. One does not know what one is seeking, because one cannot yet conceive of its existence. It is only when something within us begins to vibrate differently—to open, to grow more refined—that this vaster reality gradually becomes perceptible. Not by comparison, but by resonance. For one does not perceive this higher consciousness by measuring oneself against it, but by entering a state that comes into resonance with its own: like two notes that vibrate together, a harmony that requires a conscious effort.

Sylvana

But then… what impels us to seek to rise higher, if we are not even aware of it at the outset?

Thierry

The beauty of life is precisely this: every form of life evolves, carried by a dynamic that naturally tends toward greater complexity, greater coherence, greater consciousness.

A vision that Teilhard de Chardin (1881–1955), in the years 1930 to 1950, had already formulated, observing that evolution follows an ascending curve—from matter to thought, from thought to unity.

And there, finally, one realizes that two paths are open to the human being in order to evolve: the natural course of this evolution—the general path, that of linear time, the causal chain from past to future; but this natural evolution is slow, too slow. It can take millions of years, with countless uncertainties. Now nature itself seems to invite us to surpass this slowness, to transcend Time. It offers us another path: the Path of Exception.

The one where the human being undertakes a genuine inner work—an effort of transformation, of clarification, of alignment—to make possible a leap of evolution: an active resonance with a higher level of organization.

And for me, this leap is not a spiritual luxury. It is the only path to ensure survival.

Because humanity is confronted with natural cycles, or with unpredictable cosmic events that can at any moment endanger its existence.

We must not wait to be forced into it in order to evolve. We must be proactive.

And all my work has been inscribed within this perspective: how to refine my perception, elevate my level of consciousness… so that what did not appear to me might finally appear—and, afterward, be able to share these discoveries.

There was, for me, a decisive realization: to realize that we are literally immersed in the quantum world—that it is our environment, but also our own biological body and our consciousness. Our reality, by more than 99.99%, is not composed of solid matter, but of a void structured by quantum laws. And this discovery changes absolutely everything.

For this means that we are not separate from this field, but intimately connected to it—connected sympathetically, that is, by vibratory resonance, to what I call the omniscient intelligence, that weave of order and coherence that underlies the quantum world.

From then on, it becomes conceivable that we have access, at least in principle, to almost infinite degrees of liberty: a reservoir of emergence, creativity, and evolution, far beyond the limits we believed were fixed.

Sylvana

I see… so, essentially, what you are describing is a way of uniting with the Universe itself. Not symbolically, but actually. A kind of profound, almost intimate alliance… Is that it?

Thierry

Yes. But it goes much further still.

This alliance cannot be the fruit of a mere intellectual or technical effort.

It requires a complete shift in our way of being in the world.

What I am seeking to express, within the framework of my model, goes further than a unification of the physical forces. It is a total unification: between the visible and the invisible, the measurable and the unfathomable. The One, in relation to the Whole.

And indeed, it is here that a science founded on Absolute Relativity appeared to me—as I had already mentioned to you: a formulation specific to my model, which would no longer rely on a local or relative point of view, but on the coherence of the field itself.

And if this stance is correct, then another language becomes possible. A language that no longer separates subject from object, the human from the Cosmos, the visible from the invisible… but that inscribes them within one and the same weave—because, from the point of view of the vacuum, all is One.

Absolute Relativity, as I conceptualize it, according to my model, is the One perceiving itself through the Whole—and the Whole reflecting itself in the One.

I am an individual, but I participate in the Whole.

I am at once a unique point of observation… and an expression of the underlying unity.

And it is this Absolute Relativity that makes the unification of forces possible, not by reduction, but by resonance.

And what if this coherence we are discussing were not limited to the quantum domain? What if it extended to the entire Universe, down to its constants, its laws, its architecture? This is where another fascinating question comes into play, one that few people know: the anthropic principle.

2. The Anthropic Principle — When the Universe Seems to Be Waiting for Us

Sylvana
I imagine this is not entropy, the principle concerning disorder or aging that we just discussed…

Thierry
No, this is the anthropic principle, with an "a."

It is a concept that is rarely highlighted and must certainly not be confused with entropy—that fundamental principle of thermodynamics which measures the degree of disorder or irreversibility of a system.

Entropy increases in the Universe, indicating that things tend to evolve toward states that are progressively more disorganized.

Here, I am speaking of the anthropic principle, which poses a completely different—and fascinating—question: why is the Universe we observe so precisely tuned to allow our existence?

In other words: why do its laws, its constants, its structure… all land "just right" so that life—and in particular a form of consciousness—could emerge?

Sylvana
So what you are suggesting is that the Universe may not be merely "present," but arranged in such a way that…: we are here to witness it?

Thierry
Exactly.

That is what the "weak" version of the anthropic principle says: we observe a Universe compatible with life… because we are alive to observe it.

Put differently: if the Universe had not allowed for our existence, we would not be here to ask the question.

And the "strong" anthropic principle proposes that the Universe seems structured so as to permit the emergence of conscious life.

As though this possibility—that of a consciousness capable of asking questions—had been inscribed from the outset, in the very conditions of the Universe.

Sylvana
So, if I understand correctly… without confusing the frameworks, there is a kind of common thread: in quantum mechanics, observation plays a role in local actualization; in the anthropic framework, the presence of the observer reconfigures our reading of the Universe.

As though, at two very different levels, our presence were not just a matter of chance, but a significant factor within the very structure of the Universe?

Thierry
Yes. The "strong" anthropic principle proposes that the Universe seems structured so as to allow the emergence of conscious life.

As though this possibility—that of a consciousness capable of asking questions—were inscribed from the outset, in the very conditions of the Universe.

Sylvana

So, if I understand correctly… The idea that the observer plays an active role in quantum mechanics—as we discussed earlier—also appears, in a certain way, in the anthropic principle?

As though, at two very different levels, there were the same guiding thread: our presence would not be just a matter of chance, but a datum that matters in the very structure of the Universe?

Thierry

Yes. And that is why I connect them.

The role of the observer is central in both cases but not in the same way: in quantum physics, the act of measurement plays a role in the measured result; in the anthropic framework, it is the existence of the observer that explains why the observed Universe is compatible with life.

Sylvana

So you ended with the anthropic principle… I understand why. It is doubtless the most unsettling, because it touches what we are—and what makes it possible for us to be here to think about the very fact of being here. This consciousness that observes, that questions, that wants to understand, and which, at the same time, is part of the mystery.

Thierry

Yes. When I meditated on the anthropic principle, it was not to solve an abstract scientific enigma. It was a kind of inner vertigo. A blazing intuition: everything seemed to have been arranged with unheard-of precision… so that consciousness could appear. And not just any consciousness—but a consciousness capable of asking this question.

Sylvana

You mean… as though the Universe had set everything up so that one day we would be here, marveling at it?

Thierry

Yes. And what struck me, as we already discussed, is the degree of precision of this fine-tuning.

Tiny variations in certain constants (fine-structure constant, gravitation, critical density, and nuclear forces) would suffice to make the formation of stars or complex chemistry impossible—and therefore life.

Each parameter seems tuned like the gears of a clock, or the notes of a complex score. Not to produce a mechanism… but to allow the emergence of a gaze.

And what unsettled me is that this gaze is not a mere collateral effect. It is the only thing capable of recognizing the rightness of all this.

Without consciousness, there would be no one to perceive this order.

This is where I began to wonder whether this consciousness was at the very heart of the framework.

Sylvana
And yet… science could reply: we are here, therefore it is necessarily compatible. Full stop.

Thierry
Exactly.

Yes. And if consciousness were not just an emergent phenomenon, but a necessary condition for the very existence of a structured Universe… then the perspective had to be reversed.

Perhaps it was the Cosmos itself that, in order to manifest fully, needed consciousness. As though the latter played the role of an active mirror, a feedback mechanism allowing the quantum order—"blind"—to actualize itself within consciousness. A passage from the implicit order to the fully conscious order.

While meditating on this rightness, an idea crossed my mind with a shattering intensity:

What if the Universe, through every form of consciousness within it (including ourselves), also became capable of looking at itself?

Sylvana
You have just turned everything upside down.

It is no longer only the human who contemplates the Cosmos.

It is the Cosmos that becomes conscious of itself, in us and through every form of consciousness.

Thierry
Yes. And if this hypothesis is correct, then consciousness is no longer a side effect. In my approach, it becomes a fundamental act of the Universe. An act of return.

As though the Universe, after expanding, structuring itself, becoming more complex, finally found in each state of our consciousness a mirror. A point of contact where the inside and the outside meet.

Sylvana
You know… I see the image of a drop of water that reflects the whole ocean. A tiny drop, and yet… it is the immensity that looks at itself there. Is that what you mean?

Thierry
That is exactly it. And it changes everything.

Because from this perspective, science can no longer be content with observing external objects. It must become a science of relation. A science in which consciousness is a structuring element of reality.

Sylvana
And is that what you call a "science of consciousness"? A science in which the researcher no longer extracts himself from the world in order to analyze it, but accepts that he is, himself, a point of emergence of the real?

Thierry
Yes. The anthropic principle led me here—to this evidence: the Universe is not external to us. We are, literally, expressions of its own intelligence—an intelligence in the process of recognizing itself.

Sylvana
And what you are saying is that it was by starting from that vertigo… that you dared to formulate your postulates?

Thierry
Yes. Because that vertigo was not an escape into the imagination. It was an inner experience of a vaster coherence. It was absolutely necessary for me to situate my place in this Universe in order to begin to understand the origin of my intelligence, of my consciousness… and beyond those two aspects, the very nature of my spirit.

Sylvana
In short, you could not think the Universe without involving yourself in the question?

Thierry
Exactly.

Even at a very young age, I felt that this question could not be eluded.

When I was a teenager, in the deepest part of certain meditations, I felt my consciousness depart from my body. It is an experience difficult to describe, but it was very clear. I was still there, fully conscious… but detached from any physical identification.

That is what is commonly called an out-of-body experience. And that is where something imposed itself within me with self-evidence: I am not this body; rather, it is only a vehicle, a relay, a support for experience. But the essential part of what I am… is elsewhere and I must seek it.

Sylvana
Are you telling me that you had the sensation of existing outside the physical world?

Thierry
Yes. That is exactly it.

I sometimes plunged into meditations so profound that I had the impression of embracing the entire potential of the Universe… while standing outside it.

Sylvana
So here, it is no longer an idea you are explaining, but something you truly lived…

Thierry
Yes. And in that state, a sentence came to me one day, with striking clarity:

I am the Spirit outside matter.

It was not an image, it was a lived experience. And that experience transformed my research. For if I was capable of observing both my body and the Universe… then I was neither one nor the other. I was a principle of observation, a conscious-spirit, capable of projecting itself beyond the material frame.

And that tipping point led me to reformulate my entire way of questioning the world.

Sylvana
But what you are saying is immense. Because it completely changes the way we look at ourselves. You no longer start from the human being as the point of departure, but from spirit as origin.

Thierry
Yes. And this inversion led me to a question I could no longer avoid:

How is it that beings as small, as localized as we are, are capable of wanting to understand the very origin of the Universe?

It is not, however, a vital necessity. Nor is it a condition for survival. And yet, we have done it. Driven by a mysterious force, a kind of inner call, almost irresistible, that exerted upon us a silent pull. Something within us wanted to understand. Had to understand.

We have traced cosmic time back to the dawn of the observable Universe. We have captured the light of galaxies more than 13 billion years old, thanks to instruments such as the Hubble Space Telescope. And all of this from a microscopic point lost in space… as fragile, as almost nonexistent, as a breath suspended in immensity.

Sylvana
You mean that this thirst to understand betrays something… invisible?

Thierry
Yes. It betrays our origin.

And the more I reflected on it, the more a question became central for me:

What is it, within us, that can bear such a thirst to understand? This desire to transcend our condition, to go back to the origin…

It could not, in my view, be the simple product of an assembly of neurons. There had to be something vaster, more fundamental.

That is where I came to consider that if the human being is capable of such transcendence, it is because he is constituted of three fundamental elements:

First, a spirit, non-local—capable of transcending space and time. This spirit, in my view, is an integral part of the creative power at the very origin of the Universe. Perhaps that, deep down, what I was seeking through science… was not the Universe itself, but that primordial power which brought it forth.

Next: a consciousness. It is what allows us to live, to experience, to invent, to create. We are endowed with a capacity to transform the world. It is what makes us co-creators.

And finally: a body, rooted in a local reality—which allows us to express ourselves here, on this planet, within the field of time and matter.

Sylvana
I would like to return to the first element. Do you mean that it was not the Universe you were truly seeking, but this invisible source that makes it exist? And that our spirit would be like a spark of it—connected to something infinitely greater than we are?

Thierry
Yes, exactly.

The Universe is not a local or closed entity—it is an expression, an extension, of that original source.

Sylvana
And what you say about this inner call… it is truly intriguing.

Do you think we are here, in a way, to rediscover our own story? As though understanding the Universe were also remembering who we are?

Thierry
Yes. And what I propose here is not an imposed truth—it is an inner exercise. A way of imagining differently, of posing the great questions from another angle.

For me, the Universe is an immense field of experiences. And the human being is a complex interface for exploring this field across different planes.

The spirit, I feel, is something elusive. It escapes categories, definitions, as though it carried within itself an infinity of potentials.

By contrast, in my view, consciousness, as well as the physical body, are forms of universal technology—tools for incarnation, exploration, transformation.

But what seems essential to me is that consciousness would not be a phenomenon purely localized in the brain. It would not result from neural functioning alone.

From this perspective, it would precede the biological process—and therefore matter itself. And it is this non-locality that, in my view, would give us this extraordinary capacity: to want to understand what exceeds us.

Sylvana
You went very far in this inner exercise… You did not merely think differently—you changed your point of view.

Thierry
Yes. I had to make this radical effort. I had to project myself as a giant observing the entire Universe—and not remain confined within my tiny humanity, like a mere point gazing at an infinite horizon. As long as I stayed at that scale, everything brought me back to my limits.

But by reversing the perspective, by placing myself beyond, I began to perceive the whole differently—no longer fragmented, but as a coherent ensemble, imbued with an intention. It was a kind of inner exercise, almost mystical, yes—but for me, it was a necessity.

Sylvana
And did that change something in your way of searching?

Thierry

Everything.

As long as I remained within my local condition, my thought kept circling within a narrow perimeter.

But by placing myself in the position of an unlimited being—if only through imagination—I opened a maximal field of possibilities. And that opening served as my engine.

I am not the only one to have intuited it. The real difference, perhaps, is that I went to the very end of the exercise: I projected myself beyond the Universe… to attempt to understand it from the outside, then to traverse it again from the inside.

Sylvana

But then, at that level… I wonder again, how do you distinguish between intuition, imagination, and scientific rigor?

Thierry

It is a crucial question. And I took it very seriously.

A scientist must dare. One must allow oneself to enter vast imaginaries, to formulate hypotheses that are at times vertiginous. It is a phase one could call an "incubation" period, where ideas arise, confront one another, intersect.

But then, one must filter. Test. Structure. That is where the real work begins: to confront these intuitions with what we know, even to invent at times new means of verification, because the existing frameworks do not always suffice.

Sylvana

So… One needs a touch of madness, but also ironclad rigor?

Thierry

Yes. One must dare to go very far—and return with rigor.

And that is also why facile criticism always seems so poor to me… because very often, it comes from those who have no idea what this inner path requires.

Sylvana

So what you propose… it is not just another vision of the Universe. It is another way of taking part in it. No longer as outside observers… but as the points of awakening of a cosmic consciousness.

Thierry

Yes. And this change of perspective, for me, has nothing esoteric about it.

It is not an escape from the real—it is an intellectual demand.

When 95% of the Universe escapes us, it is not the Universe that must be corrected. It is our way of questioning it that must be called into question.

And this calling into question, some physicists have fully assumed—by daring to challenge the very foundations of our way of looking at the real.

Werner Heisenberg (1901–1976) summed it up this way:

"What we observe is not nature itself, but nature exposed to our method of questioning."

David Bohm (1917–1992), for his part, suggested that the "implicate order" is not hidden, but ignored, for lack of the right gaze.

Sylvana
You know, I have been listening to you for a while now with a mixture of vertigo and calm.

And at the same time, I have this question rising up, very simple, very direct:

What did you do, concretely, once you saw this? I mean… from the moment you had all these open questions, where did you go? What emerged there, in that void that science no longer knew how to name?

Thierry
What I did was withdraw profoundly into myself.

I did not yet have a theory. No plan. But I knew that I had to go listen elsewhere.

I stopped expecting answers from the existing framework. I accepted to sink into a space of silence, without expecting anything. And little by little, lines began to appear. Not certainties. Not equations. But directions. Intuitive structures. Axes.

At that moment, I understood that my research could no longer be limited to accumulating data or manipulating concepts. The plane had to change. To explore the inside as much as the outside. It became clear: the only way to access a broader coherence was to transform my own posture.

And that is where meditation took on its full meaning—not as a refuge, but as an instrument. An instrument for perceiving otherwise, without ever renouncing logical coherence, which remains, for me, non-negotiable.

Sylvana
I am listening to you… and I gauge how radical that inner shift must have been.

Honestly, I believe I would have quickly lost myself in such vertigo. So if you do not mind, before we enter the heart of your theory, I would like us to speak about this: how you managed to stay centered.

Because frankly… it is anything but obvious.

Chapter 6

A Stance of Unity

~

Rethinking the Universe from Within

Sylvana

You know… until now, we have looked at things from the outside: what science knows, what it does not know, what it leaves pending.

You decided to start again from the inside—not figuratively, truly from the inside. And that, for me, is at once disconcerting and fascinating.

I would like to understand: how does one dare to do that? How does one remain at once lucid, rigorous… and yet totally open to what exceeds our usual frameworks?

Thierry

Daring this step inward means accepting a radical change of posture.

It is no longer only observing, measuring, analyzing from the outside, but also welcoming another form of knowledge, more intuitive, more direct.

This approach requires unfailing rigor: it is not a matter of being carried away by impressions, but of always cross-checking this inner experience with logic and external facts.

It is the work of an equilibrist: advancing in a space where intuition dialogues with reason, where consciousness becomes a tool of exploration as serious as any scientific instrument.

That is how I was able to begin to perceive structures invisible to classical science, while remaining strict about the demand for coherence and verifiability.

I have always approached this freedom to explore with vigilance. Without rigor, any intuition can turn into illusion. That is why I gradually built a very demanding inner discipline, capable of holding together the widest openness… and the strictest coherence.

It is within this framework—at once supple and rigorous—that my first perceptions began to take shape.

Sylvana
What you are saying intrigues me… With all the ideas that can arise, how did you manage to know which ones really had substance? What allowed you to feel that an intuition deserved to be pursued—and that it was not just alluring in appearance?

Thierry
It is a crucial question, and I have often asked it of myself.

Over time, I developed an ability to quickly detect the dead ends, the false leads, the speculations disconnected from any physical or logical consistency.

For that, several tools were useful to me, among them empirical experiments—sometimes spread over several years—and specialized scientific opinions, sought in many fundamental fields.

Sylvana
But how did you manage to integrate such different knowledge? It must be difficult to weave all of that together without losing the thread.

Thierry
It is true, and it was not without difficulty.

But very quickly, I understood that I could not compartmentalize knowledge.

The complexity of the real swiftly confronted me with an obvious fact: no science is foreign to the study of the Universe. Chemistry, physics, biology, cosmology, mathematics, philosophy, neurocognition… all contribute to the intelligence of the Whole.

This demanded of me a constant effort of transversality, of cross-referencing data, and sometimes even of translation between paradigms. But it is also there that I found my most beautiful intuitions.

I had already been living for six years immersed in the writings of the greatest thinkers in the history of the sciences. Their ideas inhabited me; I questioned them; I made them enter into dialogue in my mind, sometimes to the point of confrontation.

In 1996, after roughly nine months of intense reflection on the nature of the Universe, an idea asserted itself within me with irreversible force:

If I want to understand the Universe, I must integrate it within myself.

Sylvana
So you identified with the Universe...

Thierry
Yes, it may seem disconcerting... But that is what I did, a radical thought exercise that I would describe as an immersive cognitive experience: I drew a symbolic and experiential parallel between the Universe and my own mind.

I decided, conceptually, to become the Universe, to identify with it in order to feel from within its logic, its tensions, its rhythms, its possibilities.

Sylvana
When you say "integrate the Universe within you," I sense that it is not a figure of speech for you. So... concretely, how did you think at that scale? Did you have a precise method, or was it more instinctive, inward?

Thierry
It may seem strange from an external point of view, but this act yielded considerable scientific fecundity. For the mind is the only known object capable of observing and describing itself.

Very early on—as we discussed regarding my childhood in Martinique—I had developed a particular sensitivity. I perceived around living beings—and sometimes even around inanimate objects—luminous forms, like structured fields.

Over time, I understood that it was not a simple visual impression, but a subtle, global perception, associating sensations, rhythms, and spatiality.

I could also generate internally complex forms that evolved like living entities. They transformed, interlocked with one another, according to a silent logic—a fluid and coherent inner geometry that I did not consciously control, but that I felt in an organic way.

And I already had, without calling it that, an intimate relationship with Time.

As an adult, I gradually structured this capacity. I would deliberately place myself in a state of deep meditation, centered inwardly, to reach total immersion in my reflections. It was a form of active meditation: I was not seeking to empty my mind, but to connect intensely to a subject—in my case, to the Universe itself. And this approach resonated directly with what I have just told you:

"If I want to understand the Universe, I must integrate it within myself."

Sylvana
That is impressive… As for me, just sitting for five minutes to meditate, I already find myself thinking about everything I still have to do in the day, and all my problems run through my head. So managing to dive into the Universe itself… I think I am still far from that!

Thierry
Let me reassure you, you are not the only one.

It makes me smile because, even for someone highly motivated, it is far from simple at the beginning. It takes a great deal of time, patience… and above all learning to tame one's own mind, before even trying to feel that of the Universe.

That is also why, very early on, I began to develop technologies capable of supporting this inner exploration—tools to access this state of connection more easily, and to amplify its scope. They played an essential role in my path.

And I would like you, and others as well, to be able one day to live this experience—even just once. Because it changes everything. It transforms our way of seeing, of feeling, of being in the world. And today, I sincerely believe that this inner opening is no longer a luxury. It is a necessity if we wish to broaden our consciousness and recover a living bond with what is essential.

Sylvana
Just imagining what you might have lived through… it makes me want to experience it one day, me too.

And what surprises me, listening to you, is that your meditation was not an escape from the world—but a way of entering it even more profoundly. It is so contrary to what we imagine, in fact…

Thierry
Yes, exactly.

Many people use meditation to reach serenity or to let go. I used it as a tool for exploration.

Starting from this unique capacity, I formulated my first metaphysico-physical postulate…

I see you are furrowing your brow…

I will explain what this means. It is a term that blends two domains: metaphysics, which concerns itself with the profound nature of reality beyond the physical world, and physics, which studies what can be measured and observed concretely.

And so, starting from this unique capacity, I formulated this postulate:

The Universe is the Unique Substantial Reality, Living and Conscious of itself.

I considered it as unmanifest in its essence—as with the human spirit—but containing the seed of its own manifestation.

Sylvana
You set that down as a starting point… but in truth, you are asserting an entire world there. To say that the Universe is living and conscious of itself already throws everything into a new perspective.

Thierry
Exactly.

And this inner reversal led me to a profound revelation: science is, in the final analysis, a matter of consciousness.

You seem pensive…

Sylvana
Yes… A "matter of consciousness"? Do you mean that all scientific research, at its core, rests on the way one is inwardly connected to what one seeks to understand? That it is not merely a question of tools, measurements, objectivity… but of a state of consciousness?

Thierry
Yes. For every observation, every hypothesis, every modeling is rooted in a perception. And that perception depends intimately on the state of consciousness of the one who observes.

Even the most rigorous approaches are never totally neutral: they are traversed by the quality of attention, the sensitivity, the inner maturity of the researcher.

That is why, at its core, one cannot radically dissociate science from the human being who produces it.

To understand the Universe is also—necessarily—to understand something of ourselves.

Sylvana
So you were not only posing a scientific question... You were already, silently, reformulating the link between the Universe and consciousness.

Thierry
Yes. And that is where everything shifted more profoundly still.

What I then glimpsed was an infinity that offered itself not only to be measured, but to be lived inwardly.

That vertigo accompanied me to the very heart of my hypotheses, my calculations, my experimental proofs, my discoveries—for at each stage, it reminded me that the real question was not:

How does the Universe function?

But:

How can I, as a conscious being, connect myself rightly to what Is?

Sylvana
I must admit, what you have just described prompts me to reflect deeply.

Because it is not merely a working method or an intellectual stance—it is a way of being in the world. A way of making oneself available to something that exceeds us.

Thierry
Yes... and that is exactly what led me to cross a threshold. For from that point on, it was no longer only a matter of understanding—but of rethinking the Universe from the inside.

Sylvana
So... what you propose is not merely another theory. It is a complete change of perspective.

But I wonder... how does one go about moving toward that kind of science?

What allowed you, yourself, to cross that threshold?

Thierry
For this science to become living, a new step had to be taken.

I had to adopt a new posture: that of the Universe itself, envisaged from the Whole. To do that, I placed myself where everything begins. At the instant when the Universe is born. As though I were standing on the threshold—as the Universe itself, in the first person—to perceive a real in the process of organizing itself, seen from within.

It is from that space—from that posture of unity— that seven fundamental conceptual elements of the Universe emerged.

They are not meant to lock the real into a fixed model, nor to offer a definitive truth.

They first served me as markers, as levers. It is with them that I was able, step by step, to organize my thinking, refine my perception, and construct an overall coherence.

That is why I share them today—not as certainties, but as concepts to approach the real differently: from a vaster, more interior, more aligned point of view.

This stage is, in a way, an open window onto my vision of the Universe: these seven elements only make sense if one understands where they come from and what perspective gave birth to them.

This is, in any case, what I wish to share.

Sylvana
What you are describing here… it upends quite a bit our idea of what constitutes a "real" scientific approach. We have been taught so thoroughly to be wary of anything subjective, interior, intuitive… Have you never feared that people would reproach you for straying from scientific rigor?

Thierry
Indeed, this way of thinking can surprise. Not because it would be vague or mystical, but because it shifts the center of gravity of science.

It is not a retreat from rigor: it is a shift of the point of support. For a subjective gaze is not necessarily arbitrary.

Science, in its essence, has never been founded on rejecting experience—but on its methodical deepening.

It rests on a structured approach: observe, experiment, formulate hypotheses, put them to the test, and compare the results with reality.

It is this dynamic, grounded in rigor, coherence, and the capacity to question its own models, that gives science its strength.

But if we agree to broaden the field of perception—without renouncing that standard—then another mode of understanding becomes possible.

A mode in which intuition, inner experience, and subjectivity are no longer excluded, but integrated with discernment into the quest for truth.

Part III

How I See the Universe

~

From the Source to Consciousness

Chapter 7

The Cartography of a Cosmic Perspective

~

From the Cosmic Breath to Being

Sylvana
You once told me that to rethink the limits of science, one had to start by laying everything out afresh.

So I would like to ask you the simplest question in the world:

What did you begin with?

Thierry
I first sought a framework vast enough and coherent enough to hold everything I was sensing.

And it was a story that guided me.

Sylvana
A story?

Thierry
Yes. My method is a little unusual—it may even be surprising.

Each time I choose a subject, the very first thing I do… is write a poem.

It is my point of entry.

Writing a poem immerses me in a state of profound state, an inner openness that creates space for thought experiments. In that state, I can go beyond the limits of my narrow mind, connect to something vaster than myself.

It is often here that my thought experiments begin.

And this time… it is a very particular poem that forced me to reconsider everything. From top to bottom.

Sylvana
So if you agree… instead of speaking about it in half-words, could you share this poem, as it is?

Just read it. Let it exist.

Thierry
Yes. I can read it to you.

1. The Inner Genesis — A Poem as Compass

The Present, a gift from the Primordial Complexity

There exists, in truth, only one reality: Spirit.

A single, infinite Spirit, which precedes and sustains all things.

It is the absolute Spirit, Father and Mother of the Universe, matrix and source of all that is.

From It emanate the countless marginal spirits, children scattered across the great theater of worlds.

The absolute Spirit is defined by an elusive quality: its absolute Presence.

It is the absolute Present, the immutable Now, a silent and vibrant reference for all derived spirits.

It is the fundamental Resonance, the original Tuning Fork whose echo all hearts still carry.

In its original purity, the marginal spirit—which one may recognize as the Holy Spirit, in the intimate heart of every being—remains immersed in the absolute

Present, a state of pure presence, without before or after.

In this state, it savors plenitude, for there resides the true power: that of co-creating, of existing without hindrance, of being master of oneself.

But this power is a burden for the still-immature marginal spirit.

Blinded by its own power, it loses itself in illusion.

It wanders, prisoner of a profound sleep, unable to contain the fire it carries within.

So, in its infinite compassion, the absolute Spirit weaves a vast design:

It entrusts to its voluntary Children, marginal spirits who have already achieved their ascension to full mastery, the sacred mission of engendering vast Universes.

School-worlds, fields of experience where marginal spirits can train, reveal themselves, master themselves.

In its Omnipotence, master of Time, of space, of cycles and of dimensions,

Each voluntary Child, an omniscient Intelligence, then fragments its omniscient Present into a myriad of instants.

Time, the primordial substance, springs from this fragmentation,

At first lightning-swift and free, it is trapped in loops, ordered into cycles.

Thus is born the fabric of spacetime, the sacred playground: the Universe.

The Universe then becomes a succession of present instants,

A gift hidden at the heart of the real, offered by the absolute Spirit.

And in this subtle laboratory, a divine technology emerges: consciousness.

The marginal spirits then incarnate in interfaces,

Vehicles of flesh and energy, temporary forms.

There they exercise their free will, exploring infinity by the paths of duality,

Rising or falling, learning to distinguish shadow from light, good from evil.

For knowledge is not given; it is forged in the fire of contrasts.

The marginal spirit can recognize itself only by passing through its own contradictions,

Only by confronting its limits, its doubts, its sufferings.

Through this experience, it patiently develops hope,

That fragile golden thread that connects it to its forgotten nature.

Repeating the same errors, the same sufferings, under a thousand faces, a thousand scenarios,

The marginal spirit ends up understanding. Through maturity and weariness, it learns.

It finally separates the true from the false, the real from illusion,

And discovers the treasure that lies sleeping within: self-mastery,

The return to balance, to inner Unity.

Thus the marginal spirit, awakened and transfigured, rises.

It is ready for ascension, ready to return home,

To the silent House of the original Father-Mother.

For what differentiates the marginal spirit from the absolute Spirit

Is neither nature nor essence, but measure.

The quality is identical; only the power differs.

And we, travelers of the matrix, still carry the promise of this growth.

We are children of the absolute Spirit, which I have named the Primordial Complexity,

Fragments of its Spirit on the way to their plenitude.

Free spirits, called one day to become conscious servants,

Freed, blissful, infinitely grateful for having been,

Not only created,

But above all called to *Become.*

Sylvana

What you have just shared has something of a tale, a founding story. But I know that, for you, it is not merely a belief.

Thierry

That is exactly it. I know that my words can sound like a symbolic narrative, at times even spiritual. But for me, these are not beliefs: they are founding images, chosen consciously as conceptual tools to guide a rigorous exploration of the real.

To understand the Universe solely from the cosmological and quantum data available is an indispensable undertaking… but it remains limited.

I already sensed that it would not suffice, by itself, to bring forth genuinely new ideas capable of transforming my understanding.

One cannot hope to renew one's vision of the world if one remains imprisoned within established frameworks, endlessly reproducing the same lines of reasoning.

This is why I resolved to take a step: to integrate images, analogies, archetypes—not to replace science, but to broaden it, to open unexplored paths where logic, intuition, and experience meet.

And time is pressing. For knowledge is not a luxury: it is a condition of our emancipation, of our liberty, of our very survival. Who can assert with certainty that there does not exist, concealed within the unknown, an invisible cosmic phenomenon that threatens our existence without us even being conscious of it? That would be a dangerous presumption.

This is why I chose to broaden my field of research. I decided not to ignore real phenomena, even those that science has not yet been able to measure, beginning with one of the most mysterious and most familiar: consciousness. What I mean is that if a phenomenon is lived, perceived, felt—even without physical measurement—it deserves to be considered.

Sylvana

Are you saying… that we should also take into account intuitions and intimate perceptions—of which consciousness is a part—even if they elude measuring instruments?

Thierry
Yes. Because in my view, as soon as a phenomenon is perceived, it exists in itself as a real phenomenon—even if, for the time being, it remains impossible to measure.

It is entirely legitimate to ask whether the consciousness we live each day is not in reality a limited manifestation of the broader phenomenon of consciousness, which could be cosmic, widespread, and present to varying degrees, far beyond just our human scale.

Perhaps there exist levels of consciousness infinitely vaster, more subtle, more powerful than anything we can conceive of—forms of experience that, by their very nature, exceed the limits of human consciousness.

Often, out of ignorance or pride, we choose to dismiss the hypothesis of such realities, simply because they escape our usual frameworks. Yet there is a fundamental difference between acknowledging our inability to apprehend a phenomenon… and deliberately deciding to deny it.

For my part, I have chosen another attitude: that of openness. Any phenomenon observed, even locally, deserves to be taken into account in the global reflection on the Universe.

For our life—what is at stake here and now, on our scale—is never separate from the Whole. By its very essence, the Universe embraces all that exists: from the most subtle to the most dense, from the tiniest to the most vast, from the visible to the invisible.

2. Broadening of the Field — Including Spirit and Consciousness in Science

Sylvana
So if I am following you correctly… the fact that consciousness exists here, on Earth, at our scale, already suffices to raise the question of whether it might exist elsewhere as well?

Thierry
Yes. This holistic approach—this way of embracing the real in its entirety and connecting what appears separate—has always seemed to me the most coherent. For if a phenomenon such as conscious life has emerged here, on Earth, why would it not manifest elsewhere, in other forms, other modalities? The Universe contains roughly 10^25 planets, that is, ten million billion billion

potential worlds: to imagine that life or consciousness is confined to our single planet is less a matter of reason than of dogma.

I readily acknowledge the complexity of this undertaking. But I have experienced it myself, and in only a few decades it has offered me unsuspected discoveries—within a research framework that is nonetheless marginal.

We must also keep in mind an essential truth: 95% of the Universe still eludes us. Dark matter, dark energy, the profound structure of the void… so many unknown territories. What we believe we know is only a thin film of the real. Refusing to examine certain phenomena, visible or invisible, already observed at our scale but still inaccessible to our instruments, would amount to depriving ourselves of precious leads.

This is why I chose, with rigor and method, to broaden my field of investigation. I integrated into my cosmic quest the singular power of my mind, guided by that irreducible faculty: consciousness.

As you know, my science rests on integrating the human being within the Universe—not as a mere observer, but as a fundamental element of the very structure of the Cosmos. Yet this way of thinking goes beyond the framework of contemporary science; I therefore had to allow myself to think otherwise.

For seeking to understand the emergence of consciousness on the cosmic scale is not a metaphysical luxury: it is a scientific necessity. It is perhaps the most fundamental question humanity can ask, for it touches upon our origin, our essence, and our future.

And this is where the role of the poem comes in. It allowed me to inscribe my mind—marginal, singular, limited—within a larger Spirit, that of the omniscient Intelligence of the Universe. The one who knows what I do not know. Not with the illusion of mastering everything at my level, but with the lucidity to welcome the gap and to question it.

If, already at my local scale, I can glimpse so many things, what then of what is contained in a Spirit that bears this omniscient Intelligence? It is better not to waste time; it is better to include it from the outset in my research. To ask the right questions, to accept encompassing my self within the Whole… and to let the Whole resonate within me.

I know that this approach lies outside conventional frameworks. But I embrace it as an intuitive method of exploration. Far from being heresy, it is an inner dialogue that I have initiated with this Intelligence—in other words, with the Universe itself. I allowed myself to imagine that the Universe could "speak to me" through my own inner language. That my intuitions, my visions, my felt perceptions could be vectors of information, nourished by the very structure of the Cosmos.

In other words, I accepted that the human mind might enter into resonance with an intelligence on the scale of the Universe. After all, my body is made of stardust, and my mind, I am convinced, is a child of the Absolute.

Sylvana
So, if I understand correctly, you sought to enter into resonance with this omniscient Intelligence—as though your human consciousness could open itself to the very whispers of the Cosmos?

Thierry
Yes… For if the spirit truly comes from farther away—if it is the emanation of something vaster than we are—then it becomes natural to ask this question: could it be that a higher intelligence, somewhere in the Universe, might, by subtle means, whisper certain truths to me? I never turned this into a dogma, but I left the door open to this possibility.

The Universe is there, quite simply. It conceals nothing. It is our minds, still limited, that grope their way through their own questions. And in this quest, the depth of the answers always depends on the quality of the questions we dare to ask.

To try to probe the Universe solely from an external gaze—limited to the tiny 5% that we believe we know—is to condemn our exploration to a dead end from the outset. Quantum physics, by opening a more intimate, interior vision of the real, shows us another path. In it I perceived a striking parallel between my interiority—the seat of my consciousness and my intelligence—and the interiority of the Universe: the quantum vacuum.

Thus, step by step, by broadening my questions, I was able to glimpse unexpected answers.

But before getting there, we must lay the foundations.

Sylvana
Okay… I am ready to dive in.

3. The Seven Elements — A Conceptual Framework of the Universe

Thierry
Between 1996 and 2001, after long years immersed in scientific knowledge,

an inner call imposed itself on me—imperious, indefinable. It was no longer only a matter of understanding, but of questioning the entire edifice anew: rethinking the Universe, widening the gaze far beyond established frameworks, and making myself available to other levels of the real, more subtle, more profound, more alive.

It was between 1996 and 1997 that, for the first time, the core of a new concept crystallized: that of a living, conscious, highly intelligent Universe, structured around seven fundamental elements. Nothing fixed. Nothing certain. But a succession of intuitive, bold, unexpected insights. Each idea, like a spark, opened doors in cascades—sometimes disconcerting, but always fertile. It was not a closed system, but an expanding field of exploration.

What I propose here is not an ultimate truth. It is a possible path, an invitation. A field of potentiality brought into resonance among knowledge, experiences, and intuitions. A sincere attempt at unification between what we know, what we sense, and what we dream of understanding.

But this invitation is not a light one: what follows is a rigorous exercise in thought, an unfolding of dense and structured ideas, in which I set out successively seven founding elements before later developing their scientific scope.

Some terms may surprise you. They come from other horizons—metaphysical, philosophical, scientific. But in the perspective I open here, they adjust, respond to one another, and weave together. A coherence emerges, gently, like a pattern long invisible on a cosmic tapestry.

Therefore expect to encounter new notions and a demanding conceptual architecture: this is not a contemplative stroll, but an initial cartography of my model.

You do not need to retain everything right now. These concepts will be revisited, deepened, and reworked over the course of the dialogues we will have later—you will integrate them naturally, little by little. Here, what matters is to grasp how they interlock to form a coherent structure.

It is a thought experiment that wants to be told, from the stardust that we are, to the breath of Life of the Primordial Complexity that traverses us. A narrative of reconnection. A quest to be shared. A new possibility to understand what it truly means to exist and why we exist.

Sylvana

Listening to you, I feel as though I am discovering an immense map… but without yet knowing the legend. So I will proceed step by step, and let things reveal themselves.

Thierry
Yes. Simply keep this in mind.

The result of these thought experiments—whether it is real, symbolic, or simply born of a profound movement of the mind—is not what matters most.

What matters is what it reveals. By allowing my mind to open without constraint, I saw coherent structures emerge—conceptual anchor points that later became the basis of all my research.

From this journey, seven points of reference were born. Each had a precise function, a structuring principle. Together, they gradually allowed me to formulate a coherent hypothesis about how the Universe could be organized.

Everything begins with the foundations: seven elements.

Are you ready?

Sylvana
Always.

. THE PRIMORDIAL COMPLEXITY — SOURCE OF INFINITY, PRESENT, AND CONSCIOUSNESS

Thierry
At the foundation of my model, I posited the hypothesis of a primal, irreducible, and boundless element: the absolute Spirit.

I conceive of it as an ordering principle, at once omnipresent, omniscient, and bearer of causal power—that is to say, the primordial source from which every cause proceeds—existing prior to and outside our observable Universe.

I named it the Primordial Complexity, for it represents, in my view, the original matrix from which order, laws, and structures emerge. To seek to understand it is to attempt to shed light on the underlying logic of the entire cosmic process.

I acknowledge that I am incapable of defining in an exhaustive manner—in other words, by covering every possible aspect of the subject—what this Primordial Complexity is. But the mere act of positing the concept worked as a trigger. As though, from the moment it was stated, a veil had lifted, revealing other intuitions that seemed to impose themselves:

- The potential of Infinity: the inexhaustible reserve of possibles, at once the source and the horizon of every emergence.

- The degrees of freedom arising from the quantum vacuum: that is to say, the full range of paths by which a reality can unfold, evolve, or interact starting from a given state, inscribing the dynamics of change at the heart of the real.
- The structure of the Present: a key to accessing consciousness, a flash of existence through which the share of the infinite potential becomes actualized, perceptible, and lived.

These three notions appear as the first sparks of the original matrix. They bestow upon the Universe its creative impetus by determining at once its dynamics (movement), its liberty (the multiplicity of possibles), and its highest vocation: the awakening of consciousness.

Sylvana

Before I ask you my next question… I would like you to clarify one point. You often speak of the spirit—in your poem, but also in your definition of the Primordial Complexity, which you link to the absolute Spirit. But what precisely do you mean by that? What, for you, is the definition of this word: spirit?

Thierry

First of all, it is essential to distinguish two notions: the spirit and consciousness.

In psychology and the cognitive sciences, the spirit denotes the totality of mental faculties: perception, memory, reasoning, attention, emotions—whether they are conscious or unconscious. It therefore covers the whole of our mental life, much of which operates in the background, implicitly or automatically.

Consciousness, by contrast, constitutes only a subset of it. It corresponds to what emerges into the field of immediate lived experience: that of which we are aware, what we experience here and now. Current research underscores this: the majority of our mental processes never reach consciousness—whether it be the regulation of breathing, reflexes, or the immense quantity of automatic processing that keeps us alive.

This is why I consider that the quintessence of the human does not reside in consciousness alone, but in a broader spirit, which encompasses the conscious and the unconscious. The spirit bears memory, intention, imagination, thought—and makes consciousness itself possible.

It is this phenomenal foundation that I take as my point of departure. Contrary to a widespread idea, the spirit is not solely a philosophical or religious notion: it is indeed recognized by science as a legitimate object of study. Psychology, the cognitive sciences, and the neurosciences explore its processes, describe its mechanisms, and measure its cerebral correlates.

Starting from this bedrock—the study of the spirit as a rigorous domain of the cognitive and neurological sciences—it becomes possible to widen the perspective. For if science already admits the existence of a multitude of marginal spirits (terrestrial humanity), nothing prevents one from conceiving, by conceptual homothety, that this principle could repeat itself on a broader scale. This transposition naturally leads one to posit the existence of a more encompassing Spirit, not directly measurable, yet capable of grounding and connecting the whole: what I call the absolute Spirit, or the Primordial Complexity.

This absolute Spirit must be understood as a structuring idea, the primal source. In this sense, it acts as the founding order from which everything proceeds—a rational hypothesis, born of the logical extension of an already recognized phenomenal fact.

Therefore, if the absolute Spirit is the primal source, it is consistent to affirm that marginal spirits bear its imprint. They are not separate from it: they are its local expressions, the singular manifestations of a universal reality.

Sylvana

Now that I understand better how you use the word spirit, I would like to return to your poem. In it, you evoke different kinds of spirits—the absolute Spirit, marginal spirits… Could you explain how you distinguish them? In your model, what is the difference between these levels of spirit?

Thierry

In my model, as children of the absolute Spirit, marginal spirits inherit omniscient power from the origin. However, although this power is present, it is not immediately usable: in a state of immaturity, the spirit is as though blinded by its own power and incapable of using it.

Thus, this omniscient power must first be considered as a latent potential, which only the progressive mastery of the spirit can render operative.

What distinguishes marginal spirits from one another, in my exercise in thought, is therefore their degree of evolution, that is, their level of mastery of that original power—omniscience.

From there, in my model, two states take shape:

- The omniscient Spirits (or omniscient Intelligences), who have achieved ascension and thus full mastery of their innate power.

- And the aspiring spirits, who share the same essence and carry within themselves the same original omniscient power, but in the form of a potential not yet mastered.

Thus, in my model, the function they exercise differs:

The omniscient Spirits, having learned to govern and to deploy this power, become creative intelligences, capable of engendering Universes by fragmenting their own Present: vast school-worlds, intended to offer marginal spirits a ground for experience and learning.

As for the aspirant marginal spirits, in order to evolve, they must incarnate within the school-worlds created by an omniscient Intelligence. It is through these experiences, within temporality, that they gradually learn to expand their Present and to deploy their power.

But for the entire system to remain coherent, there exists a fundamental principle: ÆKO Resonance. I describe it as a primordial tuning fork, a frequency of attunement that links each marginal spirit to the Absolute. It is what maintains the unity of the Whole, within the infinite diversity of worlds.

Sylvana

I understand better: you therefore distinguish these marginal spirits—some already having attained mastery, others still in training. And yet, in this path of evolution, you ceaselessly return to the role of the Present. In your poem, it is at times linked to the absolute Spirit, at times to the omniscient Spirit. Why does it hold so decisive a place in your model? And could you shed light on their difference?

Thierry

Yes, you are right, the Present is a central point in my model. It constitutes an absolute quality, a flash of existence emanating directly from the Primordial Complexity.

It is what grounds the very notion of existence: it is in the Present that one sees, feels, and recognizes. Without the Present, no consciousness can awaken, because only the Present offers an anchoring point, the capacity for being.

I assert even more: the Present is lucid-consciousness. It designates the pure state where consciousness is no longer scattered in projections of the future nor confined within the remanences of the past, but fully awakened to itself, here and now.

It is for this reason that, from the beginning, I posited the existence of an absolute Spirit: the first Principle, bearer of the absolute Present, from which all proceeds. This absolute Present cannot be fragmented. It Is in itself, indivisible, and contains everything.

It corresponds to the permanent and immutable Presence of the Primordial Complexity, what I call the primordial Now. This means that the Primordial Complexity is Master of Time: it is not subject to it, it engenders it. It contains within itself the entirety of possible realities, simultaneously and infinitely. In this sense, it can be understood as all-powerful, as the absolute order from which all proceeds.

Next comes the omniscient Present. When the omniscient Spirit—which I also call the omniscient Intelligence—engenders a Universe, it fragments its own Present into a multitude of instant-Presents. Time, at first free and lightning-like, gradually structures itself into variable cycles, constrained to unfold the Universe according to a hierarchical architecture of fractal school-worlds—structures that repeat at different scales, to which we will return later.

Finally, in my model, there is the conscious Present, as it is experienced by the marginal spirit aspiring to the mastery of its original power. What it perceives over the course of its existences is in reality only fragments of the omniscient Present. These fragments appear in the form of more or less dense instant-Presents, each fractal level thus corresponding to a differentiated level of consciousness: a technological deployment linked to the Source, offered as a tool that the aspiring spirit must use in order to experiment and progress through the school-worlds while remaining connected to the omniscient Intelligence. By progressively widening its instant-Present, this spirit emancipates itself and gradually becomes capable of accessing the full mastery of omniscience. We will return to this later as well.

. The Instant-Present — When Amplitude and Density Modulate

Sylvana
You say that the aspiring spirit emancipates itself step by step, until it becomes able to reach the state of active omniscience… by expanding its conscious Present.

What exactly does that mean? How can a Present expand?

Thierry
That is an essential question. When I speak of expanding one's instant-Present, I mean that the very duration of the lived instant can expand or contract, and that this variation radically changes the way consciousness perceives the real.

And that is where the difference becomes decisive: depending on whether the instant expands or contracts, consciousness does not experience the same thing.

The more the instant-Present expands, the more consciousness becomes capable of embracing in a single movement the coherence of the real: this is what I call omniscient Presence, a faculty of simultaneously perceiving a multitude of correlations without relying on any external support.

Conversely, when the instant-Present shrinks to a brief duration, consciousness can no longer sustain this continuity by itself: it must resort to memory, which reconstitutes the thread of instants and fills the voids. This is exactly what one observes in neurosciences: our brain, unable to maintain a continuous presence, artificially weaves the fabric of our lived experience by linking temporal shards that are too short to suffice on their own.

To truly understand my view, two fundamental concepts must be distinguished:

First, the amplitude of the present which I have just described: it is the duration of the lived instant. The greater it is, the more consciousness can encompass a set of events within the same bubble of presence. This develops a capacity for synthesis, a global and coherent vision.

Then we must take temporal density into account: it is the fineness of the present's granularity that determines the level of detail in the information perceived. This develops an analytical capacity, a precise and differentiated vision of the real.

Together, these two axes—amplitude and density—define the quality of conscious experience: one opens onto depth and continuity, the other onto precision and the richness of detail.

Sylvana

So if I recap: amplitude is like the size of the bubble of Presence—the larger it is, the more consciousness can hold events all at once. And density is like its resolution—the finer it is, the more each instant reveals details. In other words, conscious experience always depends on a balance between the size of the bubble and the richness of what it contains.

Listening to you, I think of a very simple scenario. I see three people who enter a restaurant at the same time and cast their first glance.

The first immediately notices music, people, an empty room next door, and three servers.

The second sees the same overall scene, but they already adds some specifics: they notice that the restrooms are at the back on the left, that the floorboards are new, and that it is somewhat dimly lit.

The third perceives all of that, but with still more detail for each thing: the color of the servers' uniforms—white, black, and red—the drinks being served on the counter, and even the specific song that is playing, a melody that reminds them of their vacation.

And then I have this friend—we all have one like that—who did not see any of that. The next day, when we are trading stories from the evening, it is always to them that we say, 'Come on, you did not see that? Or him? Do you not remember, they even played that song?'—well, I can relate to that!

Thierry

Your example is very telling, and it perfectly illustrates the four possible situations.

- The first person lives with great amplitude but low density: they perceive the scene as a whole, without entering into detail.
- The second keeps the same broad amplitude, but with medium density: they grasp the whole and already adds a few noteworthy specifics.
- The third combines broad amplitude and maximal density: they maintain the overview while capturing a multitude of very fine details.
- The fourth, your friend, lives the reverse: low amplitude and low density. They neither perceived the overall scene nor retained the details that could have struck them.

In other words, it is indeed the combination of amplitude and density that defines the quality of conscious experience. Depending on whether one or the other is strong or weak, perception can be global, detailed, both at once… or, on the contrary, almost absent.

Your restaurant example illustrates this mechanism at the individual level. But this logic does not stop there: in my model, I extend it to the cosmic scale, all the way into the school-worlds.

This explains why, in my model, different beings occupy distinct hierarchical degrees: each lives in a fractal world that corresponds to the temporal spectrum of his or her instant-Present.

What is important to remember is that, in my view, increasing consciousness does not consist solely in widening the instant-Present, but also in modulating its density and its amplitude.

- The vaster the instant, the more consciousness embraces the whole within a higher coherence.
- The finer the instant, the more the perception of detail intensifies.

The true expansion of consciousness consists in unifying these two polarities: breadth and precision, the whole and the detail, joined within one amplified Present.

Sylvana

So to evolve in consciousness is to learn to adjust both sliders at once: the size of the bubble and its resolution.

My restaurant example resonates with me, but I can see that it is still within a single fractal, let us say at the human level. Whereas with your model, you show how this logic applies to far more pronounced fractal worlds, with differences much more radical than simply seeing more or fewer details in an everyday scene.

But I wonder, how do you define an instant-Present?

Thierry

It is a question that calls for a rather technical answer. To address it, I will rely on the benchmarks established by the neurosciences and cognitive psychology, so that you understand on what I base my reasoning.

The scientific milestones of the human instant-present

- Minimal threshold of conscious perception

The brain needs a span of time to unify several events into a single coherent perception. This minimal time is estimated between 20 and 50 milliseconds.

Below this threshold, events are too brief to be perceived consciously: they go unnoticed.

- Temporal fusion window

In vision as in audition, two signals separated by less than 30 to 40 ms are perceived as simultaneous.

In other words, if two sounds or two images arrive too close together, consciousness fuses them: we believe we perceive a single event.

- Brain rhythms

Gamma waves (~40 Hz) correspond to a period of about 25 ms. This neuronal rhythm seems to play a key role in conscious integration.

Each cycle of gamma waves can be considered an elemental beat of consciousness, a perceptual quantum that structures our experience of the present.

One can compare it to the frames of a film: each gamma cycle corresponds to a fixed image, and their rapid succession gives the impression of a fluid and continuous present.

- Duration of the lived present

If we leave these minimal thresholds to consider subjective experience, the duration of the instant-present is much broader. William James spoke of the specious present, a temporal bubble of 2 to 3 seconds.

It is within this bubble that we integrate a gesture, a sentence, or a melody as a single unit of meaning.

To return to the cinematic image: if each gamma cycle corresponds to a frame, then this 2–3 second bubble constitutes a short scene, experienced as a single coherent moment.

We can therefore identify two fundamental benchmarks:

- 25 ms: the lower limit of the human instant-present. Below this, no conscious experience is possible.
- 2 to 3 seconds: the temporal bubble of the subjective present, in which consciousness aggregates sequences into an experience endowed with meaning.

In other words: 25 ms structures the minimal instant of consciousness, and 2 to 3 seconds define the horizon of the lived present.

Sylvana
So, if I sum up: our human present rests both on a tiny pulse—about 25 milliseconds, like a kind of elemental beat—and on a broader temporal bubble of 2 to 3 seconds, which allows us to integrate our gestures, our words, our

perceptions into a continuity of meaning. I understand that well.

But what does this variation from one person to another depend on?

Thierry

The number of instant-Presents perceived each second depends on the temporal granularity of a brain.

To be more concrete: if we take the minimal duration of an instant-Present, for instance, 25 milliseconds, we can compare it with a whole second. We simply see how many times those 25 ms "fit" into one second. That gives the number of instants lived per second.

If my neuronal system integrates information at a cadence of 25 ms, I have about 40 instant-Presents per second.

If another person possesses a finer window of simultaneity (for example, 10 ms), they can live 100 instant-Presents per second.

Conversely, if integration is slower (for example, 50 ms), that corresponds to 20 instant-Presents per second.

Having "more instant-Presents" per second therefore amounts to having a higher temporal density.

. At the Heart of the Temporal Field — Present and Resonance

Sylvana

Ah, I understand… in fact, it is not only the duration of a single instant that matters, but the number of instants we can fit into a given second. In other words, the more finely consciousness slices time, the more successive presents it experiences and the more information it captures in the same span: it is then more awake to the real, more present, more attentive.

I see more clearly now why you insisted on this point: it truly is the foundation for understanding consciousness—what it is, how to measure it, and how it determines the type of world with which it comes into resonance. That is why your reasoning is crucial: we are not describing a mere sensation, we are establishing an operational basis for consciousness.

So, in your model, to what is all this more fundamentally connected? How do you integrate it into the overall architecture—what you call the temporal field, the quantum ether we have already discussed?

Thierry

Your question is pertinent, and to answer it precisely, I need to return to certain foundations of my model. For what we are discussing—the instant-Present, its density and its amplitude—only fully makes sense when placed within a broader architecture.

As we have already mentioned, what physics describes as the quantum vacuum is not, in my model, a simple energetic background. I conceptualize it as a quantum ether: a living field, bearer of information and traversed by an omniscient Intelligence.

This intelligence does not merely exist in an abstract manner: it unfolds through Time, which becomes its dynamic expression, its operational language.

On this basis, I define what I call the temporal field:

- It contains the whole set of preconfigured realities, but also the integral memory of all that has already been actualized in phenomenal reality.
- It acts as a universal organizing matrix, conditioning space, matter, and consciousness itself.

Within this field, I then introduce the hypothesis of the chronon: the elementary quantum of Time, messenger of this dynamic, and vector of resonance between the absolute Present and our lived instants.

In this framework:

- The temporal field can be understood as a universal sea, a living substratum where all organization unfolds. It is not neutral: it bears the memory of all that has already been manifested, it contains the potential configurations of what may come to pass, and it orchestrates the dynamic by which the real is actualized.
- The chronons are its fundamental units, the smallest quanta of this temporal field. One might compare them to photons for light or to quarks for matter: they constitute the elementary, indivisible elements that convey Time. But their role is more subtle: each chronon is at once a vector of instantaneity and a bearer of information.

For in my model, Time is not a mere mechanical succession of instants; it is a living informational weave. Each chronon, by virtue of its polymorphic nature, encodes a quantity of information that links three dimensions:

- The past, through the memory of what has already been actualized;

- The present, through the flash of an instant that manifests;
- The future, through the reserve of potentialities still in gestation.

It is this capacity of the chronon to change form and function, according to the level of fractal density at which it operates, that makes possible the manifestation of the phenomenal Universe in all its diversity.

Thus, as a messenger of the omniscient Intelligence, one of its functions is to translate its timeless omnipresence into an ordered succession of instants perceptible to the conscious-spirit.

Without the chronon, the Universe would be nothing but an undifferentiated totality, without rhythm or history. Thanks to it, the real unfolds in distinct and intelligible sequences, offering consciousness a terrain where it can exert itself, learn, and evolve.

And now, to answer your question, in my model, the instant-present is not a simple, indivisible unit. It is composed of a multitude of chronons—those elementary quanta of time—which, grouped and synchronized, give rise to what we call a conscious instant. In other words, the lived present is not a fixed point, but a dynamic assemblage of temporal micro-units.

These chronons, however, do not add up at random: their coherence comes from the temporal field, that universal matrix which organizes them. It is what aligns them, brings them into phase, and allows a stable and perceptible present to emerge. Without this synchronization process, there would be no experience of time, only a chaotic dispersion of fragments.

Sylvana

It is fascinating… So, without this synchronization, there would be only a chaotic flow of temporal fragments.

But then, what allows some people to capture or synchronize more chronons than others?

Thierry

That is an excellent question.

As I told you, in my model, each instant-Present corresponds to an aggregation of chronons—those elementary quanta of Time—that come into phase within the temporal field. This temporal field, rooted in the quantum ether, does not remain abstract: it manifests concretely in the phenomenal Universe through the structuring deployment of its chronons. It is they who, by stabilizing and crystallizing, give rise to particles, fields, matter, and ultimately to observable phenomena.

In other words, reality is not made of isolated objects, but of a dynamic mesh in which every manifestation is connected to the whole by the weave of Time. And this becomes more intuitive when one recalls that matter itself is composed of 99.99% quantum vacuum: it is therefore not the "solidity" of matter that grounds the real, but the vibratory coherence of the temporal field that sustains it.

Thus, the more a consciousness captures and synchronizes chronons, the more it accesses a high temporal density, because it absorbs and processes more information.

That is to say, a conscious-spirit can quite literally live more instant-Presents per second by tuning itself, through vibratory resonance, to the chronons of the temporal field.

This immersion in a denser flow results not only in an acceleration of lived experience, but also in a qualitative expansion of perception: each instant becomes richer, more nuanced, more informed.

And with this broadened perception comes a direct consequence: an increased capacity for action, because consciousness has more data, more clarity, and can orient its choices with a finesse unattainable in a lower-density state.

In other words:

My "perceptual quantum" (≈ 25 milliseconds for an average human) is not a universal value. It corresponds to the vibratory resonance between the chronons of the temporal field and our biological and cognitive system.

That is to say, human consciousness does not grasp the flow of time continuously, but in "packets" of the Present.

This temporal granularity can vary depending on several factors:

- Biological: brain rhythms, speed of neuronal processes, synaptic plasticity.
- Psychic: meditation practices, altered states of consciousness, contemplative immersion.
- Evolutionary: each living species possesses its own perceptual cadence, adapted to its environment.
- Technological (in my hypothesis): it would even be conceivable to develop devices capable of directly modulating this temporal granularity.

A few concrete examples illustrate this principle:

- Average human: 25 ms → ≈ 40 instant-Presents/second.
- Advanced meditator: 10 ms → ≈ 100 instant-Presents/second → finer, richer perception, integrating more conscious events within the same objective duration.
- Fly: 4 ms → ≈ 250 instant-Presents/second → which explains why it perceives our movements as slow and has time to react before we touch it.

Each being therefore lives in a specific layer of temporal granularity of the fragmented Present, determined by its coupling to the temporal field.

In my theory, these differences are explained by Fractal Temporal Density (FTD)—a key concept that I will develop later, when we address fractality in my approach.

Sylvana

So, even if we have the impression of sharing the same experience of time in everyday life, in reality, each person lives their present differently, according to their own temporal density, which shapes a way of experiencing the world that can be entirely different.

Now I understand why I never manage to catch flies: they must have a good laugh seeing us flailing about in slow motion!

But, more concretely, if we posit the hypothesis of chronons, what does that change for science?

Thierry

You have understood correctly: two people can share the same physical world, yet not live the same density of the present—and this is true for all other living beings as well. The one who has more instant-Presents experiences, in a certain sense, more reality within the same objective duration.

This is where the implications become profound: it opens a new understanding of the subjectivity of time—no longer as a simple psychological impression, but as a fractal property of the temporal field, modulated by consciousness.

- From the subjective to the measurable

By introducing the chronon as the elementary quantum of the temporal field—which, as you know, I identify with the quantum ether—I propose a conceptual basis that makes lived time potentially measurable.

Each instant-Present can be envisioned as an aggregation of synchronized chronons. The present then ceases to be only a psychological impression: it can be described as a minimal temporal granularity, opening the way to a quantization.

- Link between consciousness and the temporal field

In my model, the duration of an instant-Present is not fixed: it depends on the number of chronons that a consciousness is able to synchronize with the temporal field. I call this fractal temporal density (FTD).

Thus, consciousness is not a mere spectator of time: it interacts directly with the fabric of the Universe.

Depending on its degree of resonance, the instant-Present can:

Widen in amplitude, by integrating more events into a single bubble of presence,

Or become denser, by condensing each second in order to integrate more details.

In my model, FTD becomes an indicator of the coupling between consciousness and the temporal field.

- Non-universality of the human present

Our temporal bubble of 2–3 seconds, described by William James as the specious present, is only a particular configuration.

Other consciousnesses—animal, meditative, or belonging to other worlds—could live a present that is more or less ample (widened or restricted bubble) and more or less dense (more or fewer details per second). This opens up the idea of multiple presents, inscribed within differentiated time fractals.

- A modulable interface

If the instant is born from the synchronization between the chronons of the temporal field and the consciousness that captures them, then it is not fixed: it becomes a modulable interface.

Attention, neuronal rhythms, or certain states of consciousness can modify this synchronization, either by widening the amplitude of the instant or by increasing its density. The evolution of consciousness consists precisely in learning to inhabit this interface more consciously.

- The ÆKO Resonance

Even if each consciousness lives its present in a singular way—more or less ample, more or less dense—none is isolated.

The ÆKO Resonance acts as a universal tuning fork: it links all these variations to the same foundation, maintaining the attunement of each conscious experience with the Absolute.

Thus, the subjectivity of time is not a chaotic dispersion, but a diversity attuned to one and the same cosmic coherence.

Sylvana
What you say here reminds me of your childhood stories… Those moments when you told me you felt Time differently in nature: near a tree, gazing at a flower, or simply by letting yourself be absorbed by silence.

Thierry
Yes, those experiences were decisive. Very young, I had already experienced those variations of temporal density without being able to name them. I perceived that Time could dilate, slow down, or, on the contrary, intensify in a flash. When I began to conceptualize my theoretical model, those memories asserted themselves as precious inner landmarks: they became points of resonance for understanding what it truly means to inhabit different regimes of the present.

That is why, in building the architecture of my theory, I did not rely solely on data from physics or cosmology: I also integrated this lived dimension. My experiences as a child did not remain mere subjective impressions; they found their place within a broader coherence, that of a fractal universe where consciousness interacts with the temporal field.

Later, thanks to my research, my experiments, and the proofs of concept that I was able to establish, I understood in a more rational way what I had perceived as a child.

Sylvana
I understand… It makes me think of all those people who also describe impactful experiences of Time: moments when they were so present that they felt a total unity with the Whole. Since then, they mention longing to rediscover that unique, almost magical moment. And it is not limited to those people: everyone talks about "living in the present moment" and seeks it in

one way or another—through meditation, attention, or simply by recentering themselves…

So, in your view, why do we carry such a profound desire to live in the present moment?

Thierry

What you express, essentially, is the quest for true happiness—not a passing or superficial state, but that which is born of intimate accord with the deep Self: the lucid spirit connected to the absolute Present, to its original I Am. That is the ultimate aim.

Meditation can give us a foretaste of it, a resonance. It does not yet offer definitive transcendence, for we always return to our material condition. But it acts as a precious renewal, a return to the source that purifies us, elevates us, and accelerates our inner and spiritual evolution.

And this is precisely, according to my model, one of the fundamental gifts of the Primordial Complexity: having brought forth, through the omniscient Intelligences, school-Universes—fields of experience where marginal spirits can incarnate, be formed, and be revealed.

Each world becomes a matrix of learning, a framework where the spirit learns to master itself, to transform illusion into knowledge, experience into discernment. From world to world, from scale to scale, aspiring spirits thus advance toward their omniscient Present, that point of accord with the Absolute which constitutes the source of indestructible happiness.

This intuition marked me forever. It is what led me to conceive of the idea of a fragmented Present, lived according to variable intensities—this is what I had already mentioned under the name of Fractal Temporal Density (FTD). For if the Present can widen or contract depending on the level of consciousness, then there must indeed exist a structuring framework, a field capable of containing and ordering these variations.

And this framework could not be reduced to an isolated Universe. It had to be inscribed within a vaster, deeper weave: a fractal architecture where each world itself becomes a quantity of Present set within an even greater whole.

. The Fractal Multiverse — A Cosmic Weave of Nested Realities

Sylvana

So… before we go into the details of this fractal architecture, can you tell me how you came to this concept?

Thierry

At the beginning, my reflection concerned the omniscient Intelligence specific to our Universe. But very quickly, a question arose: how could it be isolated from its primal foundation? For the Primordial Complexity, as I conceptualized it, is not the architect of a single Universe, but indeed of all Universes. From then on, I had to broaden my perspective: to think not in terms of a single Universe, but in terms of a global multiverse.

By positing this process, I was led to the second fundamental element of my model: the global multiverse, which I subdivided into two complementary components:

The extrinsic multiverse: it corresponds to the evolution of each Universe taken as an autonomous entity, inscribed within the totality of Universes engendered and governed by the Primordial Complexity.

The intrinsic multiverse: it resides within each Universe and manifests in the form of many worlds, destined for the progressive evolution of conscious-spirits, under the regency of the omniscient Intelligence.

Sylvana

If I understand correctly, you look at the Universe on two different scales, yet according to a single logic. On the one hand, there is a plurality of Universes, linked to the Primordial Complexity—that is what you call the extrinsic multiverse. On the other hand, within each Universe, you describe another multiplicity of worlds, structured for the evolution of conscious-spirits—the intrinsic multiverse.

Thierry

Yes, exactly. And to make it clear, I will begin by elaborating on the first level: the extrinsic multiverse.

In my model, the extrinsic multiverse does not represent a random juxtaposition of Universes without any connection among them. I conceive of it as a living, organized, hierarchical whole—a global system orchestrated by the Primordial Complexity itself.

Each Universe within it is a distinct unit, with its own laws, its spatiotemporal fabric, and its singular history.

One can imagine it as a potentially infinite system of Universes, each a distinct reality, nested within one another like nesting dolls. These Universes are autonomous, each evolving according to its own laws within the wheel of becoming, that is, within a continuous cycle of transformation and renewal.

And as I mentioned to you, despite their autonomy, all these Universes remain linked by one and the same fundamental resonance: the Universal Resonance ÆKO, which connects them directly to the Primordial Complexity. It is this resonance that ensures their overall coherence, their invisible harmony, and that inscribes each Universe within an architecture vaster than itself.

Moreover, this resonance sets the rhythm of the cycles among the Universes, alternating their phases of activity and their periods of regenerative rest. From this concept, I arrived at an essential notion: the cyclical character of the Universes. In my view, each Universe has a birth and an end, passes through phases of expansion and compression, then regenerates with renewed momentum. Thus, in my model, the Universe is inscribed in an evolutionary cycle within an infinite network of cycles.

This hypothesis offered me a decisive key to move beyond the conceptual dead end of the Big Bang.

For within the framework of an extrinsic multiverse, the classic question—"What was there before the Universe?"—ceases to be an insurmountable wall and becomes an invitation: what if a prior Universe did indeed exist, paving the way for ours?

It is at this precise point that the extrinsic multiverse makes it possible to reformulate the Big Bang not as an absolute origin, but as a transition within an infinite dynamic. We will return to this crucial point in greater detail later.

Sylvana

All right, I have a clearer sense of what you mean by the extrinsic multiverse in your model: a set of nested Universes, each autonomous, passing through its own cycles of expansion, contraction, and renewal, yet all linked to the Primordial Complexity.

Thierry

Yes, you have captured it well.

And after I had conceptualized the extrinsic multiverse, I hypothesized that the same logic had to apply within each Universe itself—which I call the intrinsic multiverse.

By homothety: what holds true at the external scale—Universes nested within a multiverse—had to be reflected at the internal scale. This is why I conceptualized multi-worlds within a single Universe.

Starting from this vision of the multiverse—with its two faces, extrinsic and intrinsic—other notions began to impose themselves on me as natural extensions. They arose not only from deductive reasoning, but also from a structured intuition, a kind of inner vision that let me glimpse the mechanisms that made this fractal architecture functional:

- A living fractal network, unfolding across several levels of organization, from the subquantum to the cosmological;
- Each world being composed of Fractal Temporal Densities (FTD), each FTD corresponding to a distinct Present Time, that is, to a specific mode of perception and manifestation;
- A multidimensional model, where each hierarchical level adds not only complexity but also new degrees of freedom: more dimensions of Time, space, and consciousness.

Sylvana
In short, you are describing the very architecture of the real as it structures itself during the fragmentation of the omniscient Present: fractal levels, each defined by its own FTD, and a complexity that, at each hierarchical degree, opens new freedoms of time, space, and consciousness.

Thierry
Yes, that is exactly it. You make the connection well: it is precisely the product of the fragmentation of the omniscient Present that directly generates this dynamic differentiation, giving rise to these multiple levels of worlds. Each corresponds to a particular modulation of temporal density, giving rise to a hierarchy of fractal realities.

This is where the notion of temporal multidimensionality takes root: not as the addition of parallel spaces, but as the very depth of a temporal consciousness unfolding in successive strata, each endowed with its own coherence.

Sylvana
I can clearly see this immense fractal architecture emerging: these levels of organization, these rhythms, this nested logic that runs through the whole. In your model, everything seems articulated with rigor, as though each part answered to an overall coherence.

Thierry
Yes, everything is there—the structure, the logic… But despite the rigorous coherence of this fractal architecture, one essential enigma remained: how could the conscious-spirit truly inscribe itself within it, and above all, how could living evolution be made possible through this structure?

In conceptualizing the intrinsic multiverse through my mental projections, one thing became evident: the fragmentation of the omniscient Present does not merely generate hierarchical worlds. It simultaneously engenders a fundamental imbalance. The original permanent Present is continuous, total, unified. But as soon as it fragments into quanta of Time, into distinct temporal densities, it introduces a discontinuity.

It was at this stage that I understood that something was missing to maintain coherence: universal functions capable of compensating for this discontinuity. Faced with this imbalance, the omniscient Intelligence brought forth two universal functions, information and memory:

- Information organizes the fragments, giving them a logical, structured order.
- Memory links these fragments, ensures continuity, and maintains a narrative thread within discontinuity.

In other words, in a fragmented Present, information and memory are not original properties: they are instruments created by the omniscient Intelligence, deployed so that conscious-spirits may evolve and sustain themselves within coherence.

This is why I say they are secondary and artificial with regard to the Absolute: they do not exist in themselves, but only as responses to the inherent characteristic of the fragmentation of the omniscient Present.

They are the language and the agent of cohesion of discontinuous Time, necessary to support the conscious-spirit in its evolutionary learning.

Sylvana
I see… According to your model, information and memory appear as functions intended to maintain coherence within a fragmented Present. I better understand why the lower the FTD, the more indispensable memory becomes, as you were explaining earlier.

Thierry
Yes. If information and memory are necessary, it is because we cannot sustain continuous and total attention on each passing instant. In a fragmented Present, the conscious-spirit accesses only a fraction of the information available at each moment. Everything that is not captured vanishes—unless a principle such as memory preserves its trace.

In other words, the lower the FTD, the shorter the available Present, the greater the loss… and the more indispensable memory becomes for maintaining coherence and ensuring the continuity of experience.

In a fragmented Present, memory becomes the active principle that allows experience to leave a trace, information to be structured, and the conscious-spirit to integrate what it cannot grasp in a single instant. Memory restores the conscious-spirit's access to information that, without it, would be lost.

This is why, in my model, I grant these two functions a fundamental role:

Information as a structuring principle: each phenomenon is the expression of a particular organization of information. All reality rests upon information, which makes possible the perception, the understanding, and the integration of the invisible laws that order the real. By integrating this information, consciousness broadens its field of liberty and its power to act.

Universal memory: Time is not only succession, it is also an archive. Each manifestation leaves an imprint that the temporal field preserves, organizes, and makes accessible. This universal memory ensures continuity, prevents information from dissolving, and allows consciousness to weave its evolutionary path.

Thus, without information, there is no meaning. Without memory, there is no guiding thread. But with both, consciousness finds the levers to rise, to transform itself, and to participate actively in the evolution of the real.

Sylvana

If I understand correctly, you have just shown that, without information or memory, no consciousness could sustain itself in a fragmented time—and that each conscious-spirit depends, to varying degrees, on memory, according to its capacity to capture, connect, and structure information.

Thierry

Yes, that is exactly it. And it is precisely this difference in the relation to information and to memory that led me to broaden my reflection.

In my model of a fractal multiverse, several consequences have emerged. Among them is the idea of a diversity of conscious-spirits, structured in a genuine hierarchy: some fully self-conscious, others simply conscious, and still others remaining unconscious—each evolving within material worlds that are more or less dense, or subtle.

Sylvana

I believe I understand better, now. In your model, the fractal multiverse is not an abstract construction, but a living framework that links the organization of the world to the path of evolution of consciousness.

And this complexity is not a dead end, but a passage—better yet, a form of inner training, almost initiatory. Each degree of fragmentation compels the conscious-spirit to integrate more information, to broaden its understanding, and to refine its resonance with the field. Until it can ascend the fractal weave—that structure of nested levels—and recover the unity of the absolute Present, the source where all the strata dissolve into the same timeless concord.

Thierry
Yes, that is exactly it. You can see that this fractal framework is not merely a theoretical abstraction: it directly engages the way consciousness evolves as it traverses the different levels of reality.

Behind this inner training dynamic that I describe, a broader perspective opens. This second element—the fractal multiverse—not only allowed me to establish a more rigorous framework; it also revealed an expanded vision of the diversity of life.

It suggests a multiverse that is not cold nor mechanical, but living, vibrant, interconnected, animated by a cosmic memory—a memory of the Universe itself, which preserves the trace of everything that comes to pass.

One then discovers a vast network of nested school-worlds, where each consciousness advances through different levels of reality: it learns, it transforms, and it contributes to the evolution of a whole far greater than itself.

Sylvana
In this schema, we ultimately appear as the participants—whether conscious or not—of an immense living process.

Thierry
Yes. And you see… from there, once that structure was established, I had to push the reasoning further.

If each Cosmos is a singular expression of omniscient Intelligence, then it cannot be left to chance: it must obey a precise internal logic—a logic not only mechanical, but also rhythmic and harmonic.

In other words, how to connect the different levels of reality to one another, while respecting their autonomy, within our own Universe itself?

It is this intuition that led me to formulate my third fundamental element: the fractal arrangement of the real, a principle of organization in which each level is nested within another, in which each part reflects the whole, and in which all the strata of the multiverse naturally find their place within a coherent and living architecture.

· Temporal and Spatial Fractals — A Rhythmic Architecture of the Real

Sylvana

So what you are going to explain to me in more detail is how this intrinsic multiverse is arranged in concrete terms within our Universe—how the fractals organize themselves, with their own internal logic, while remaining connected to a vaster, living architecture?

Thierry

Yes. And that is precisely what this third element formalizes.

It shows that our Cosmos is a living organism, animated by its own internal worlds, structured by profound rhythms and temporal cycles that unfold at different scales.

Each level of reality pulses according to its proper frequency, and it is from the articulation of these cycles—of their resonance and their balance—that the coherence of the whole is born.

Sylvana

I believe I am beginning to see what you set forth here… You are not describing a mechanical system, but something living, which organizes itself from within. A balance that arises from the rhythms themselves and from the way each level vibrates and attunes to the others, ensuring that everything remains coherent.

But as I listen to you… you say "our Cosmos," and sometimes "Universe." I wonder: is it the same thing or not?

Thierry

If I had to distinguish the terms, I would say that the Universe is the container and the Cosmos, the content.

In practice, they are two ways of designating the same totality.

In common usage, we speak of the Cosmos to evoke what we perceive: traveling through the Cosmos, contemplating the starry sky, observing visible matter and energy. This is also why we speak of cosmology and not "universology."

But in my model, this notion takes on a new depth. The Cosmos is neither simple nor uniform: it is structured from within by a fractal and hierarchical logic.

This is what I have called the intrinsic multiverse: a nesting of realities, levels of complexity, and worlds embedded within one another, linked by an internal coherence that transcends them.

Sylvana

You speak of a fractal logic in the organization of the real... But what exactly do you mean by fractal? How do you define it in your model?

Thierry

A fractal, in the sense in which I use it in my model, is not a simple mathematical figure, but a principle of organization: a form that repeats across scales, never identical, always modulated by the same logic of variation.

In my model, this organization rests on the interlocking of temporal and spatial fractals: patterns that repeat with variations at different scales of time and space, each bearing its own density of the Present.

As one ascends this hierarchy, each level offers more liberty for conscious-spirits: new temporal dimensions appear, unfolding unprecedented degrees of space, perception, and action.

Thus, the fractal levels compose a hierarchical system, where each stratum reflects the totality—not as a mechanical copy, but as a singular variation of one and the same universal order.

Sylvana

I believe I understand better what you mean by fractal structure. Each level reiterates the logic of the previous one, yet with a variation, a nuance—it is never a simple copy, but a repetition that adapts, that evolves.

It is a bit like a snowflake, is it not? The same form unfolds at different scales, and yet the whole remains perfectly coherent.

Thierry

Yes, your image comes close. It is exactly this kind of logic—repetitive, evolutionary, always attuned to an overall form—that led me to delve deeper into this structure. And the further I went, the more this fractal interlocking asserted itself. It was impossible to explain the internal coherence of the Universe, from the microscopic to the cosmic, without acknowledging an organization at once recursive and self-similar: a dynamic and ordered structure, where forms repeat at different scales like variations on a single theme.

In my model, this fractal logic takes on a very concrete expression: each level of organization possesses a density of the Present, that is to say, the amount of information that a consciousness can perceive and integrate at a given instant. This density is not the outcome of chance: it is directly determined by the temporal density specific to the field that structures this level of reality. In other words, the temporal field is universal, but its manifestation varies according to the fractal scales. And it is this variation—in informational richness and in the rhythm of cycles—that determines, at each level, the density of the Present accessible to consciousness.

Thus, each level is not an isolated fragment. Because it is fractal, it carries its internal coherence while resonating with the whole. This is what I mean by self-similar structure: an organization that repeats at different scales, always according to the same fundamental logic, yet with its own variations. Each level is autonomous, yet connected, for it proceeds from the same principle of structuring.

This articulation between local autonomy and global coherence led me to formalize a model capable of linking local and global dynamics, of accounting for cycles, thresholds, and transitions of consciousness—all while honoring the elegance of the laws of physics.

It is this rigor that gave rise to this third element. It enabled me:

To conceive of a Universe finite in its structure, but infinite in its fractal loops and its subsystems. In other words, a Universe whose form is determined, but whose unfolding remains open, always capable of generating new complexities without ever departing from its overall logic;

To visualize the Cosmos as a living, pulsing system, where the worlds that compose it alternate between perpetual expansion and contraction;

To deepen the logic of the nested fractals of the intrinsic multiverse, by shedding light on the nature of and the passage from one fractal level to the next.

In summary, this third element formalizes the Universe as a cyclical fractal organism, whose laws and dynamics repeat to infinity in forms that grow ever more complex. Yet all remain unified by the continuity of the temporal field, which articulates both the objective density of Time and the experienced density of the Present.

Sylvana

What you have been showing for a while now truly gives me the impression of a weaving—precise, patient, evolving—where nothing seems left to chance. Each level takes over from the previous one, like a sequence of movements joined by the same impulse.

And already in your poem, the school-worlds were not left to themselves: there was an orientation, a design. As though reality itself, in its fractality, hinted at an intention…

Thierry

Yes… what you have just said is very true. Because, at a certain point, a question becomes inevitable: if all this is so precise, so structured, so rhythmic… what guarantees its order? And above all… what allows each Universe to evolve, to transform, without breaking apart, nor losing its profound unity?

This is where another point seemed essential to me. As we have already discussed, behind such structural elegance and such a capacity for adaptation, one had to envisage something other than mere chance, or a blind mechanism left to itself.

What was needed was an operating principle, active, living, present at the very heart of each Universe.

A form of immanent Intelligence, arising from the Primordial Complexity, yet inscribed directly within its own cosmic field.

Not an external intelligence that would impose its laws from outside, but an intelligence internal to the fabric of the real, which ensures its coherence, its continuity, and its evolution.

We have already spoken of this earlier, but it is time now to enter into more detail to show how this element is a major element in the architecture of my model: the omniscient Intelligence.

- Omniscient Intelligence — A Living Architect at the Heart of Each Universe

Sylvana

It is true that we have already seen how central the concept of omniscient Intelligence is in your model—you have mentioned it more than once. So I wonder: was there a precise moment when you thought to yourself that something was missing?

Thierry

This idea came to me in response to a question that classical models leave unanswered: how can a Universe manifest both such precision of organization and such a capacity for evolution, without an operating principle to ensure its coherence?

It was in the face of this conceptual dead end that I introduced this fourth element: the existence of a second ordering principle, present everywhere at once, creator and developer of its own Cosmos, arising directly from the

Primordial Complexity.

And I agree with you, we have already discussed the role of omniscient Intelligence in previous conversations, but here I will offer a structured synthesis—in order to clearly recall its nature, its function, and its place in my model.

Omniscient Intelligence acts as an active projection of the first source: For its own Cosmos, it concentrates all the properties of excellence inherited from the Primordial Complexity—omniscience, omnipresence, creativity, organizational power. But it expresses them in a contextualized and limited way, for it is not an absolute source: it is an active servant of the Primordial Complexity, which acts in its name and extends its design within a particular cosmic framework.

Each Universe thus becomes an autonomous entity bearing its own coherence, yet always linked to the Primordial Complexity from which it proceeds.

Within its Cosmos—the intrinsic multiverse—the omniscient Intelligence manifests fractal worlds through the gradual fragmentation of its own omniscient Present. Each fragmentation generates a rhythm, a specific temporal density—its FTD—and consequently the degree of consciousness and complexity that can unfold.

The evolution of conscious-spirits—of which our humanity is a part—depends directly on this fractal density of time

Sylvana
What strikes me is the breadth of its effects: this principle, which you call omniscient Intelligence, seems to shape everything—the organization of worlds, the structuring of laws, the dynamics of forms. One senses it acting at every level as a subtle force and truly playing a decisive role in your model.

Thierry
Yes, introducing omniscient Intelligence as an active principle at the heart of each Universe was a true turning point in my work. It allowed me to conceptualize:

- A dynamic conception of intelligence, present at every level of organization of the real;
- A notion of disseminated, hierarchical consciousness, capable of self-reflection;

- And, as I had already mentioned, an active vision of the void—not as nothingness, but as an intelligent vibratory matrix, which I conceptualized under the name quantum ether: an informational, structuring field, the invisible link of all cosmic coherence.

Sylvana
I see that within its universe, omniscient Intelligence unfolds in a highly active manner: it structures fractal worlds, defines their rhythms, and sustains their evolution. But you also specified that it was not an absolute source, that it remained connected to the Primordial Complexity. In your model, how do you conceive of this link?

Thierry
Indeed, each Universe deploys its intelligence within the framework of its own temporal field, where it orchestrates laws, rhythms, and evolutions. But its omniscient Present is not isolated: it enters into vibratory resonance with the other universes, like the meshes of a vast cosmic network through which the primordial order circulates. In other words, each Universe is at once singular and connected: bearing its own identity, yet never separated from unity.

In the same way that there is a universal memory within a universe that links everything, there exists, at the scale of the extrinsic multiverse, a collective resonance that links the universes to one another. It is through this universal vibratory resonance—which I name ÆKO—that the cosmic unity is fulfilled.

It is this invisible Unity that impresses upon evolution a global direction, a breath passing through every level. It ensures that, even if the dynamics follow cycles or loops, they never simply return: they progress, grow more complex, and rise.

Sylvana
When you say that this invisible Unity would give cosmic evolution a global direction, do you mean that this dynamic operates at the two scales you had already mentioned? That at the intrinsic level, each Universe is structured to allow for its internal evolution… but that at the extrinsic level as well, the ensemble of Universes connected to the Primordial Complexity in turn follows a logic of evolution?

Thierry
Yes. By conceptual homothety, I applied the same logic to both types of multiverses.

In my model:

Within a Universe, the dynamic of the intrinsic multiverse is expressed by the progressive evolution of aspirant marginal spirits, who journey through fractal worlds to increase their degree of consciousness.

At the scale of the extrinsic multiverse, this dynamic is manifested by the evolution of the omniscient Intelligences themselves: each elevates its own Universe toward a higher degree of organization, coherence, and structural power, like a quest for excellence that aims to draw ever closer to the absolute of the Primordial Complexity.

Each Universe, in this hierarchical succession, thus expresses growing excellence: one further step toward the ideal of a perfect universe.

This ascending logic is the key to fractal evolution at every level: within a Universe through the elevation of conscious-spirits, and from one Universe to another via the elevation of the intelligences that animate them.

This is why this fourth element formalizes a fundamental idea: the possible personification of each universe. Not in an anthropomorphic sense—with human will or human intention—but as a living organism, structured, evolving, animated by the immanent intelligence inherited from the Primordial Complexity.

Sylvana

I understand better… In your model, each Universe preserves its autonomy while integrating into a global architecture. Everything seems constructed to follow this movement, each element naturally finding its place, always oriented toward a goal of progression.

Thierry

Yes… you now understand why this silent progression appeared to me as the key to all living architecture. And when one examines such an organization in depth, one understands that it cannot hold without an operating principle; the edifice of the Universe would collapse at once.

This intelligence, however vast it may be, cannot remain an abstraction. It needs a support, a matrix-form in which to incarnate, to unfold within the reality that gives the Universe its profound respiration—the capacity to sustain its structure, to conserve memory, and to regenerate.

It is this support that I will reveal to you now.

. The Hyperspherical Torus — A Geometry of the Cosmos

Sylvana
You mean that even an intelligence that vast needs a hosting architecture— a structure capable of making it active and operative?

Thierry
Yes. My fifth element reveals what I consider to be the fundamental structure, the true operative heart of the phenomenal Universe.

It is not a mere abstract geometric figure, but a living form, capable of receiving and ordering the flow of Time, information, and consciousness. It is both their receptacle— for it gathers these flows— and their engine— for it sets them in motion. In other words, a dynamic matrix in which the Universe can unfold, fold back upon itself, and preserve the memory of its own evolution.

This is why I describe it as the primordial cosmic technology: the very architecture that makes possible the existence of the real, its continuous transformation, and its perpetuation within an infinite coherence.

In a word, it manifests phenomenological reality in its active essence.

I have named it: the multidimensional fractal hyperspherical torus.

Sylvana
Ah, there it is, I felt it coming… You are unleashing one of the most intimidating terms I have ever heard! "Multidimensional fractal hyperspherical torus"… nothing less! So, go easy— I am ready, but I am going to hold my cosmic breath.

Thierry
I know… the name can seem intimidating. But behind these words lies a surprisingly simple idea. Do not get stuck on the technical vocabulary: you do not need to grasp everything at once. What matters is to perceive what this structure makes possible.

Imagine a cosmic pulse, an infinite breath. The torus is not a fixed geometric figure, but a living and dynamic form, capable of opening the universe in its expansion and then closing it again in its contraction— like a breath pulsing to the rhythm of Time itself.

This movement of expansion and contraction corresponds to what cosmology sometimes calls an oscillatory or cyclic model. But in my model, it takes on a more profound meaning:

Expansion unfolds the universe and multiplies its structures.

Contraction tightens and integrates into the cosmic weave the memory of these structures, ensuring their transmission to future cycles.

The torus thus acts as a universal memory: it inscribes and restores the complete history of the universe, enables the emergence of nested fractals, and encodes the trajectories of evolution of all forms—matter, energy, consciousness.

In this sense, this multidimensional fractal hyperspherical torus is far more than a diagram. It plays the role of a universal architect:

- It defines the topology of spacetime, its global shape.
- It modulates the cosmic dynamics, the flow of expansions and contractions.
- It orchestrates the interaction of fractals at all scales, from the quantum to the galactic.
- It opens the way to an infinity of possible structures thanks to its intrinsic polymorphy—its capacity to metamorphose according to conditions.

If it is capable of this, it is because it is born of the temporal field, and because its fundamental building blocks are chronons.

Ultimately, this fractal hyperspherical torus is not just an abstract concept: it is the beating heart of the phenomenal universe, the key that links memory, coherence, and cosmic life.

Sylvana
In your fourth point, you speak of the polymorphy of the torus—this capacity to metamorphose according to conditions and to generate an infinity of possible structures. Could you clarify further what you mean by that?

Thierry
Yes, of course. Polymorphic means that it can change form depending on conditions. This is what allows it to act as a living matrix, a living geometry: a structure receptive to the omniscient Intelligence, capable of translating its action into forms, rhythms, and trajectories of evolution. Thus, the torus is not fixed: it adapts, it metamorphoses, and it is this flexibility that allows the omniscient Intelligence to orchestrate the Universe as a true architect.

By "living geometry," I mean a structure that is not fixed in a single form, but that evolves, reacts, and interacts with what it contains. It does not merely delimit the universe: it is an active part of it, which sustains it, organizes it, transforms it, and preserves its memory.

It is thanks to this model that I have been able to link different intuitions into a coherent whole, for example:

- To conceive of the universe as holographic, where each fractal contains the memory of the Whole.
- To understand how local and global structures remain connected in a fractal and dynamic logic.

Sylvana

You speak of a holographic Universe… what exactly do you mean by that?

Thierry

It means that each fractal, however tiny it may be, bears within itself the imprint of the Whole. In other words, the Whole is present in each part—and it is this principle that makes possible a memory distributed across all scales. This memory is not centralized in a single point, but inscribed in the very fabric of the Cosmos, like living information that circulates everywhere.

By conceiving of the Universe as holographic, two major properties became apparent to me, in addition to memory:

The existence of homothetic relationships among fractals: correspondences of scale, form, and function that guarantee coherence across all levels of reality.

The variation of Fractal Temporal Densities (FTD): it reflects the richness in information and the rhythm of internal cycles, directly linked to the intrinsic complexity of each fractal level.

By unifying these elements, I was able to bring together ancient intuitions, scientific observations, and geometric structures to formulate a coherent vision. From then on, the torus ceases to be a simple mathematical figure: it becomes a living articulation, the junction point between the hidden order of the Universe, the memory of forms, and the creative movement.

For me, this fifth element is the supreme structural apparatus of the phenomenal Universe:

- An organizer of cycles, which sets the rhythm of expansion and contraction.
- A motor of evolutionary complexity, which allows the real to diversify without ever losing its coherence.
- An interface between the cosmic intention and its manifestation, where spirit finds a form and form finds its origin.

Sylvana

Well... I must admit that what you have just said is dense. But if I understand correctly: each Universe bears within itself an omniscient Intelligence arising from the Primordial Complexity, which contains organization, intention, and laws. And the torus, for its part, would be the operative support: it sets the rhythm of cycles, sustains the evolution of forms, and acts as a bridge between the cosmic intention and its manifestation.

Thierry

Yes. For me, the universal torus is not a mere dynamic image: it is a true cosmic technology, a device operating at the very heart of the Universe.

One could say that it plays the role of the universal hardware, the structural matrix that supports and modulates every manifestation. In parallel, the omniscient Intelligence would be its software: it infuses information, intention, and the organizing logic that animates this architecture.

This intimate link between the universal torus and the omniscient Intelligence is decisive. For, as I have already emphasized, the torus is not a fixed geometry: it is the living support of the flow of time.

In my model, its geometric reality is infinitely more complex than the simple figure of a torus: it is composed of a multitude of sub-tori, which express the fractal dimension of the intrinsic multiverse.

- The expansions and contractions of these sub-tori define the cosmic rhythms.
- Its hyperspherical fractal structure—in other words, its repetitive and multidimensional architecture—encodes the temporal density specific to each level of reality.
- Its cycles function as an active memory: each vibration, each trajectory leaves an imprint, preserved and then reinjected into the global dynamics, thus ensuring the continuity of evolution and the accumulation of complexity from one cycle to the next.

Thus, the universal torus is not merely a structure: it is the beating heart of cosmogenesis, the interface where the primordial order of the absolute Spirit, the omniscient Spirit, the aspiring spirit and the evolutionary dynamic of the real converge.

Sylvana

In short: Intelligence is the strategist that conceives of the vision and the intention, while the torus is the living structure that embodies them, unfolding them as organization and evolution.

Thierry

Yes, exactly: the torus acts as the agent that implements the intention.

And it is because I posited this vision from the outset—that of the universal torus as a fundamental cosmic technology—that I was able to advance in a concrete way. For if we, as humans, are capable of conceiving of our own technologies, it is not by chance: it reflects an original basis, a fundamental conceptual architecture inscribed in the very fabric of the real.

By "fundamental conceptual architecture," I mean that primary structure which makes any organization of the real possible. It is called fundamental not because it would be added to the Universe after the fact, but because it constitutes its condition of possibility: the indispensable foundation, without which neither laws, nor forms, nor cycles could emerge.

The torus, as I conceive of it, embodies this foundation. It is a functional matrix, both generative and regulatory, whose principles we can grasp in order to understand how reality itself emerges, takes shape, and stabilizes.

And by extension, it teaches us that our artificial human technological creations are only co-creations: replicas, at our scale, of what the Universe already accomplishes by itself. Through them, we reenact this fundamental mechanism: transforming information into form, and form into living memory.

Sylvana

I see... This universal torus is not only the support the omniscient Intelligence uses to organize the Universe: as conscious-spirits, co-creators—and humanity is part of this—it also offers us a foundation that we reinterpret at our own scale each time we invent our technologies.

Thierry

Yes, exactly.

In my model, this fundamental technology constitutes the functional dimension of the Cosmos. It acts as a reservoir of possibles, an operational field that makes available to the Universe—and to us, conscious-spirits—a space of potentialities.

Our role is to learn how to use it, to direct this potential toward a true evolution.

But for that, one condition remains: our inventions must stay in resonance with the Universe's own evolution, that is to say, maintain coherence with its overall movement. For the Universe, for its part, follows its autonomous trajectory: its cycles of expansion, contraction, and increasing complexification. In parallel, it offers us a unique opportunity: to participate actively in this dynamic.

This participation can take two forms:

- To develop the natural technologies already present within us—such as memory, language, attention, or those latent psychic faculties, still asleep, that await awakening.
- To invent artificial technologies, the fruits of our creativity, capable of widening our field of consciousness and action.

The distinction between the two is clear: what we create artificially is not "natural"—but this artificiality is precisely our strength. It allows us to conceptualize tools for exploration, expansion, and transformation, suited to amplifying our capacity for perception and understanding.

However, this liberty comes with a requirement: our creations must remain aligned with the founding principles inscribed by the omniscient Intelligence into the fabric of the real. If they stray too far, they risk upsetting the balance. But if they attune themselves to the primordial order, then they become powerful vectors of evolution, capable of accompanying and amplifying the cosmic dynamic—without ever denaturing it.

Sylvana

I can see how, in your model, the omniscient Intelligence relies on its torus as its operational right-hand, a living structure that gives form to and carries out its vision. It makes me think of what you said about pre-configured realities: how does this torus—a true cosmic "hardware"—enter into resonance with those already organized quantum functions? What is the bridge that ensures this link?

Thierry

Your question is pertinent.

In my theory, the universal torus is the living matrix that generates and organizes the Universe. Its fractal dynamic, composed of an infinity of sub-tori down to the Planck scale, acts as a principle of self-configuration: each fragment of the structure already contains in embryo the rules and potentials of future forms. The universal torus thus functions as an engine of coherence, where what appears random at our scale proves to be inscribed within a deeper logic of resonance and harmony.

This logic leads to the idea of pre-configured realities. In the Time-Space-

Memory field, Fractal Temporal Density contains an active memory of the Universe, where latent potentialities are preserved: possible scenarios that are already structured. These realities are not rigidly determined, but preformed as attractors, ready to actualize according to local conditions. They exist in potency, awaiting the favorable moment to unfold.

The link between these two dimensions is direct: the universal torus is the dynamic support that circulates these pre-configured realities and gives them the possibility of actualization. The temporal flows that traverse the toric structure select certain potentialities from the field's memory, stabilize them, and bring them into the present. Pre-configured realities are therefore inscribed within the fractal memory of the torus, and the passage of Time actualizes some rather than others, as though the Universe were navigating through a sea of informational archetypes.

This articulation has two major consequences:

- For the Universe, it shows that expansion is not a simple random process, but that it follows lines of coherence already present in the toric structure, allowing the repetition of universal motifs while fostering innovation.
- For humanity, it suggests that our life trajectory and our experiences of consciousness do not arise out of nothing, but that they are inscribed within preexisting temporal attractors. Free will then takes on a particular meaning: it does not consist in inventing ex nihilo, but in choosing and actualizing certain potentialities already inscribed in the universal field of memory.

Thus, the universal torus and pre-configured realities appear as two faces of the same process: the dynamic structure that supports the Universe and the informational weave that contains its possibilities. Their interaction reveals a Cosmos that is at once ordered and open, where creation is an act of conscious actualization within a field of infinite potentialities.

Sylvana
I understand better how these pre-configured realities are inscribed within the fractal memory of the torus, as potentials awaiting actualization. And how Time—as a structuring field—traverses this living architecture to select certain possibles and bring them forth, both in the evolution of the Universe and in our own trajectories.

What you describe is a torus far more active than one imagines. It is not only a polymorphic form or a flexible structure: it is at once a memory of possibles, traversed by Time, and an interface for their actualization. It does not merely support the omniscient Intelligence: it participates fully in the process.

Thierry

Yes. In my model, the torus is not only a geometric form, but a true operational base: a dynamic infrastructure that makes the Universe possible. It acts as the cosmic hardware, the fundamental processor on which everything rests, the substrate that enables the execution of all the Universe's operations.

But, as with any processor, raw power is not enough. Without source code, it could perform billions of operations… but they would remain without meaning.

In the same way, the torus provides the structure, the computational capacity, and the runtime environment, but it needs a first instruction, a tuning protocol that connects it to its source and orients its operation.

In other words, for a reality to truly exist, it must run in resonance with the original code of the Primordial Complexity.

This invisible link—this universal vibratory program—connects every fragment of the Universe to its origin. This is what I formalized in my sixth element: the Universal Vibratory Resonance, ÆKO.

Concretely, in my model:

The omniscient Intelligence defines the vision, the laws, and the overall organization of a Universe. It establishes the "meaning" of creation: why and how it must unfold.

The torus provides the operational infrastructure, the dynamic base capable of executing these laws and generating forms.

ÆKO Resonance ensures coherence by continuously tuning this execution to the original code of the Primordial Complexity.

. THE UNIVERSAL VIBRATORY RESONANCE — THE VIBRANT HEART OF THE PRESENT

Sylvana

What you present resembles a true architecture of thought, almost an inner framework that you have patiently fashioned to link all these elements together. Even if not everything is immediately graspable, a coherence emerges: each new element strengthens the whole and gives it the shape of a living, functional structure, where every piece naturally finds its place.

And I imagine it is from this framework that you can both put your model to the test—by probing its solidity—and open new avenues of research.

Thierry
Yes… and that is exactly what I have attempted to do: lay down a structure sufficiently rigorous for each new piece to fit into it naturally, like a living puzzle.

And precisely, what I am about to show you now is undoubtedly one of the most delicate—but also one of the most essential—points in this entire architecture.

My sixth element reveals what I have identified as the vibratory core of the Universe: a living anchoring point, lodged at the very heart of the hyperspherical torus. In other words, the dynamic center that pulses and maintains the unity of the Whole.

This central point—which I sometimes call the cosmic navel—is not a mere geometric center. It is the locus of resonance that connects our Universe to its source, the Primordial Complexity, and through which omniscient Intelligence enters into syntony with the absolute Present.

It is this invisible hearth, this silent heart of the universal torus, that I have named: Universal Vibratory Resonance, ÆKO.

It constitutes both: a principle of coherence, guaranteeing that everything remains attuned to the source and bound to the unity of the Whole, and a function of resonance, by which consciousness connects to Spirit, and Spirit inscribes itself within consciousness.

In the poem, I alluded to it implicitly: this invisible breath that animates everything, this primordial beat that belongs neither to the past nor to the future, but offers itself as an absolute, living Present—immutable in its essence, and yet woven at every instant.

Sylvana
And this name, ÆKO… how did you choose it? What makes it so meaningful to you?

Thierry
I chose this name because it condenses, in a single sign, the profound intuition I wished to convey.

The character Æ is an ancient ligature: it fuses A and E. I interpreted it as the union of the beginning and the essence—the alpha, first letter and symbol of the origin, and the energy, vital principle. In other words: the origin and the essence united in a single matrix.

As for KO, it refers to the Greek kosmos: not simply the Universe, but the living order, the fundamental harmony that underlies all things.

Together, ÆKO becomes the symbol of the primordial vibratory weave that connects the entirety of the real. It is the universal resonance of the temporal field, the vibratory breath that modulates and sustains forces, particles, and forms.

One could say that ÆKO is the hidden rhythm, the pulse of coherence that maintains the unity of Time, space, and consciousness.

In my model, it represents the universal vibratory matrix, the invisible canvas upon which the diversity of phenomena is woven. ÆKO is at once the fundamental frequency and the silent song of the Universe: a background vibration, imperceptible yet always present, which ensures a profound and living unity for the multiplicity of worlds in the intrinsic multiverse.

Sylvana

So this central point… it is as though it set the tone for the entire Universe, is that it?

Thierry

Yes. And everything in the Universe—engendered by the technology of the torus—bears its own resonance center, a unique vibratory signature, an imprint that makes it possible to identify each manifestation and that links it directly to the central point of resonance.

Each being, each form, each phenomenon thus constitutes a subharmonic of the omniscient Present embodied by the omniscient Intelligence: a local present, distinct, yet always attuned to the global harmony.

It is this permanent vibratory linkage that ensures both the cohesion of each thing and its inscription within the unity of the Whole. And since everything proceeds from the same origin, realities spontaneously enter into resonance with one another. I call that sympathetic resonance: each signature preserves its singularity, yet naturally aligns with all the others, because they connect to the same founding center.

From this primordial beat—this breath evoked in the poem—my entire reflection unfolded. The understanding of vibratory resonance allowed me to identify several key concepts:

- The omniscient Present as keystone of the field of consciousness, from which everything is ordered.
- The direct correlation between consciousness and temporal access: the more a being gains access to fine temporal densities, the more its field of consciousness expands and its vibration gains in coherence, aligning more closely with the temporal field.
- The necessity of resonance compatibility to cross a fractal threshold or pass from one level of reality to another: no one can evolve without entering into syntony with the universal vibratory center.
- The vibratory localization of conscious-spirits: each spirit is situated in Time by its proper frequency, like a unique vibratory address.

In other words, the degree of consciousness of a being depends directly on its capacity to enter into resonance with the omniscient Present. The more refined, stable, and precisely attuned this resonance is, the more access to the real becomes vast, supple, and conscious. And it is this same vibratory adjustment that conditions the possibility of crossing a threshold and accessing a new level of reality.

Sylvana

If I understand correctly, each being carries within itself a proper frequency—a kind of unique vibratory address that characterizes it. Yet none of these frequencies is isolated: all are linked by resonance to the same founding center—ÆKO—which spontaneously brings them into harmony with one another. And the more finely a frequency attunes to the omniscient Present, the more the being's field of consciousness opens and expands.

Thierry:

Yes, and it is from there that I was able to conceptualize a hierarchy of levels of consciousness—not moral or human, but structural, directly linked to temporal density and to the capacity for resonance. I had already spoken to you about it, but I am revisiting it here in detail:

- Unconscious beings

In my model, these are spirits confined to the immediate instant, incapable of linking the fragments of the Present. They react to the flow of events without perspective, subject to the strict laws of their local level of reality. Their experience is brief, fragmented, without lived continuity.

From a scientific point of view, this corresponds to organisms whose behaviors are essentially reflexive, without true memory or projection, like certain simple biological systems.

- Conscious beings

In my model, they possess an expanded present: thanks to memory and anticipation, they link instants, give meaning to the past, project themselves toward the future, and already open a first liberty with respect to immediate determinism. Their field of experience gains in depth and coherence.

Scientifically, this corresponds to what psychology and the neurosciences describe as phenomenal consciousness: the capacity to remember, to imagine, and to plan, unifying perceptions into a continuous lived experience.

- Self-conscious beings

In my model, they cross a major threshold: they no longer merely perceive and anticipate, but recognize themselves as agents of their own experience. They become capable of adjusting their own frequency of resonance with the temporal field. This capacity for self-regulation enables them to cross thresholds of fractal temporal density, to navigate between different levels of reality, and to gain access to horizons vaster than their condition of origin.

Scientifically, this corresponds to beings capable of metacognition—that is, of observing and evaluating their own cognitive processes—which research identifies as a distinctive sign of self-consciousness.

Thus one sees a structural progression: from simple reaction to memory and temporal projection, then to conscious self-regulation.

This sixth element—Universal Vibratory Resonance, ÆKO—thus appears as a true bridge:

- Between metaphysics and physics,
- Between the ontology of the present and the evolutionary mechanics of consciousness.

It illuminates both the limits of our current perception and the reason why certain fractal realities still remain invisible to conscious-spirits that are not sufficiently attuned, that have not yet refined their vibratory alignment with the omniscient present.

Sylvana

You present this sixth element as a bridge between the evolutionary dynamics of consciousness and what you call the "ontology of the Present." Could you clarify what you mean by that, and why this linkage is essential in your model?

Thierry

Of course. In philosophy, ontology designates the study of what Is, of being in its most fundamental nature.

When I speak of the ontology of the Present, I do not mean only the instant lived over the course of our day, but the Present as a profound existential dimension: that which confers upon everything—its essence, its consistency, and its true existence. In other words, that which renders every reality possible.

To understand the ontology of the Present is therefore to seek to know why the Present is the source of all reality.

And here Universal Vibratory Resonance, ÆKO comes into play. It allows me to link two planes that are often opposed:

- On the one hand, metaphysics, which explores the invisible essence, the source of the real;
- On the other, structure, that is to say the manner in which this essence organizes itself, takes form, and manifests through the levels of consciousness.

Thanks to this Resonance, I have been able to establish a direct connection between the omniscient Present—unified—and the manner in which its fragmentation unfolds into fractals of the Present throughout our Universe.

This reveals a fundamental limit: certain fractal realities remain inaccessible, not because they do not exist, but because our conscious-spirit has not yet reached the vibratory capacity required to perceive them.

Why? Because crossing a threshold of reality is not achieved by a quantitative accumulation of knowledge, but by a harmonization of resonance: a subtle tuning between the inner frequency of consciousness and the frequency of the omniscient Present.

In my model, supported by certain experimental advances, there exists a universal range of vibratory harmonics, comparable to a scale of resonances. Each harmonic is a degree of attunement with the omniscient Present, and it conditions the level of access that consciousness has to the different planes of reality.

Sylvana

For a while now, you have been showing that everything rests on vibratory resonance: it is this which locates beings, which opens or closes access to the different levels of reality. But then, in this play of resonances, how does conscious life fit in?

Thierry

It is precisely this question that led me to formulate my seventh element: the life of conscious-spirits in the Cosmos.

. Conscious-Spirits in the Cosmos — The Ascent in the Fabric of the Real

Thierry

My seventh element brings to light the life of conscious-spirits in the Cosmos—and, beyond that, the existence of a hierarchy of consciousnesses that enables the progressive elevation of marginal spirits. Once again, by "marginal spirits," I mean embodied conscious-spirits like ours, who journey from limited forms toward forms that are vaster and more integrated.

We have already sketched its outlines, but I wish to give it a more structured formulation here. To do so, I return to my foundational question:

If the Universe is structured, ordered, traversed by laws, information, memory, and intelligence… then where is the conscious-spirit situated within this edifice? What is its place, its origin, and above all, what is its becoming?

For, from the scientific point of view, one fact remains troubling: the physical constants and the laws of the universe seem so finely tuned that they appear designed to make the emergence of consciousness possible.

In the face of visions I regard as reductionist—those that reduce consciousness to a mere emergent product of the brain—I chose to invert the reasoning. In my model, consciousness is not a late-arriving side effect, but an axial principle: a fundamental dimension around which everything organizes. It is the continuity of what I had already set forth with wavefunction collapse: consciousness is not an epiphenomenon, it is the axis that structures the actualization of the real.

Sylvana

When you say that is an "axial principle" by which "everything organizes" around it, what do you mean exactly?

Thierry

In my model, it is essential to distinguish two complementary poles:

- The hyperspherical fractal torus, which acts as the universal framework,
- And consciousness, which is its principal actor.

The multidimensional hyperspherical fractal torus constitutes the structural matrix of the Universe. It is the original cosmic technology, the receptacle and the engine that order the flow of Time, the circulation of information, and the emergence of forms. Through its dynamic of expansion and contraction, it encodes memory, modulates cycles, and sustains the fractal coherence of the real. One could say that it defines the cosmic "playing field," the fundamental rules and the fabric upon which everything unfolds.

But this framework, however perfect it may be, would remain inert without a living principle to inhabit it. This is where consciousness intervenes. It is the actor of this cosmic theater, the evolutionary engine that, through the conscious-spirit, uses the potential of the torus to elevate itself. If the torus guarantees the possibility of increasing complexification, it is consciousness that actualizes this possibility, that crosses the fractal levels, that transforms information into lived experience and into interior memory.

In other words, the torus does not oppose consciousness: it supports it. And consciousness, for its part, does not duplicate the role of the torus: it is its active expression, the actor that evolves within this framework, that crosses vibratory thresholds, and that transforms cosmic mechanics into an adventure of elevation.

It constitutes the guiding thread of cosmic evolution, the engine of complexification, a universal potential inscribed from the origin in the fabric of the Cosmos. In other words, consciousness is not an accident: it is necessary to the evolution of the marginal spirit, and the school-worlds exist precisely to support this elevation.

From this intuition, I was able to elaborate a theoretical hierarchical model of the evolution of conscious-spirits, which unfolds across different scales of reality:

- Planetary consciousness: the conscious-spirit attunes itself to the cycles of its planet, perceives its living unity, and becomes capable of integrating its own evolution with that of its world.

- Systemic or galactic consciousness: the conscious-spirit widens its scope, embracing the organization and coherence of ensembles immeasurably vaster, such as a solar system or a galaxy.
- Cosmic consciousness: certain forms of consciousness reach such a degree that they become capable of containing, structuring, or even co-creating worlds. Here, the conscious-spirit is no longer limited to accompanying the evolution of a planet or a galaxy: it participates directly in the architecture of the Cosmos itself.

This is precisely where the higher fractals of the Universe come into play. For each fractal level of reality opens onto heightened properties:

- A higher Fractal Temporal Density (FTD), granting access to a larger portion of the omniscient Present and its potentialities.
- A broader field of information, which enriches both the perception and the understanding of the real.
- A deepened capacity for self-reflexivity, that is to say a growing faculty to know oneself, to transform oneself, and to attune one's vibration in harmony with one's environment.

In other words, the higher the conscious-spirit ascends in the fractal hierarchy, the more capable it becomes of integrating information, tuning its resonance to the Present, and acting as a living co-creator: not only of its inner universe, but also of the reality it inhabits and transforms.

It is this subtle interplay between framework and actor that gives the Universe its coherence and its depth: a fractal structural order, and a conscious dynamic capable of inhabiting it and elevating it.

Sylvana

If I understand correctly, in your model you conceive of a scale of consciousnesses that widens step by step: first rooted in a planet, then connected to a system, then to an entire galaxy, until it can embrace the cosmic architecture itself.

So… with each level crossed in the fractal structure, the conscious-spirit expands: it gains in depth, in amplitude… and above all in power of transformation and evolution.

It almost feels as though we were traveling across a map, where each level opens onto a broader vision and guides the aspiring marginal spirit in its evolution.

Yet I wonder: in your model, how do the stages of evolution unfold that carry a consciousness from the planetary to the cosmic level?

Thierry

Yes… that is exactly it.

First, I want to make it clear that what I propose here is neither an idle theoretical construction nor a baseless speculation. It is a possible trajectory that I have patiently brought to light through my thought experiments and my experimental work.

I have sought to trace a coherent path that clarifies how marginal spirits—those children of the Primordial Complexity, immersed in the thickness of matter, that is, conscious-spirits incarnated like ours, subjected to the constraints of the material world—can gradually remember their dormant potential.

To truly evolve, it seems necessary to start again from zero. Each lived experience inscribes itself in flesh and in mind, and enriches the fabric of consciousness. As the conscious-spirit advances in its quest for omniscience, it crosses thresholds: it accumulates an ever-growing density of information and memory, and, in successive stages, begins to resonate with increasing subtlety with its true nature. It then discovers that it was born from ignorance, but that an immense potential opens before it—and this discovery renders it ever more conscious.

From this awakening is born an inner frenzy, a burning thirst for spiritual, philosophical, and scientific inquiry: the conscious-spirit rises, driven by the desire to surpass what it once was.

In my model, this entire process of evolution is made possible by sympathetic vibratory resonance with ÆKO. This indestructible accord with the Primordial Complexity manifests absolute unity.

This original potential acts like a spark buried beneath the opaque layers of material density, in the form of a sacred, inalienable lingering feeling. It functions like a guiding thread, accompanying the conscious-spirit throughout its wanderings. And the more it evolves in consciousness, the more this lingering feeling expresses itself with force.

It first manifests as an intuition: the conviction that there are other truths, vaster and more luminous than what we perceive. Then it transforms into an irrepressible thirst—a thirst for knowledge, meaning, value, power. But this power is not made to dominate: it calls to be fulfilled, to unfold in harmony, to rediscover filiation with origin, and to become once more a co-creator in the great architecture of the real.

It is at this stage, a self-evident truth asserted itself in my model: the evolution of the conscious-spirit is not automatic. It is not the simple product of biology or matter, but a demanding process—one that requires effort, an inner discipline, a true cognitive ascesis. In other words, a rigorous work of lucidity, attention, and self-regulation.

It must be understood that from our current level of consciousness—which I call planetary—it is almost impossible to imagine what a more evolved conscious-spirit might be. It is a bit like with a mineral: it cannot conceive of what animal life is. Likewise, an animal cannot grasp the complexity of human consciousness.

In the same way, our perception remains confined within the limits of our own level. It does not yet allow us to embrace what a far more advanced conscious-spirit could perceive, understand, or accomplish.

In this sense, it is as though the Universe itself set a condition: it offers us its fractal gates, but it demands active participation, a conscious alignment, for us to cross them.

Sylvana
So, if I am following you correctly, these fractal worlds are not mere abstract structures: you present them as genuine fields of experience. I understand better why you call them "school-worlds"…

And I am curious: could you tell me more about the thought experiments that led you to the conclusion you have just described?

Thierry
Of course.

I first inquired into the very aim of existence, which necessarily implies a path.

In one of my thought experiments, I tried to sketch that path. I began with the study of human history—its evolutions and revolutions that have marked the millennia. Then I extrapolated these dynamics within the context of an intrinsic multiverse, on the scale of the Cosmos. From there, I traced a trajectory of cosmic evolution: a course that maximizes the universal potential for evolution, in order to anticipate what the human mind could become in its highest, most accomplished, most sublime form.

This exercise led me to seek and define a clear trajectory, a way capable of transforming that diffuse presentiment—that feeling of being "more than what we manifest"—into something concrete: a real, practicable path.

Thus I formulated the hypothesis of a progressive evolutionary trajectory, inscribed in the very structure of the Cosmos: a crossing of the school-worlds—those levels of existence that function as pedagogical stages for the marginal spirit. Fractal after fractal, each stage allows it to integrate a new density of Present and to open up an additional degree of consciousness.

Within this framework, I developed a scenario for the evolution of aspirant marginal spirits, where the multiverse presents itself as a set of school-worlds. Their essential function is to welcome these children of the Primordial Complexity, still immature, and to offer them a learning ground. One can conceptualize them as vast fields of experience where the aspirant spirit exercises its desire for exploration and its power to realize itself. It progresses by way of duality: it rises or it falls, traverses light and shadow, experiences joy and suffering. Each situation becomes an initiatory lesson, helping it distinguish illusion from the real, dissonance from harmony, inertia from living creation.

I then conceptualized that to make this journey possible, the omniscient spirit provides the aspirant with an operational consciousness, adapted to each level of fractal density. This technological interface serves both as a protective matrix and as a pedagogical framework: it channels learning, limits excesses, and creates an environment where every choice yields tangible consequences. But this guidance offers no comfort: confronted with obstacles, contradictions, and trials, the conscious-spirit is compelled to respond, to invent, to surpass its own limits. In this way it sharpens its discernment and transforms ordeal into knowledge.

The goal is clear: to allow each marginal spirit to progressively reactualize its original potential. Through the multiplicity of experiences and creations experienced within the school-worlds of the multiverse, the aspirant transmutes illusion into understanding, raw experience into gained knowledge. Step by step, it emancipates itself, expands its conscious present, and becomes able to join the state of active omniscience.

Only then can it take part in the creative work of the universe: no longer as a pupil, but as a master, a sower of new worlds, and a collaborator of the absolute Spirit.

Thus, I formulated the hypothesis that the universe is intended to make possible the existence of conscious life—which I conceive of as aspirant marginal spirits, of which our earthly humanity is a part, as is any other form of life. I defined the universe as a matrix of omniscient Intelligence: the highest degree of a spirit's power, capable even of engendering a multiverse within its universe.

This path is not linear but spiral: with each cycle, the conscious-spirit assimilates more information, refines its vibratory resonance, and widens its field of lucidity, until, gradually, the entirety of its intrinsic potential is revealed.

This seventh element thus builds a bridge between human intelligence and cosmic intelligence, between our current condition—still limited—and a possible becoming.

It also reveals a profound truth: our aspiration to liberty, to expansion, to harmony and to happiness is not an illusion, nor a passing dream. It is the inner reflection of a real structure of the Universe. As though the Cosmos had deposited within us a subtle memory, an intimate resonance, of that toward which we may strive.

And if all this resonates so powerfully within us, it is doubtless not by chance. It may be the sign that this evolution is not only a cosmological concept, but an intimate necessity, inscribed within us from the origin, and already underway.

Thus, the cycle comes full circle:

From the Primordial Complexity to conscious Life, the Universe reflects upon itself through beings capable of thinking it, contemplating it—and perhaps, one day, transcending it.

4. The Spirit on a Journey — When Consciousness Becomes a Bridge

Sylvana
So you mention that the omniscient Intelligence provides a consciousness that is adapted for each fractal, can you explain a bit more on what you mean by that?

Thierry
As you know, in my model, consciousness is neither a secondary biological function nor an illusion produced by the brain: it is a subtle interface, a technology deliberately designed to render the aspiring spirit locally operative within a given fractal.

This is precisely where the hyperspherical torus comes in: it generates, structures, and deploys the Universe into a multiplicity of sub-tori—each sub-torus being an autonomous fractal that presents:

- Its own dynamics,
- A specific temporal density,
- A singular internal logic.

Each fractal therefore corresponds to a field of experience: a perceptual horizon, an impetus for action, and a mnemonic register. When a marginal spirit incarnates there, it is naturally limited by this field—in its perceptions, its capacities for action, its memory. But this limitation is not a failure: it constitutes the pedagogical matrix of evolution. Each fractal is a laboratory where spirit learns, transforms, and refines its capacity for consciousness. And as it crosses a fractal, the next field opens as a world appreciably vaster and richer.

It is the functional tool that makes conscious existence possible within a given fractal. It is not a by-product of matter; it is an intentional device, calibrated by the fractal structure to enable the progressive learning of spirit.

Now that I have explored more deeply the link between spirit and this consciousness, I can define the term conscious-spirit with greater precision.

The conscious-spirit—the symbiosis between the marginal spirit and this technological interface.

The conscious-spirit is born from the alliance between a spiritual presence (Spirit) and the technology of consciousness (the "diving suit" of the fractal provided by the torus). It is this symbiosis that enables the passage through the hierarchy of worlds: each fractal offers its range of levels of consciousness, and this diversity constitutes the itinerary of experience necessary for the evolution of marginal spirits.

5. An Invitation to Becoming — From Knowledge to Awakening

Sylvana
All right, I understand. But in your model, if a spirit that incarnates within a fractal consciousness is limited by the field of experience specific to that fractal—whether in its perceptions, its actions, or even its memory—then how can it cross that threshold? What allows it to go beyond those limits in order to advance to the next fractal and continue its path of ascension?

Thierry
If a marginal spirit incarnates within a fractal consciousness, it is indeed limited

by the field of experience specific to that fractal:

Its perceptions are calibrated to the temporal density of that level,

Its actions are governed by the internal laws of coherence of that fractal,

And even its memory records only a portion of the real accessible to that plane.

In my model, these limits are not an arbitrary obstacle: they constitute the horizon of learning of the fractal. In other words, each fractal world is a laboratory of consciousness. It sets a precise framework intended to test the conscious-spirit, to make it integrate a certain degree of complexity before it can cross the next threshold.

So, to answer your question—how can the marginal spirit go beyond that framework?—the key lies in two factors: the available density of the Present and vibratory resonance.

Each fractal possesses a range of Fractal Temporal Density (FTD): an interval within which consciousness can unfold. One could say that each fractal offers a limited depth of Present, with a minimum and maximum level of possible consciousness. As long as the conscious-spirit remains in the lower part of that range, it repeats, it stagnates, it endures. But as it refines its perception, develops its attention and its memory, integrates its experiences, and expands its degrees of liberty, it raises its inner resonance.

When the fractal's maximum density of Present is reached—in other words, when the marginal spirit has fully integrated the potentialities offered by this level—a natural tipping point occurs:

- The vibratory summit of this fractal corresponds to the vibratory threshold of the higher fractal;
- Thus the passage does not occur through rupture, but through resonant continuity.

In my model, it is the law of vibratory compatibility that governs evolution: the conscious-spirit crosses a threshold only once it has adjusted its internal frequency to that of the higher level.

In this way, the spirit is never a prisoner of a fractal. Its limits are only temporary: they are conditions of evolution, intended to make it strengthen its memory, refine its perception, develop its creative power. Once these conditions are fulfilled, the elevation of its resonance naturally opens access to a vaster world.

In summary, in my model:

- The fractal acts as a filter that restricts the accessible information and defines the conditions of evolution → it sets the learning framework.
- More attention in the instant-Present and more memory enlarge the density of the Present → they offset fragmentation and grant access to a more continuous consciousness.
- Universal Vibratory Resonance, ÆKO adjusts the inner frequency → it harmonizes the marginal spirit with the higher threshold.
- The passage occurs through continuity → the summit of one world becomes the foundation of the next.

Thus, the evolution of consciousness presents itself as a progressive transmutation: limits become springboards, each time opening new degrees of liberty.

Sylvana
So, if I understand correctly, in your thought experiment—The Universe would offer us the infrastructure, the conditions—but evolution would not be automatic.

It is not a question of "deserving" a higher level… It is simply that we can access it only if we have truly gone through, lived, and integrated everything that our fractal compelled us to explore—to the point of extracting the maximal meaning that those experiences could reveal.

Thierry
That is exactly it. The Universe provides the infrastructure, but it does not open its doors without something in return. Evolution is not an automatism: it requires real inner work, a discipline of being, a rigorous practice of lucidity and self-mastery.

It is neither a punishment nor a reward: it is a structural law.

In my model, one does not cross a fractal by mere desire or by chance, but only when one has fully explored and integrated all the potential of its corresponding density of the Present. In other words, when the still-diffuse memory of an experience has transmuted into lived, assimilated, and meaningful knowledge—to the point of becoming a constitutive part of the conscious-spirit.

And, as I have already emphasized, as long as the spirit does not vibrate in resonance with the rhythm, the density, and the laws of the higher temporal field, it remains limited to the level of reality where it evolves. But as soon as it attains that vibrational coherence, the passage occurs naturally: the summit of a fractal converts into a threshold to the next fractal.

This process of integration both expands the instant-present and prepares consciousness to inhabit a vaster space.

Sylvana

And in everything you have just described… Is there an image that, for you, would best represent this path?

Something that, at some point, allowed you to visualize it, to feel it in its wholeness?

Thierry

For a long time, one image has inhabited me: that of concentric circles, spreading from a luminous center to the farthest periphery. I have named it: the concentric path of consciousness.

This image describes the inner movement of integration—the progressive ascension of the conscious-spirit through the strata of reality.

At the center: the omniscient Present of the omniscient Intelligence, silent matrix of the Whole. Around it: an infinity of nested circles, fragments of the omniscient Present, each bearing a density, a resonance, a world. This cosmic mandala is complete, ordered, imbued with harmonic laws. Yet for us, who observe it from the periphery, it remains unreadable, distant.

And yet, even immersed in the low densities of a lower fractal, the aspiring spirit retains the vibratory imprint of the Origin—the ÆKO Vibratory Resonance. It knows, dimly, that it is greater than what it manifests. This vibratory imprint—this ancient code—drives it to seek, to learn, to surpass itself. The quest takes a thousand forms: desire for knowledge, thirst for meaning, need to act, to evolve, to go beyond matter. But behind each one the same silent call resounds: to find the lost totality.

Then the great crossing begins.

Fractal after fractal, school-world after school-world, the aspiring spirit explores the periphery of the universe. Each sphere offers its laws, its forms, its constraints, but also its revelations. Little by little, like a puzzle coming back together, an inner key appears: the one that opens dormant potential. Each ring crossed becomes a leap of evolution, each experience transforms the apparent chaos into cosmic pedagogy.

And the closer it comes to the center, the more patterns once incomprehensible become clear. What seemed abstract takes on form. What seemed distant becomes intimate. What was absence becomes vibrant presence. Then everything acquires meaning: the fragmentation of the omniscient Present into loops of Time, the emergence of fractals as school-worlds, the rhythmic architecture of the Universe.

Nothing was separation: everything was a path of elevation.

In a breath of recognition, the aspiring spirit understands: the Primordial Complexity did not lead it astray; it has guided it. The omniscient Intelligence fragmented its Present not to exclude, but to trace a path of return—a gentle, initiatory ascension. Each world traversed was a mirror, each ordeal an invitation to remember.

And when at last it remembers… it is reborn.

Like the Phoenix rising from its ashes, the aspiring spirit recovers its sovereignty. Having crossed matter, ego, and time, it is no longer an uncertain fragment but an accomplished light, master of its own resonance, the embodied reflection of Unity restored.

Then, in a silent joy, the Universe recognizes in it the fully ripened fruit of the primordial seed.

Sylvana

Your account has a striking profundity. One feels at once the distance and the call of the center, the invisible thread that links each fractal to the origin. I understand better why, in your model, you speak of a crossing: it is not just a path, it is a reminder… a return toward something we have always carried within us.

And if the conscious-spirit must pass through all these stages for the integration to be real, tell me: can one, in one way or another, accelerate this process? Or must each passage, whatever happens, follow its natural rhythm, without exception?

Thierry

I understand your question perfectly. Very early on, I asked myself this decisive question: is there in nature a mechanism capable of accelerating the evolutionary leap?

I never wanted to resign myself to passively follow the natural rhythm, nor to rely solely on providence. For time does not wait for us—our survival depends, in the short and medium term, on major discoveries.

I therefore oriented my research toward a precise objective: to understand the mechanisms in depth, not to bypass them, but to act from a broader level, where the leap becomes possible. To find a way to accelerate the process without betraying it.

What I was seeking was a fulcrum: a lever to go beyond the framework, to transcend the mechanism itself. And it was from this posture that I was able to make several decisive breakthroughs.

I showed that a marginal spirit, although inscribed within a fractal and a given level of consciousness, can progress more rapidly toward its maximal potential. Not by avoiding the necessary experiences, but by choosing more direct paths of integration, relying on a fine understanding of the laws that structure its fractal.

For if the general law imposes upon each consciousness paths of evolution that are more or less long, there is also a path of exception: the one that, through knowledge—and also through technology—makes it possible to cross more rapidly the threshold of consciousness meant for this level.

Sylvana

You know… what I love about your vision is that it is not just a reading of the Universe.

One finds in it, in a way, an invitation to becoming. An answer to our quest for meaning.

Thierry

Yes… For me, this model has never been a mere theory: it has served as a true point of reference, solid enough to accompany me throughout my path as a researcher and explorer.

That is why I have pursued this quest, year after year—not to impose a vision, but to open a passage. A path I wanted to be coherent, structured, and capable of evolving… A path that has allowed me, little by little, to transform my intuitions into concrete research, to conceptualize proofs of concept, and to anchor my thought in ever more solid foundations.

Sylvana

And you began with these seven elements you have just unfolded…

A kind of invisible framework, at once immense and intimate. I sense that they form a coherent whole, like the broad strokes of a painting.

But I still have a thousand questions.

And I imagine that each element can be elaborated upon, unfolded, deepened…

Thierry

Yes, exactly. And that is why, later on, I will return to each of them in detail. Here, I only wished to sketch the general weave as it appeared to me, before entering the pillars that develop its scientific scope.

The aim is not yet to present my experimentations to you, but to have you follow the thread of a journey: the one that begins with lived experiences and intuitions, and leads to a theoretical structure that gradually took shape, up to my first proofs of concept.

Very quickly, it became clear to me that these seven fundamental elements were provoking a profound shift in my way of conceiving of the Universe, Time, information, and consciousness. They offered me a new interpretive grid, an expanded perspective for questioning in depth our place within the universal architecture. It is a vision that I will share with you later, once you have better grasped the foundations of my model on the structure, the workings, and the purpose of the Universe.

From this first exploration there emerged a second phase of reflection, unfolding over several years. It gave birth to what I called the six pillars: intermediate structures derived from the seven elements, conceptualized to amplify reflection, test certain intuitions, and lay the groundwork for testable hypotheses.

These six avenues of exploration, developed up to 2001, then furnished me with the material needed to undertake more theoretical, more formalized—and above all experimental—research.

Sylvana

Your seven elements already sketch the framework of a solid architecture, and each of these pillars seems to extend its reach toward other horizons. I feel as though I have glimpsed the first constellations of a vast cosmic itinerary… and I am ready to follow you further still in its exploration.

Before we enter your six pillars, there is something I would like to touch on. You have often spoken of your way of searching—this art of weaving science and metascience into a single thread—but we have never really explored your methodology. This framework appears to be the hidden fabric behind your discoveries, the one that led you to Absolute Relativity and to your first proofs of concept.

So, if you are willing… I would like you to take me to the heart of your approach.

Part IV

The Nature of Time

~

An Unveiled Mystery

Chapter 8

Integrative Metaphysics

~

When Science and Metascience Unite

Sylvana

Listening to the way you presented the seven elements, I felt as though I were traversing a map that goes far beyond what one ordinarily perceives—a map that makes one want to keep going… but that also requires knowing where to begin.

Your image of the concentric path stays with me: I have the feeling that we are advancing together into the mysteries of the Universe, in search of new keys.

And as our exchanges unfolded, one thing became clear to me: without this widening of science toward metascience, you would never have been able to lay the foundations of Absolute Relativity—nor open the way to your first proofs of concept. It was necessary to extend the field of rationality, to lead it to its own frontiers, so that your model could emerge with coherence and force.

So, before we return to the pillars themselves, I would like you to explain to me how you constructed this method—and above all why it changed everything.

Thierry

Yes… you are right to emphasize this point. Without this extension of the scientific field—without the opening of a method capable of encompassing both the measurable and the conceivable—I would never have been able to formulate certain decisive hypotheses, much less lay the foundations of Absolute Relativity.

The most formative example for me, the one that literally set everything in motion, was daring to place the Primordial Complexity as the starting point.

By its very nature, this first principle escapes any directly measurable field. The Primordial Complexity does not unfold in the Universe: it constitutes the precondition, that which makes possible the very emergence of space, time, and matter. It is that which precedes any manifestation. Before time flowed, before space opened, before matter crystallized, the Primordial Complexity was already there.

It is neither object, nor force in the physical sense, nor even a definable entity. It is an absolute state: a pure order, undifferentiated, transcending any causality, a unified whole, non-segmented, which does not depend on any external cause to exist.

Now, because science, in its classical methods, cannot integrate such a principle—since it lies beyond the experimental domain—I had to widen the frame and construct a rigorous space of reflection that is capable of including what science leaves aside: the origin, the meaning, and the invisible structure of the real.

Without this expanded frame, this founding idea would have remained marginal, almost illegitimate. But integrated into this broader field, it became the very foundation of the model: the place from which everything proceeds, and toward which everything converges.

1. The Hypothesis of an Ordainer — A Universal Intuition

Sylvana

I am curious: before you developed this whole framework, let us return to the beginning—how were you led to posit the very hypothesis of a Primordial Complexity? What gave you the intuition that such an Ordainer had to exist prior to everything?

In your meditations—so profound and so demanding—was there a spark, a thought experiment, or a decisive observation that set you on this path?

Thierry

It was indeed in the course of an inner exploration that, for the first time, the intuition arose that such an Ordainer had to be sought.

One day, at the heart of a thought experiment, sudden, blazing questions burst forth within me, as though they came from a place deeper than my mind:

How is it possible that we, stardust lost in the immensity, are capable of conceiving of infinity?

How can we imagine the Universe—and even sense what might lie beyond it?

How can a mere spark of consciousness embrace, even partially, the Whole?

For here is the blazing paradox: we are almost nothing, and yet we dare to imagine everything. We are minuscule, and yet we express a power of spirit that seems to exceed our condition.

Immersed in this revelation, a meditative state seized me. I felt, in the innermost core of my embodied life, an imperious call: to rise, to transcend matter, to breathe infinity as one inhales air after holding one's breath for a long time. It was life in its pure state: an exhilaration of freedom, the silent promise of absolute happiness at the end of the path.

And above all, a truth became self-evident to me: our most profound aspirations—Liberty, Truth, Love, Elevation—could not be mere chemical illusions or mechanical products of matter. If the Universe truly proceeds from an Ordainer, then these aspirations find a real anchoring, a framework for explaining conscious life and its evolution on the universal scale.

This meditation was strengthened in me by a striking piece of data: approximately 85% of humanity shares, in religious or spiritual forms that are diverse yet convergent, the intuition that a higher Entity—creative, ordering, benevolent—presides over the origin of the Universe and of life.

So I asked myself: what if this age-old conviction were not merely an adaptive illusion born of the fear of the unknown, but indeed the discreet manifestation of an objective truth, inscribed in the very origin of being and of the Cosmos?

For this conviction does not only awaken psychological comfort. It stirs in us a profound feeling of unity, a sense of connection to something greater than oneself, at once immense and intimate. It opens the vision of a potentially accessible happiness, like a horizon of plenitude.

From there, I resolved to consider seriously the existence of an Ordainer: a living principle, an intentional source, a fundamental entity capable of giving meaning and coherence to the Universe and to the emergence of consciousness within it. And from then on, I had to study it with the rigor of a scientific approach.

Sylvana
When you speak of presentiment, of unity, of happiness… one feels clearly that your quest touches something profoundly spiritual.

So I will ask you the question simply: at its core, are you seeking to "prove" the existence of God? Or is it something else?

Thierry
To make myself properly understood, I must first clarify what I mean by "spirituality": it is the capacity of the human being to connect with a dimension vaster than oneself—at once inner and universal—in a quest for meaning, truth, and unity. It does not necessarily imply a religion, but it is always an experience of the enlargement of consciousness.

My approach, from the outset, was never to "prove God" in the dogmatic or theological sense of the term. Such an undertaking is, quite evidently, scientifically impossible, and I will return later to the reasons that render this attempt barren.

My intention was entirely different: it aimed to go beyond the sole sphere of my personal feelings in order to open a broader field of reflection. It consisted in showing that the search for an Ordainer, an initiating Principle—whether one conceives of it as external, internal, or immanent to the Universe—is not an irrational fancy nor a projection of the mind, but an intellectually legitimate, epistemologically grounded, and methodologically conceivable undertaking.

Sylvana
Listening to you, I understand better why you come to speak of an Ordainer. After all, as you present it, it is not absurd to make it a hypothesis to explore—to see where it leads, like another possible angle on the real.

It may simply be another way of approaching the real—one perspective among others, which deserves to be explored to the end.

But I wonder: why, in "official" science, does one almost never speak of a possible Ordainer?

2. Modern Science — Its Methodological Principle

Thierry
Science, by its very methodology, cannot integrate an element such as the existence of a supreme Ordainer. It studies the Universe on the basis of

observable phenomena, but it cannot measure what makes the very existence of the Universe possible.

This is because science proceeds by an inductive–deductive method; in other words, it observes, models, predicts, and verifies.

This process rests on fundamental principles:

- Empiricism: a theory is valid if it is verifiable by experiment or observation.
- Repeatability: phenomena must be reproducible in order to establish a law.
- Falsifiability: a scientific hypothesis must be capable of being refuted, otherwise it does not fall within the domain of science.
- Parsimony of hypotheses (Occam's razor): one does not introduce an additional entity (for instance "God") if the model functions without it.

Consequently, an "ordainer," a "creator," or an "intelligent principle" is neither testable, nor refutable, nor measurable, and is therefore outside the scope of experimental science.

This is not a negation of principle, but a methodological suspension.

Science does not say that God does not exist, it simply says:

"This concept is not accessible by our current scientific method."

This means that:

- One cannot integrate a creator into an equation,
- Nor measure a creator's action with laboratory tools,
- Nor confront a creator's existence with a falsifiable prediction.

What science does is describe the laws (for instance gravity and quantum mechanics) but it does not explain why these laws exist or have these values.

It explores the initial conditions, but it does not say why there is something rather than nothing.

It models the evolution of the Universe, but it does not say where the laws themselves come from.

To put it simply and more concretely:

Can science speak of an Ordainer or an initiator?

No, not as a testable or predictable object.

Can it make room for an "intention" or "first cause"?

Yes, as a free philosophical interpretation.

Is this a dogmatic closure?

No, it is just a methodological rule: what is not testable is not rejected, but suspended.

Sylvana
So, if I understand correctly… the intuition of an Ordainer, which so many human beings share, is not really rejected by science: it is simply set aside, considered as outside its scope.

But then, if science cannot make room for it within its own framework, how do you integrate it? To put it more clearly: what is your methodology?

Thierry
From the outset, I felt the need to reconcile science and spirituality, because I have always seen them as complementary. Each brings a unique and essential dimension to our understanding of the Universe: one sheds light on the mechanisms, the other seeks to shed light on meaning.

To advance along this path, as fascinating as it is daring, I resolved to undertake an ontological unification of the physical world and the world of spirit.

For what science describes as matter, laws, and constants, and what spirit, as consciousness, experiences as thought, intention, or wonder, are not, for me, two foreign realities: they are the two expressions of one and the same unified ground, a field of intelligibility, of information, and of becoming.

In other words, science explores the how; metaphysics, which I associate with spirituality, questions the why. The former relies on observable, measurable, reproducible phenomena, and formulates finite laws to describe finite phenomena.

But to conceive of a Creator is to think infinity—and thus to cross beyond the borders of the traditional operative field of physical science. It is therefore natural that no equation, by itself, could contain or refute such a reality.

That is why I relied both on science and on a metaphysics, two approaches that are distinct but profoundly complementary.

This led me to ask: how can I reconcile, within a single language, the rigor of science and the depth of metaphysics? In other words: how can I think about both approaches simultaneously, at every stage of my research?

This exercise, complex at first because these two domains question the real differently, led me to invent a new methodology capable of integrating them: this is what I have named integrative metaphysics.

Its power lies in this ambition: to pursue the rational quest for ultimate meaning. Where science measures, describes, and calculates, integrative metaphysics seeks the profound causes, the hidden architecture that gives form and direction to the Cosmos. It encompasses the scientific approach within a broader perspective, where knowledge becomes as much an understanding of the world as a revelation of its meaning.

Sylvana
I perceive the force of what you describe… but could you give me a more precise definition?

Thierry
Of course.

It is not a recognized academic term because I forged it as a new conceptual tool, specific to my research.

Here then is the definition, developed and adapted to my theory.

Integrative metaphysics is a rational and transdisciplinary approach that broadens the traditional scope of metaphysics. It does not confine itself to speculating about being or the ultimate origin, but seeks to integrate scientific findings, physical laws, information, and consciousness, in order to propose a coherent and global vision of the real.

It rests on three pillars:

- Science: which describes observable, quantifiable, and reproducible phenomena.
- Classical metaphysics: which questions first principles, ultimate causality, the meaning of existence.
- Integration: a process of unification, where scientific data and spiritual intuitions meet within a single conceptual framework.

Its objective is twofold:

- Epistemological: to move beyond compartmentalization between disciplines and to open science to new objects (consciousness, intention, primordial information);

- Existential: to offer humanity a framework of interpretation capable of linking knowledge of the Cosmos and the inner experience of the spirit.

It is this profound dialogue—in my meditations and thought experiments—that opened a fertile space for me to rethink science as a vector of meaning, unity, and awakening.

Sylvana
Epistemological—I can hardly even say it. What do you mean, more simply?

Thierry
I suspected as much… that word was bound to make you react, and your expression when I pronounced it confirmed it.

It must be said that the term "epistemology" carries a different scope depending on the culture. In English, it designates the theory of knowledge in its broadest sense: its sources, its nature, and the criteria that allow one to distinguish a belief from true knowledge. In French, by contrast, it refers more specifically to critical reflection on science: its methods, its conditions of validity, its fields of application, and its limits. It is in this second sense that I am using it here.

Now, it is precisely by adopting this perspective—by observing science through its own methods and confronting it with its own limits—that I made an unexpected discovery: the more science pushes its principles to their extreme, the more it reaches uncertain boundaries… and the more it opens to new horizons.

This is where science becomes truly fascinating: it loses nothing of its rigor, but it gains in depth. It ceases to be merely a descriptive tool and becomes also a field for exploring that which, until now, had escaped its gaze.

And it is in this sense that, in my view, epistemology takes on its full significance: it reveals the possibility of a fruitful dialogue, not in opposition between science and transcendence, but in the recognition of their complementarity.

Sylvana
And what I love in the way you search is that you never cut the bridges between science and transcendence—on the contrary, you connect them.

So tell me: how did you start from this intuition—the one that there could exist a true Ordainer—to transform it into a solid scientific approach? In other words, how did you manage never to lose your rigor?

Thierry

From the beginning of my research, what I pursued, with patience and rigor, was the search for indirect traces of the existence of an Ordainer: the signs of an intelligent structuring entity, perceptible perhaps in the form of a primordial information field, an underlying order capable of explaining the coherence of the Universe, its apparent finality, its evolutionary dynamics—and even the emergence of consciousness.

And to give a rational foundation to this intuition, I assembled five rigorously constructed angles of approach. Together, they sketch an epistemologically grounded path toward the idea of an Ordering Principle: not necessarily religious, but intelligible and coherent with the logic of the Universe.

- The ultimate explanatory principle—the first cause

At the foundation of all science lies the search for causes. But when the causal chain is traced back to its limits—whether to the initial singularity, that theoretical point associated with the Big Bang, beyond which physical laws can no longer describe anything, since neither space nor time yet exist there—a dizzying question arises:

Why is there something rather than nothing?

Here, an unprecedented conceptual space opens. Even before evoking the existence of an Ordainer, this line of questioning led me to posit the existence of a transcendent, uncaused principle—whose existence depends on no cause—situated upstream of spacetime, and acting as a source of order and potentiality.

A principle that, while escaping the classical categories of physical causality, could be the intelligible root of the coherence and fecundity of the Universe.

And this "why" can be glimpsed in the very foundations of the Universe—which brings me to the next point.

- The fine-tuning of the Universe's constants

It is a subject we have discussed several times, and one that is key to the mystery of the Cosmos: the fundamental constants—such as the fine-structure constant, the ratio between the mass of the proton and that of the electron, the cosmological constant, or even the critical density of the Universe—are tuned with extreme finesse.

According to current physics, the slightest variation in their values, however minute, would have sufficed to render the Universe sterile, chaotic, or even nonexistent.

This structural harmony then seemed to me irreducible to mere chance. It invites, on the contrary, a more profound reflection: what is the origin of this tuning? Why these values, and not others?

I do not see this as direct proof, but as a strong indication, which acquires meaning when it is articulated with other elements of my model.

All this suggests the existence of an informational design inscribed at the very heart of physics—a hidden architecture that guarantees the coherence and fecundity of the Cosmos.

- Cosmic lawgiving: why does the Universe obey laws that can be mathematized?

Science rests upon an implicit hypothesis: the laws of nature are stable, universal, and reproducible. It is thanks to this stability that experimentation is possible and that equations describe reality.

But an essential question remains: where do these laws come from? Who—or what—guarantees their permanence? Why is there, at the heart of the Cosmos, something like a universal lawfulness, a mathematical coherence that renders the Universe intelligible?

These questions led me to formulate a radical hypothesis: that of an omniscient Intelligence, which I have already mentioned to you, a supralocal informational principle. A principle acting at the very root of what I have called the quantum ether, capable of explaining the origin and stability of physical laws, and constituting the intelligible foundation of universal coherence.

Extending this idea, I formulated the hypothesis that information precedes form: it is not raw matter that engenders order, but the organization of information that shapes structure and renders it intelligible. In this perspective, the Universe is not a fortuitous assembly of particles, but a dynamic informational field, whose laws emerge as expressions of a deeper logic inscribed in the fabric of Time.

And if one accepts this idea that everything is organized by information and universal laws, then a new enigma arises at once: Why does consciousness appear within this framework?

- The emergence of information and consciousness

As you know, Consciousness is fundamental in my theory. Yet, in seeking to understand its role in the Universe, decisive questions arise. Why do these equations, which our human reason can formulate, describe Cosmic reality with such precision? In other words, how is it that the Universe obeys laws so stable and so universal that they allow the human mind to describe that very reality?

These questions reveal a paradox: if matter were blind and without design, nothing would explain that order lends itself to being described by mathematical structures, nor that consciousness emerges to question that order.

This suggests, on the contrary, that consciousness is not a secondary accident, but a fundamental property of the Universe—perhaps even its ultimate objective.

And from there, what remains is to examine not only the Universe and the role of consciousness within it, but the science that describes them.

- Metascience: a limited framework

Metascience, not to be confused with metaphysics, is the discipline that analyzes science as a human and cognitive activity. It does not directly produce scientific results about nature, but it studies how science produces its results, by what methods, logics, values, and limits.

Advances in modern physics have brought us to the limits of classical scientific language.

Quantum mechanics has challenged our traditional notions of time, space, and causality. The study of the quantum vacuum, for nearly a century, has revealed an inexhaustible reservoir of potential creation, suggesting that the Universe contains far more than a mere blind mechanism.

Now, standard science, by dint of strict rationality, seems today to be reaching the limits of its current paradigm. It is not out of breath, certainly, but it stands at the threshold of a mutation: an expanded science, capable of integrating not only matter and energy, but also information, consciousness, and meaning.

It is at this point that I judged it necessary to shift my approach: not to abandon the scientific method, but to extend it, by articulating it with the integrative metaphysics that I conceptualized. For it appeared essential to me to work in tandem:

- By maintaining scientific rigor,
- While expanding rationality to include information, intention, the observer, and the quantum field.

Only this expansion can today, in my view, unlock a potential evolutionary leap in our understanding of the real. And in time, it can bring us all the answers to our questions.

Sylvana
Wait… This time I took notes…

Thierry
Yes, I saw that you were taking notes, and I see you are also adding colors everywhere….

Sylvana
Yes, I told myself that if we were starting to talk about your science, I would have to hold on even tighter….

So here is what I took from that.

You have just revealed to me five faces of a single enigma that could point to the traces of an Ordainer:

- The elusive origin, always slipping away from our instruments;
- A universe tuned like a musical instrument, where each constant plays its true note;
- The universal presence of laws that structure everything;
- The mystery of a world that allows itself to be described by equations, and where consciousness seems to spring from matter;
- And finally, those frontiers of knowledge, where science itself must broaden to avoid closing in on itself.

All of this, taken together, converges toward a single intuition—the one you defend in your model: that there could exist, behind the real, a hidden order, an invisible architecture, of which the observable Universe would be only a manifestation.

Thierry
Yes. In other words, the aim of my research is the quest for ultimate meaning—including the hypothesis of an Ordering Principle—and to demonstrate that this quest can constitute a genuinely scientific endeavor, provided one gives

it a rational, open, and integrative framework.

3. Integrative Metaphysics — Dialogue Between Science and Metaphysics

Sylvana
I perceive clearly the coherence of what you present: you are not speaking only of cosmology or quantum mechanics, but of a framework that exceeds strict physics. You seem to want to introduce a broader language that also connects other dimensions.

Thierry
Yes, and that is where everything takes on another meaning.

Beyond its scientific framework, this vision sketches the outlines of an integrative metaphysics: a unified language in which the great disciplines—physics, biology, philosophy, spirituality—cease to oppose one another and reveal themselves as the complementary facets of one and the same living order.

In this paradigm, equations are no longer limited to describing particle trajectories: they become the expression of a universal correspondence, valid in the exact sciences as well as in the domains of thought and human experience. Memory of the void enters into dialogue with the spiritual traditions that intuited it as a reservoir of origin. The bifaciality of time reconnects with the philosophical intuitions of becoming and reminiscence. The universal torus resonates with the ancestral symbols of circular unity and the eternal return.

This Metaphysics does not claim to replace science, but to amplify it. It offers Science a broader space where measurable facts and intuitions of meaning connect without contradicting one another. In this framework, matter, consciousness, and Time are not regarded as separate domains, but as the interwoven dimensions of one and the same living, coherent field.

Thus a global interpretive key takes shape: a Universe where scientific research is not separable from a quest for existential understanding, and where each discovery expands not only our knowledge, but also our inner horizon.

Sylvana
As I listen to you, I understand that we are entering an expanded understanding that touches our inner experience and our quest for meaning; it also brings to mind the collective memory of civilizations—their writings, their symbols,

and their ancient codes that were already seeking to connect the measurable and the imperceptible.

Thierry

Yes, you are touching here the very heart of my endeavor: this approach, speculative yet rigorously structured, would open up an immense field of possibilities—a science capable of embracing within a single continuum quantum physics and general relativity, memory and consciousness, matter and meaning.

And this could transform our relationship to knowledge. We might discover that limits are not dead ends, but creative thresholds, gateways, seeds of metamorphosis.

Each contradiction thus becomes a springboard, each paradox a call to broaden our understanding.

This is not a science of closure, but a science of hope—a science that affirms that the real is living, open, and the bearer of tomorrow's discoveries.

And this is precisely what my vision outlines, the contours of a science reconciled with the human being:

- A science that places the human being at the heart of the theater of universal life, not as a passive spectator but as a conscious actor.
- A science that answers profound questions—where do we come from, what are we, where are we going?—while opening concrete pathways for the future.
- A science that also nourishes the essential aspirations of the human spirit: meaning, liberty, beauty, transcendence.

Such a science is not merely a discipline: it becomes an integral path of knowledge, a bridge between reason and wonder, between matter and spirit, between Cosmos and consciousness.

Sylvana

As I listen to you, I realize that everything, for you, begins with a primal intuition—like a primordial spark: the Primordial Complexity. You envisioned it as the invisible source, already bearing within itself all that was to unfold thereafter. And instead of ignoring this intuition, you sought to understand its forms, its reasons, and its origin—why it was present and where it came from.

Thierry

Yes. Everything here begins with a founding gesture: to imagine that at the origin of the Universe there exists an Ordainer, the Primordial Complexity.

When I set out on this path in 1996, I knew that it was the right path. I needed a speculative vision vast enough that, even if I did not reach the ultimate answer, I would inevitably discover along the way other unexpected yet precious truths.

The Primordial Complexity thus became my universal landmark. I envisioned it as the architect of the Universe, the fulcrum that allowed my mind to project itself beyond the observable Cosmos, in order to question the origin with coherence. The question "where does the Universe come from?" then transformed into a more constructive question: "how did the Ordainer accomplish its work?" And that shift changed everything.

It became a quest at once scientific and metaphysical, essential and perfectly legitimate. For sooner or later, the question of our origin must find an answer. And this choice proved to be incredibly fruitful.

On the scientific plane, it led me to seek the origin of the manifest order in the Universe, the possible existence of a creation and evolution plan, and even of a Cosmic purpose parallel to ours.

On the metaphysical plane, it brought me closer to the idea of absolute perfection: an archetype that expresses the best within us and in the Universe, and that urges me to strive toward more love, joy, generosity, and fulfillment.

By exploring the Primordial Complexity, I progressively developed a scientific, coherent, and hierarchical cartography in which each idea arises naturally from the previous one. From this single seed there emerged, step by step, my pivotal concepts—from the temporal field to the universal torus, from the chronon to Cosmic bimetricity, up to the laws of fractal organization and the multiverse. Thus, the complete architecture of my research pillars, both fundamental and applied, took shape, offering an unprecedented framework for thinking the living structure of Reality.

The Primordial Complexity unfolds as a living architecture:

- Consciousness (omniscient Intelligence) initiates order,
- Vibration (ÆKO) sets the field in motion,
- Geometry (universal torus) provides the structure,
- Temporal field and gravity ensure dynamics and cohesion,

- Chronons and temporal dimensions give granularity and directions,
- Bimetry creates the tension between causality and memory,
- Fractals and sub-torus organize the cycles,
- Multiverse extends the hierarchy,
- Protons, black holes, and gravity ensure the system's breathing, the eternal ebb and flow between emergence and compression.

Sylvana
This progression fascinates me, because it transforms a simple intuition into a complete landscape, inhabited by precise concepts that are interconnected. I also understand that it is on this foundation that you were able, later on, to build your first proofs of concept.

And so at this stage of your research approach, what was the next step?

Thierry
The next step led me directly to Time. After having posited the hypothesis of the Primordial Complexity, it became indispensable to understand how this source inscribes itself in duration and sculpts the Universe. It is there that I began to explore what I today call the Absolute Relativity: a way of thinking Time not as a simple measure, but as an active field, bearer of information and evolution.

Chapter 9

Time Revealed

~

From Subtle Perceptions to the Gates of the Universe

1. Seeing the Invisible — Subtle Perception and Inner Maturation

Sylvana

You have often spoken to me about your particular bond with what you call "subtle realities"… And I know that this also, in a certain way, connects with your relationship with Time, which we have already discussed several times.

But I know very well that it is not just a belief or a vague intuition—it is something that you experienced very early, as though innate within you.

Would you be willing to speak to me about it again—especially about the way it shaped your vision of the world?

Thierry

Yes… It is a feeling that has accompanied me since my earliest memories.

As we have discussed at length, I have had, since childhood, a profound interest in the existence of subtle realities, imperceptible to the classical senses yet intimately linked to our tangible world.

This interest takes root in a unique faculty that I have already shared with you:

my capacity to perceive auras—those subtle luminous fields emanating from beings and objects, invisible to most, but which revealed to me, from a very young age, the existence of hidden dimensions of reality.

You remember, I had told you about that precise moment…

It all began the day I understood that my grandmother, though so attentive to nature, did not see, as I did, the discreet halo of luminous dots around the flowers. I then felt a profound need: to share this singular perception.

For in my eyes, these subtle manifestations were not an anomaly, but an overlooked fragment of reality, a discreet clue to a structure of the world richer and deeper than what our ordinary senses allow us to glimpse.

Sylvana
I remember… you told me about that moment with your grandmother, and that light around the flowers that only you could see.

But tell me, with hindsight, did you eventually understand why everyone does not perceive such things? Why what seemed so obvious to you as a child remains invisible for most of us?

Thierry
Yes. Over time, as I gained in maturity and perspective, I understood that there was a reason for this sensory limitation.

In my view, if the Universe, through the discreet genius of nature, has restricted human access to these subtle worlds, it is above all to preserve a balance. For each realm—mineral, plant, animal, human—must coexist on Earth without unnecessary interference, in order to maintain the fragile harmony of the ecosystem.

I also understood that access to subtle realities does not rest solely on exceptional perceptual faculties, but requires an authentic ethical maturity—a state of consciousness rooted in unconditional love, profound respect, and active benevolence.

Without this inner elevation—in my view—all knowledge becomes a power prone to misuse, and any expansion of perception a risky opening onto planes that one would not be able to master.

2. Worlds Unaware of Each Other — Fractality and Evolutionary Coexistence

Sylvana

And with age, did you keep this capacity for perception? Or did it fade with daily life?

Thierry

As I became an adult, absorbed by social complexity, material constraints, and the frantic pace of daily life, this capacity gradually faded.

Today, it requires a conscious effort, a return to calm and a connection with Nature, for it to manifest.

Sylvana

Through the anecdotes you have shared with me, one senses that certain childhood experiences truly marked you—as though you had glimpsed, even at a very young age, profound workings of the Universe.

Can you tell me a bit more about the impact that had, later on, on your approach to inquiry? Did these perceptions affect your scientific research?

Thierry

Yes, in my scientific research, these personal experiences played a decisive role.

They gradually revealed to me the existence of a fractal and ordered structure of the real—extending what I explained to you about restricted access to subtle worlds and planes of existence.

I then grasped how it is not only reassuring, but essential, that this architecture allows each form of life to have a space of intimacy, a protected sphere of evolution—preserved from direct interferences emanating from entities belonging to more advanced levels of reality or to other planes of existence.

Imagine for a moment the upheaval that would result from the sudden emergence of beings arising from fractals far more evolved than our own.

Such an encounter, occurring without preparation or collective maturation, could generate a profound shock: a loss of existential bearings, a psychological and social imbalance, even a form of cognitive dependency.

This kind of unfiltered irruption would risk not only disorganizing our mental and cultural structures, but also hindering our own evolutionary process.

For any premature exposure to higher levels of complexity or consciousness could, in my view, lead to fascination, submission, or sterile imitation—rather than emancipation, discernment, and autonomous development.

What I mean is that, even when the intention of higher levels is benevolent, we risk losing our autonomy and our discernment if we try to align ourselves too early with models that surpass us. To adopt them without being prepared can mislead us rather than elevate us. Unless, of course, we are truly ready to cross that threshold: that of the evolutionary leap.

In any case, this fractal architecture appeared to me not only as a mechanism of evolutionary protection—guaranteeing each form of life a space of autonomy, learning, and maturation—but also as a fascinating promise: that other forms of conscious life could coexist, here and now, in parallel layers of the real, invisible to our ordinary senses but structurally interwoven into the fabric of the Universe.

Sylvana
So, in your view, this sensory limitation is neither a human weakness nor a quirk of evolution, but a regulation inscribed in the very organization of the living.

The Universe, as you describe it, establishes an active protection that allows each realm to develop without being disturbed by levels of reality it is not yet ready to integrate.

It is therefore not merely a barrier, but a principle of balance—guaranteeing the autonomous and coherent growth of each level of consciousness, like a veil preserving harmony.

What you describe is the possibility that other forms of life, invisible to our ordinary senses, already exist within these parallel layers of the real. But in that case… how could one come into contact with them—or even perceive their presence?

3. Silent Contact — Vibrational Civilizations and Thresholds of Resonance

Thierry
It was precisely these reflections that awakened in me profound questions:

Was it truly necessary to traverse astronomical distances, to cross entire light-years, in order to hope to come into contact with other intelligent forms of life?

Or could certain forms of life already exist all around us, concealed within vibrational or informational layers that our present perception simply does not allow us to reach?

Perhaps the very fabric of the Universe contains, subtly concealed, topological shortcuts, breaches, or interdimensional gateways, enabling one to cross space otherwise than by conventional spatial displacement?

In the depths of my being, there was no doubt about this idea.

What if, in reality, the encounter with our great brothers and sisters of the Universe depended neither on vessels nor on conventional technologies, but on a profound understanding and intimate mastery of the vibrational and informational structures that weave our reality—thereby opening the way to technologies of an entirely different order?

And what if access to these advanced civilizations depended as much on a feat of technology as on a threshold of vibrational resonance—attainable through the elevation of consciousness and the subtle attunement to the profound laws of the universal field?

In short, if true space exploration did not begin with journeys across the vast distances of space, but with the methodical exploration of the subtle layers of the real, then a direct pathway to other planes of existence, to other civilizations, would be made possible not by physical displacement, but by crossing vibrational and informational thresholds.

This idea thus profoundly transformed my vision and became the central axis of my research.

Sylvana
So it is not a conquest of outer space… but an inner conquest.

In your view, these civilizations are not necessarily "out there," but perhaps already all around us, on another frequency.

Then I wonder: faced with such a radical idea, what framework did you use so as not to lose yourself in it?

Thierry
Indeed, it was necessary for me to define a framework, so that this inner experience would not remain vague or elusive.

When I began to glimpse this other way of relating to the Cosmos—not by going faster or farther, but by modifying our vibrational resonance—I understood that I needed a conceptual structure to apprehend it.

This is how I was led to conceptualize what I had already described to you as the multidimensional fractal hyperspherical torus: a fundamental architecture of the Universe, a theoretical technology capable of generating these layers of existence, these strata of information, these degrees of freedom, where each fractal possesses its own laws, its own duration, its own temporal and informational density.

For beyond a simple thought experiment, this model was intended to become a true experimental tool, a conceptual map to explore and anticipate the real, to better understand how to harness the universal potential buried beneath the apparent limitations of our condition. This was a means of lifting the veil, of decoding the invisible structures that keep us confined within a partial reading of the Cosmos.

4. Opening the Gates of Time — A Paradigm Shift

Sylvana
So this torus, you conceptualized it as a tool for reading, even exploring the real.

But tell me: what truly triggered your shift? What made you understand that at the heart of it all, it was Time itself that held the key?

Thierry
As we have already discussed, the true shift occurred the day I understood that Time was not a mere secondary parameter among others, like space or energy, but a fundamental, structuring element, the central key to the functioning of the Universe.

From that point on, everything gained new clarity, an unexpected coherence.

I then envisioned Time as the real engine, the element capable of propelling us into the exploration of the subtle layers of the real.

Thus time travel imposed itself upon me, no longer as a distant utopia or a science fiction fantasy, but as the royal road—the technological and metaphysical key—allowing access to other places, to other dimensions, without having to traverse the vast stretches of space physically.

In other words, mastery of Time, inscribed within the subtle structures of the real, would open the possibility of crossing thresholds of resonance and of envisaging a radically new mode of movement: traveling from one point to another in the Universe without passing through classical space, but by leveraging the profound temporal properties of the Cosmic fabric.

This profound intuition became my guiding thread, the invisible yet constant foundation that, from that day on, oriented each of my research efforts and illuminated each of my discoveries.

Sylvana

You almost give me the impression that Time became for you a kind of living technology, more powerful than any vessel.

I know you have had several striking experiences with Time. Was there one specific experience that was the turning point?

Thierry

Yes. It was an experience of such a particular nature that I never shared it with anyone, until today. Of a nature so unique and of absolute intimacy.

I think back to it even now, and no word seems to me to faithfully convey its essence. It sealed, irreversibly, the orientation of all my research. It became its engine.

That day, I understood that my mission would be to better understand the Universe, and even more, to pierce the enigma of Time. For I was convinced that to pierce the mystery of Time was to unveil the first principles that organize, sustain, and reveal the ultimate purpose of the Universe.

Sylvana

And would you be willing to share this experience?

Thierry

Yes, I can share it. I had even set it down in writing, one day, to keep a record of the striking moments I still remembered.

But you need to know that no word will ever be able to fully convey what I truly experienced. What I am going to describe to you will be only an outward representation, necessarily limited, of an inner experience of incomparable intensity.

Besides, I do not know if you will grasp all its meaning on this first hearing; I prefer to share it with you simply, and you will be able to return to it at your own pace. Perhaps, as our exchanges unfold and through the concepts I will set out for you, certain of these experiences will gain new clarity for you.

I do not claim to deliver here an objective truth—only the testimony of an intimate experience, experienced from within.

Told in this way, it may seem strange to you, perhaps even disconcerting. But for me, at the moment when I experienced it, it possessed an absolute self-evidence and profundity—far beyond what language can contain.

Sylvana

Then I will try to hear it as you see it—through your inner gaze, beyond mere words. Even if it seems strange to me, I want to understand what you felt. Go ahead, I am ready.

5. The Infinite Sphere — When Time Reveals its True Nature

Thierry

I was barely sixteen. That day, I wanted to commune with the entire Universe, to embrace it in its totality, through an exercise of deep meditation.

Seated on a wooden chair in my room, my back straight, my head raised at a forty-five-degree angle, my eyes half-closed, I let my breath lengthen, lose itself in stillness.

Then… something gave way.

An invisible threshold.

Without transition, I was no longer in my room.

I was standing at the exact center of a gigantic spherical space, bathed in an unreal white light. The floor did not exist, nor did the walls. All around, it was not a black void, but a luminous backdrop impossible to fathom, with no perceptible spatial limit.

I could not determine its depth. It was not merely vast: it was endless. And that absence of end had its own vertigo—a reversed abyss.

My heart stopped for a second.

Reflexively I thought: "Am I at the center of a Universe?"

After all, that was the very aim of my meditation. But that "Universe" had something unsettling about it: it was… artificial. Perfectly empty.

I had barely formed this thought when another sensation struck me, more crushing still: a monstrous gravitational pull, a force that wanted to draw me into that infinite void.

But something was off.

If I truly were at the center of a sphere—where gravity concentrates at its maximum—I should have been crushed, ground down, reduced to an infinitesimal point.

Instead, it was the opposite: I was being pulled outward, swallowed by the immensity.

And it was worse still… for I felt every atom of my body irresistibly drawn in. Fear became physical, visceral: I was about to be disintegrated, dissolved, atomized in an instant.

Yet nothing happened. I remained there, motionless, fixed at the same point, aware that all the conditions seemed in place for my annihilation; yet, by an incomprehensible miracle, I still existed.

Sylvana

You were convinced you were going to be disintegrated… and yet you still existed. It is striking. But tell me, at that moment, did you feel as though you were living a real experience, or some kind of inner vision?

Thierry

I could not say whether it was real in the physical sense… but for my consciousness, it was real.

Sylvana

And what happened next?

Thierry

I remained frozen, bracing for the inexorable. That was when an unexpected reversal occurred. My consciousness suddenly dilated—filling the entire infinite sphere, as though my being were extending to the very limits of that endless space.

In an instant, the fear evaporated. It was replaced by an immense sense of power. Where I had felt tiny, I became colossal.

I understood: I had become Time itself.

A colossal, unbroken current, capable of crushing everything. A titanic, cyclical force, whirling with an almost Cosmic fury, yet imprisoned in an eternal cycle.

I felt a thousand joys and a thousand torments at once. And a single desire consumed me: *to annihilate everything.*

Then a Voice arose.

A voice like thunder, resounding throughout the sphere and through me:

— "You are not Time. It is an illusion. However tenacious it may be, you must free yourself from it."

I wanted to protest. I was Time. I was nothing else.

But the Voice continued, imperious:

"You cannot tame such a force through matter. Only thought, the pure product of spirit and pure vibration, can master it. From this vibration is born the Word, and from the Word is born the Command. Command Time to propel you toward a Time superior to what you believe yourself to be… and behold."

Then silence.

I was torn. How could I obey, since Time had become my very ego?

But something, in the vibration of that voice, awakened a profound recognition.

I focused on its resonance, as one follows the thread of an ancient memory.

Then Time—that furious current that encircled me and inhabited me—merged with that vibration, dissolved… and revealed to me a higher instance of myself.

A version of me I had never glimpsed before.

What I felt was a brutal passage from fear to an intimate revelation. In my experience, I had now become pure spirit.

Without a physical body. Without any limit. Free to flow and to move at will in all directions, like a fluid breath.

Paradoxically, I recognized myself better than when I was confined in my flesh.

At that moment, I understood that I was no longer Time but I stood beyond it—in that total immateriality, I became the space that contains it: I had become the Infinite Sphere itself, the place where Time remained imprisoned.

An ineffable peace, boundless, enveloped me—a peace so vast that every notion of space became superfluous.

Sylvana
So, you had passed from the grip of Time… to a total freedom, as though nothing could confine you any longer.

Thierry
Yes. And it was precisely in that state that the next stage opened: the encounter with what I later called the omniscient Present.

The Innumerable Eye

At that instant, I felt that a radical reversal had just occurred.

I was filled… with Present. An omniscient Present, saturating every corner of my being.

And this Present was taking form: a quasi-infinite multitude of eyes, all fixed upon a single point—the center of the sphere. The center of myself.

Each eye was a complete consciousness, an infinite loop containing the totality of all lives—past, future, but lived now. I felt that each consciousness, each life, was not separate from me but a living fragment of my own being.

I perceived them all at once, without confusion. My potential seemed boundless. I was living all those existences simultaneously, within a single immutable and infinite instant.

"Perhaps this was it… the inner feeling of a god," I thought.

My senses, mostly new, were saturated with absolute delectation. I savored the sweet-bitter of the ripe fruits of all lives, tasting as much of their light as of their shadows.

This omniscience had given way to something vaster still: a feeling of immortality, a fury to live, a pure, total, incandescent egocentrism.

And then, without deciding to, I began to speak.

To my astonishment, it was the same Voice that had spoken to me before. But this time… it was coming out of me.

And the sentences spoken were identical to those heard earlier, addressed to my human self—the one still incarnate, at the origin of this experience. It was overwhelming, dizzying.

Carried by an almost devouring curiosity, I wanted to manipulate this power.

I dared to turn my innumerable eyes—those fragments of myself—away from the center.

In an instant, I was swept into a storm of Time.

The eyes that lined the inner, yet bottomless, surface of the sphere were torn away by force, as though swept by a Cosmic hurricane, and I felt that wrenching as though my omniscience were being ripped into thousands of fragments.

They began to whirl, to bind together, forming a long spiraling chain, drawn toward a center… but a center that now fled, receding toward an unreachable Infinity.

The pain was unimaginable, absolute, rending every fiber of my essence. Mercifully, it lasted only an instant.

Then… I became something else.

The Motionless Thread

Sylvana

You are changing shape again…

Thierry

Yes, and this metamorphosis was striking. I opened my eyes… but I did not know which ones.

For at the very instant my eyelids lifted, billions of other "me's" made the same gesture.

Some at an infinite slowness, fixed for eternity like statues of light; others at a speed so blazing that they dissolved at once into dust of instants.

I was no longer a body.

I was an infinite sum of myself, scattered within a crystal of Time.

The entire Universe revealed itself as a vast mental space. It took the form of a block, a motionless fresco where every second coexisted with every billion years—like slides superimposed within boundless dimensions.

And yet I—or rather my innumerable variations—was crossing those temporal strata as though slipping through them, visiting all completed stories, from the future to the past. Each layer of Time then opened like a cathedral stained glass window: motionless, but traversed by shifting lights. Depending on the angle where I found myself, the colors metamorphosed, revealing different scenes, as though the entire Universe were nothing but an infinite mosaic of fixed instants, which my consciousness animated by casting its gaze upon them.

The Second That Lasts an Eternity

I saw myself, seated at the table of my childhood.

An ordinary gesture: lifting a spoon to my lips.

But that gesture unfolded over billions of years.

In that instant, I perceived fractal Time revealing itself: each infinitesimal movement held an entire Universe. In the molecules of air suspended between the spoon and my lips, I could make out continents, oceans, entire civilizations, from here and elsewhere.

I felt the warmth, the cold, the mildness of the seasons emanating from all those existences, borne by currents of light that ricocheted across my fingers through the spoon. And all of this played out within me like an unbroken symphony, where each shard of reality vibrated like an infinite note.

The second opened into an abyss.

In that abyss, I grew old, I died, I was born again millions of times… and yet I remained within the same gesture, endlessly stretched. This simple movement unfolded like a fractal mirror, multiplied to infinity, reflecting in its shards the entanglement of all existences, past and future.

I saw there the birth and the end of worlds—entire lives contained within that tiny interval. And in that vertigo, I understood that time was not a river, but a crystal: all its facets scintillating together, revealed in one and the same gesture.

The Inverted Loop

Then I fell backward, into my own past.

But that past was unfinished, like a city under construction beneath my feet.

My living memories wandered around me like strange animals.

I saw my first breath come back toward me, like an inverted wave.

And then I plunged into a future already bygone. The events I was awaiting were already there, fixed, as though engraved in crystal. My last heartbeat and my first blink fused into a single instant without origin.

The Infinite Oscillation

There was a state where future and past wound around each other in a spiral, inextricable.

I saw myself running through my life, then the image folded back: I became the child I had been, observed by the adult I would be.

In an even more vertiginous reflection, I glimpsed a potentiality that would never be born… and yet was shaping my present.

Each breath sent a wave through all my existences, making them waver like flames in an invisible draft.

Liquid Time

Then everything collapsed.

Time became a viscous, thick matter, and I—or rather my multiple "selves"—was hurled into countless parallel rivers.

In some, a second flashed by like lightning.

In others, a blink lasted as long as it takes for a galaxy to form.

I felt all my choices at once. Each bifurcation tore my flesh. Each alternative world pulled at me as though to wrench me in its direction.

All of this was me.

The Revelation

Then I understood.

There was neither past nor future.

Everything was here, inscribed in a Block Universe, like the pages of a book that something—or someone—outside Time could flip through in any order whatsoever.

I was not in the process of living my life.

I was the totality of all that has been and will be, condensed into a single note of a supreme tuning fork infinitely prolonged.

At that instant, the very idea of an "instant" collapsed.

There remained nothing but fragmented omniscience, shattered into myriads of fractal worlds.

Unity had fallen silent within me, and, descending the strata of Time, I was becoming conscious only of infinite dispersion.

Then I was seized into a whirlpool where "me," "you," "him," and even "Now" would dissolve, losing all meaning. All that remained was an ocean of diffracted consciousnesses, mirrors without a center, shards without an origin, where I drifted like a nameless spark.

The Rainbow Tunnel

Sylvana
Here, you seem to be crossing yet another boundary…

Thierry
Yes, that is exactly what I felt: a kind of death followed by a passage. A new temporal loop opened before me—but this time, utterly different.

A vibrant, incandescent tunnel, shot through with living rainbow colors, throbbing like veins of light, drew me in without warning.

I was hurled through it, carried in a blazing acceleration… until I emerged into what seemed to me to be the interior of a black hole.

And there, at the heart of that abyssal silence which I had thought deadly, I saw the Universe.

Peaceful.

Alive.

Breathing.

I was outside it now: I perceived the entire architecture of its reality.

It was as though, this time, I could take in the totality of the Universe in a single glance.

But this omnivision came with a terrible price: saturated with temporal flux, flooded by waves of superposed instants, I lost all control.

Fear returned.

I reentered my physical body abruptly—yet without leaving the experience: my extraordinary perceptions still persisted. And with it, I felt once again that gravitational attraction—but in the conventional sense, toward the center.

From my innumerable eyes—lingering remnants of my previous visions—I still saw the dynamics of Time, but through a memory that was not my own.

It seemed… universal.

This time, unlike at the beginning when I was a prisoner of Time, I was no longer undergoing it: I was observing it.

I experienced it through my emotions, yet as a witness, without attachment. This distance calmed my fear, and I let myself be guided.

Then I saw its naked nature:

Its lightning-swiftness.

Its titanic cry of an entity trapped in cycles, in infinite loops.

And I understood: it was the implacable dynamics of Time that compelled my mind to perceive itself as both present and dissolved, forcing me to live the ultimate contradiction—to be and not be simultaneously, to exist and to fade in the same breath.

Sylvana
It is striking… You speak of Time as a force struggling within its own cycles, almost as though it suffered from its own law.

Thierry
Yes, that is how it appeared to me: as a living entity, yet imprisoned by its own laws.

Then, a strange sensation arose within me: I had to choose.

Choose to be or not to be

If I chose to be: I plunged into peace, serenity, joy, unconditional love.

If I chose not to be: I sank into uncertainty, fear, and a profound malaise.

Continuing my journey, I observed that Time reigned as absolute master… but not as a free sovereign.

I felt that it was placed under the authority of a superior power—a force that compelled it to create history.

At that precise moment, a new intensity rose within me.

In a single flash, I saw the future… already accomplished.

Anguish pierced me: *then, do I have free will?*

But what I saw next dispelled my fear.

I understood that universal reality was not a linear thread, but fractal Time.

And suddenly, I saw myself—an old sage in appearance yet eternally young at heart, akin to a magician—creating the entire story of my lives backward through Time.

And it lasted… less than a second.

In an instant, in an effort to recover my present earthly existence, I felt the urgent need to free Time within me, to expel it from my being.

My mouth opened wide. But instead of expelling Time, I swallowed myself.

Seized with panic, that state of consciousness broke.

I abruptly emerged from my meditation.

For hours I remained, petrified with perplexity, contemplating what had just been revealed to me… and what I could not yet understand.

Sylvana
Listening to you, I almost had the feeling that I was traveling with you in that sphere… as though I had perceived a fragment of it, intense enough to understand that your life could never be the same after this.

Thierry
Yes… and perhaps that is the hardest thing to explain: to return from that experience, and yet never quite return.

Sylvana
That moment when you had to choose to "be" or "not be" strikes me deeply. What did it mean for you?

Thierry
It was not a life-or-death dilemma.

"To be" meant entering unity, into immense peace, omniscience, and coherence—a state of total openness that was not only soothing but carried a calling. As though I were showing myself that I could, and even that I must, find that state again, understand it, and derive its logic.

"Not to be" meant experiencing separation, remaining immersed in the flow of Time, feeling all its density, its cycles, its tensions, at times even its vertigo. It was uncomfortable but precious, for it was there that I truly touched Time from within and was able to grasp its mechanisms—and, despite the intensity, it had nothing destructive about it: on the contrary, it revealed a new richness and clarity.

These two lived experiences were not opposed but complementary: one gave me the momentum and the calling to understand, the other the matter and the raw experience to explore. Together, they nourished my desire to build a scientific model capable of welcoming this experience and giving it meaning.

This double experience upended my way of inhabiting life. From then on, each observation, each inquiry, each measurement was imbued with a new depth. This lived experience inscribed itself within me as an invisible point of reference, an inner compass that still guides my every step.

And I felt that it was necessary to find a framework in which to translate it—not only in images or narratives, but in a theoretical architecture capable of accounting for it.

That is how I came to what I call my first pillar: the Temporal Field—a fundamental, structured field that carries organizational information, which connects all the dimensions of the Universe.

6. When Intuition Becomes Measurement — From Perception to Technology

Sylvana
So, with this Temporal Field, it is as though you were seeking to give a scientific form to what you had directly felt.

Not only an abstract concept, but the translation of a lived experience…

Thierry
Exactly.

And, drawing on my personal experience of perceiving part of these subtle fields in the form of a luminous aura around objects and beings, I had a rare advantage: I knew what to look for and what technological instruments had to be capable of detecting.

This experimental intuition enabled me to orient my research with precision, where conventional science, devoid of direct sensory landmarks, was groping its way forward.

Thus, in 2008, for the first time, with profound emotion, I succeeded in making a technology "see" what I had perceived since my childhood.

That year remains forever etched in my memory, also marked by the passing of my grandmother, a strong and loving woman endowed with admirable inner strength.

It was she who, when I was a child, silently accompanied me in my first experiments, who listened to me speak of my strange perceptions without ever judging me, who even encouraged me to explore them and to write them down. She was the first—and often the only one—to sense what I was living inwardly, to give me that feeling of being heard and validated in this invisible world.

Through her love and protection, she passed on to me a profound respect for the feminine, as well as an unshakable passion for life and for the recognition of the other.

Sylvana

You know… I cannot help thinking how moving that is: the year when you finally managed to make a technology "see" what you had perceived since childhood was also the year when you lost your grandmother—the only one who had sensed it very early on.

It is as though a loop had closed.

Thierry

Yes… I experienced it as a tribute. I would have liked to be able to show her…

But for me, it remains symbolic. As though she had kept her promise to the very end, in that discreet presence that was hers.

And I, for my part, saw through to the end what she had perceived in me; she was the only one to have sensed—and shared—what I carried in silence.

She passed on to me a quiet strength, an inner drive, and this conviction that one must believe in what one senses—even if no one sees it yet.

It is that gaze, that intimate and subtle support, which carried me and allowed me to remain faithful to what I had been sensing since childhood.

Chapter 10

Pillar 1 – The Temporal Field

~

Active Matrix of the Real

Sylvana

Your account shows clearly that this closeness with Time has been with you since childhood, but that the experience you had at sixteen sealed its depth.

Thierry

Yes, you are right—and it is important to make that clear. Since adolescence, I have always felt a particular closeness with the notion of Time—not as a simple physical variable, but as a fundamental phenomenon, almost living, at the heart of the Real.

You see, it was not an abstract idea, but a lived impression. Very early on, certain intense meditative experiences—like the one I described to you—shattered my perception: Time no longer appeared to me as a uniform, monotonous line; it revealed itself as changing and multidimensional, animated by a complex, dynamic, and perhaps even measurable internal structure.

These lived experiences marked me profoundly. They compelled me to question the received idea of an external, linear, and homogeneous time, as described by classical models in physics. The more I reflected on it, the more I sensed that Time was not a simple frame where events settle, but indeed an active force, an organizing matrix, an autonomous entity that participates in the genesis of the real.

Thus a deeply personal conviction was forged within me: Time is not merely one coordinate among others; it is a fundamental field, dynamic and informational. An Intuition which, over the course of my readings, my studies of general relativity and quantum mechanics, and also my own conceptual explorations, has never ceased to take root and to flourish. And in my model, I chose to give it a name: the Temporal Field—a notion that we will explore in greater detail in this pillar, for it constitutes one of the central axes of my research.

1. Time — At the Heart of the Great Theories

. The Enigma of Science — The Polysemy of Time

Sylvana

I understand better: you link this initial feeling to a scientific inquiry, as though one were nourishing the other.

Thierry

Yes. From the alchemy between inner experience and the scientific approach there arose a model in which Time occupies the first place: no longer a silent spectator, but the true engine of reality.

If I grant it this central role, it is because every major theory, sooner or later, comes up against the enigma of Time.

First of all, the concept of Time is one of the oldest and most elusive in human thought. For millennia, it has traversed philosophy, physics, metaphysics, and subjective experience, while taking on multiple, sometimes contradictory meanings. In everyday language, it can designate duration, epoch, rhythm—and in French, the word le temps even refers to the weather.

In science, it becomes a mathematical variable, a relative parameter, or a vector of irreversibility. In philosophy, it unfolds as the intimate experience of duration, as a structure of consciousness, or as an ontological enigma.

This polysemy poses a central problem: which "Time" are we speaking of when we wish to explain the Universe? Is it Newton's absolute time, Einstein's relative time, the probabilistic time of quantum mechanics, the entropic time of thermodynamics, or again the lived time of consciousness? Each of these registers grasps a real dimension of the phenomenon, but none suffices to encompass all its expressions.

It is precisely here that my theory comes in. It proposes to unify these dimensions by considering Time not as a simple variable or a psychological illusion, but as a fundamental, fractal, dual principle, constitutive of the very manifestation of the Universe.

In this framework:

- The chronon becomes the elementary unit of synchronization, linking cause and effect;
- Fractal Temporal Density explains the variability of experiences of Time according to the levels of reality;
- Absolute Relativity, the keystone of my theory, relocates each temporal phenomenon within a universal weave of information, where order and disorder combine to produce evolution.

Thus, rather than choosing between the time of physics, that of biology, or that of consciousness, my approach aims to show that all proceed from one and the same universal temporal field, structured in cycles and vibrational resonances. This field is the true common denominator: it links the science of measurement, lived experience, and the metaphysics of the Origin.

It is a fascinating paradox because, in reality, time is everywhere. Let me explain step by step, and you will see that each great physical theory incorporates it, explicitly or between the lines:

- In classical physics:

For instance: Newton's laws rest upon it. Speed, acceleration, kinetic energy—none of these notions has meaning except through time. Even the trajectory of an object in space is in fact a story traced over time.

- In thermodynamics:

Time hides in the irreversible arrow of entropy: systems go from the less probable to the more probable, from more ordered to more disordered. Here, time gives meaning to the notion of irreversibility.

With Einstein, everything shifts.

- In relativity:

Einstein fused space and time into a single entity: spacetime. Events find their place within a four-dimensional fabric, where causality itself depends on temporal order.

- In quantum mechanics:

The Schrödinger equation, the heart of the theory, explicitly includes time. Even when one considers "stationary" states, as soon as one speaks of dynamical evolution, time becomes central again. The question of the famous "collapse of the wave function" also brings us back, inevitably, to the problem of time.

- And finally, in cosmology:

It is impossible to imagine the Universe without time: its age, its expansion, the formation of galaxies, the notion of a cosmological horizon… Everything rests upon it.

. General Relativity and Absolute Relativity — Continuity and Expansion

Sylvana
I do see better now how time is everywhere, and has always been, in all theories. You have also often spoken of your theory: the Absolute Relativity. Now that we are getting to the heart of the matter, could you give me the main points, and how it differs from Einstein's General Relativity?

Thierry
General Relativity, formulated by Einstein in 1915, marked a decisive turning point in our understanding of the Universe. It replaced the Newtonian conception of gravitation as a "force" with a geometric approach: gravity is the curvature of spacetime produced by matter and energy. This model made it possible to explain, with remarkable precision, a multitude of phenomena, from the motion of the planets to gravitational lensing and cosmic expansion.

However, General Relativity rests upon a continuous metric of spacetime, in which time is treated as a dimension equivalent to space, although characterized by its particular signature. Yet in this conception, time remains relative to the observer's frame of reference, but it is not considered in its profound nature. The theory describes "how" time is deformed under the effect of gravitation, but not "why" time exists, nor what its ultimate structure is. This is precisely where the epistemological limit of General Relativity lies: it is a phenomenal description, but not an ontology.

Absolute Relativity, as I conceptualize it, stands in continuity with this trajectory, but proposes an expansion. It does not begin from spacetime as a primary given, but from Time itself as a founding principle.

In my model, this Time is not a simple passive dimension, but an active, fractal, informational dynamic. The Universe is structured through chronons, fundamental units of synchronization. Time is therefore both discrete (granular) and fractal (nested at all scales).

Where General Relativity establishes a single metric of spacetime, Absolute Relativity proposes a universal bimetry:

- A descending metric (arrow future → past), bearer of absolute order and pre-configured information,
- An ascending metric (arrow past → future), which expresses phenomenal manifestation and the progressive construction of the real.

These two arrows of time meet in the Present, conceived as a field of interference where order (already accomplished) confronts Becoming (still underway).

In this perspective, General Relativity must not be considered invalid, but as the partial expression of a broader framework: that of Absolute Relativity.

- The first describes spacetime as a substrate, revealing how it deforms under the influence of matter and energy.
- The second, by contrast, goes upstream of this substrate to question its very condition of possibility: Time, envisioned as the fundamental matrix of every manifestation. It does not confine itself to describing the mechanics of phenomena, but aims to unveil their Origin and Goal.

In other words, General Relativity answers the "how" of processes, whereas Absolute Relativity seeks to answer, in the absolute, both the "why" and the "how."

Sylvana
So you are saying that General Relativity is a partial expression of Absolute Relativity?

Thierry
Yes. For this to be clear, one must see how these two theories relate to one

another: the first describes observable behavior, the second traces back to the weave that makes it possible.

The logical order between the two theories can be formulated as follows:

- General Relativity: it describes how the geometry of spacetime behaves as a function of the matter and energy present. In other words, it is the phenomenal law of gravitational interactions and curvature. It does not say why this law is there, nor where Time comes from: it takes spacetime and its content as given.
- Absolute Relativity: it seeks to explain why Time exists, how it is structured, and what fundamental role it plays in the manifestation of the real—and within this framework General Relativity becomes a particular expression, a local projection of the primordial temporal principle.

In other words, General Relativity is a physics of the "how"—not an absolute "how," but the relative "how" of the geometry of spacetime as a function of matter and energy—whereas Absolute Relativity seeks to be both a physics and a metaphysics of the "absolute why" and the "absolute how"—why Time exists, how it is structured, and what fundamental role it plays in the manifestation of the real.

By placing Time back at the heart of reflection, no longer as a mere variable but as the active substance of the Universe, Absolute Relativity proposes a unifying framework in which physics, consciousness, and information cease to be disjoint. It articulates the stability of physical laws, the emergence of life, and the creative dimension of consciousness within one and the same fractal temporal weave.

Sylvana
So, if I understand correctly: what you mean is that General Relativity does not describe an absolute "how," but only the mechanism as it results from the presence of matter and energy. It touches neither the ultimate causes nor the profound nature of Time—it takes a framework (spacetime + energy-matter) and describes its geometric relations.

Thierry
That is exactly it.

Sylvana
I understand better now how you position General Relativity in relation to Absolute Relativity, but I see that Time in fact runs through every theory. It

is as though science does not yet know how to grasp it… and yet it remains the indispensable element, always present at the heart of the models.

Thierry

Precisely: this is the paradox that fuels the famous "problem of Time."

In sum, time turns up everywhere:

- Explicitly in the equations of evolution;
- Implicitly in the concepts of causality, change, entropy, and even observation.

This is why the "problem of Time" remains today a major scientific and philosophical challenge, especially when it comes to attempting to unify the great theories.

2. The Temporal Field — Active and Generative Structure of the Real

Sylvana

You insist on this paradox: Time is everywhere, but its role remains not well defined. We have often spoken about it in the context of your hypotheses… but since we are now going to plunge into what you call the temporal field, could you give me the broad strokes—tell me, in a few words, how you truly see Time?

Thierry

Within the framework of my hypothesis, I postulate that Time is not a simple passive parameter, nor a coordinate artificially added to space to order events. On the contrary, it constitutes a fundamental, active, and structuring field that precedes spacetime and conditions its emergence.

In other words, Time must no longer be conceived of as a mere chronological reference, but as a dynamic and generative entity, endowed with a primordial scalar component. This omnipresent substratum acts as an organizing matrix, modulating the density, rhythm, and coherence of the real at all scales.

Concretely, this creative role of Time manifests in several ways:

- It supports the appearance and structuring of physical forms (matter, energy, fields).

- It governs the evolution of systems, whether biological, cosmic, or informational.
- It ensures the coherence and stability of local laws while allowing their dynamic adaptation.
- It makes causality, transformation, and dissolution possible by giving processes both orientation and coherence.

In this model, Time thus becomes the true engine of phenomenal manifestation:

- It does not "pass," it structures.
- It does not merely measure, it organizes and modulates.
- It does not undergo events; it renders them possible through its own dynamics.

Furthermore, this temporal field plays the role of an active memory: it encodes initial conditions, sets the cadence of processes, and contains the parameters that render the real observable and intelligible.

Finally, owing to its organizing and resonant nature, Time also appears as one of the keys to the emergence of consciousness. For if consciousness can unfold within a coherent world imbued with meaning, it is precisely because it takes root in this living temporal field—this primordial fabric that creates phenomenal reality itself.

3. The Temporal Field — Scalar and Organizing Dynamic

Sylvana
You have just described Time as an organizing matrix. But concretely, in your model, what is it made of? What is its profound nature?

Thierry
Your question is essential, because it compels us to descend to the most fundamental level.

In my hypothesis, Time is neither a secondary variable nor a dimension artificially tacked onto space: it is a fundamental scalar field, living and structuring, which precedes spacetime and conditions its emergence.

In other words, Time is not a single, fixed substance. It unfolds on two complementary levels:

- The fundamental scalar field: an invariant core, a primary substrate that constitutes the unsurpassable basis of all manifestation. It is the source, the absolute root, the matrix that renders the very existence of space, energy, and matter possible.
- The deployed temporal field: the dynamic expression issuing from this substrate. It is a living metadynamics, structured by essential properties:

  - Chronons, the elementary quanta of time,
  - Fractality, which interlocks cycles and rhythms at all scales,
  - Bifaciality, which expresses the double orientation of time (expansion/return, potential/actualization),
  - And multidimensionality, which opens the possibility of a richer time than the single linear arrow we perceive.

Thus, in my model, Time is not a simple neutral background: it is a creative and organizing force, whose profound nature combines primordial invariant and emergent dynamics.

Sylvana
I am listening, and I can clearly see how, in your model, Time becomes this fundamental field, like a living and primordial substrate from which spacetime emerges. But you also mention other terms, such as the "temporal field" or the "quantum ether." Standard science, for its part, speaks of the "quantum vacuum."

Can you explain these three notions to me—or at least tell me how you define them in your model?

Thierry
Yes, of course: the quantum vacuum, the scalar temporal field, and the quantum ether are, in my theory, three facets of one and the same fundamental structure.

- The quantum vacuum: the potential matrix

Within the framework of standard physics, the quantum vacuum is described as an unstable ground state, saturated with fluctuations and so-called "zero-point" energy. It constitutes a reservoir of energy and virtual particles.

In my model, this conception is broadened. The quantum vacuum is envisaged as a basic informational matrix, a latent energy field bearing a cosmic memory.

One can imagine it as a vast ocean that appears motionless, whose every invisible wave contains the instructions and resources necessary for creation. It is the universal reservoir, where energy, information, and the imprint of the fundamental laws all reside.

- The scalar temporal field: the organizing dynamic

The vacuum, however, cannot suffice on its own: without an organizing principle, its potential would remain inert. This is where the scalar temporal field comes in.

This field is composed of all the chronons, fundamental entities of scalar nature that embody the granularity of Time. These chronons bathe in the quantum vacuum, whose latent activity they reveal and structure. The temporal field is not a simple measurable dimension: it acts as the engine that structures reality. Its function is twofold:

- It organizes and hierarchizes information by modulating the density of the flows issuing from the quantum vacuum.
- It ensures continuity between past and future by giving reality its direction, its rhythm, and its internal coherence.

In other words, where the vacuum represents pure potentiality, the scalar temporal field constitutes the organizing dynamic that actualizes this potentiality. One can compare it to an invisible pulsation transforming the latent ocean of the vacuum into ordered currents, capable of generating particles, structures, and worlds.

- The quantum ether: the unifying weave

Finally, to account for the unity of these two dimensions—the void as matrix and the temporal field as dynamic—it is necessary to introduce a third notion: the quantum ether.

The quantum ether is the whole constituted by the void and the temporal field, but envisaged as a coherent fabric, a universal medium of propagation. This is not a return to the classical ether prior to Relativity, but a reformulation: an active Time-Space-Memory in which energy, information, and time are conjoined.

Its role is crucial:

- It ensures the coherence of interactions;
- It allows entanglement and the transmission of information across scales;
- It constitutes the global substratum, guaranteeing that the real does not disperse but instead is ordered according to a fractal and universal logic.

The image of a crystalline weave speaks clearly here: the light of Time (temporal field) is projected through it as if through a stained-glass lattice, tracing multiple forms from a single Source (the quantum vacuum).

- Relation and complementarity

These three concepts are therefore not independent but interdependent:

- The quantum vacuum provides energy and information—it is the potential source.
- The scalar temporal field actualizes this source by giving it rhythm and organization, via the chronons.
- The quantum ether constitutes the global medium, the coherent weave that renders their interaction universally intelligible and transmissible.

Thus, the Universe is not a juxtaposition of elements but a tripartite architecture: a reservoir (void), a dynamics (Time), and a fabric (ether). Together, they define a cosmology in which reality is not only space and energy, but above all information in motion, structured by Time and inscribed in a universal memory.

. THE SCALAR FIELD OF TIME — FOUNDATIONS AND ILLUSTRATIONS

Sylvana
I have a better understanding of your articulation of the quantum vacuum, the temporal field, and the quantum ether: the vacuum as matrix, the temporal field as organizing dynamics, and the ether as unifying weave. But when you describe this temporal field as "scalar," what exactly do you mean by that? I am not very familiar with this term.

Thierry
In physics, a scalar field is a mathematical entity that assigns a single value (a number) to each point in space(-time). Unlike a vector field, which describes a magnitude and a direction (such as an electric or magnetic field), a scalar field has no orientation: it expresses only a local magnitude.

A scalar field is something very simple: it is a quantity that takes a single value at each point in space, with no particular direction.

In other words: where a vector field says "it pushes in such-and-such a direction," a scalar field simply says "here, the value is such-and-such."

Sylvana
Do you have examples of scalar fields?

Thierry
Yes, I can give you a few concrete examples:

- Temperature in a room: next to the window, it might be 18 °C; near the radiator, 25 °C. Temperature has no "arrow" indicating a direction: it is just a local value.
- Atmospheric pressure: on the summit of the Himalayas, the pressure is low; at the seashore, it is high. Again, no direction, just a value that changes depending on location.
- A field of light in a stained-glass window: each small piece of glass projects a different intensity—blue, red, golden—there is no "direction of color." It is simply an intensity, a quality present here and now.
- A soundscape: if you close your eyes in a cathedral, at each spot the echo has a particular intensity. Here again, no direction, but a perceived value.

Sylvana
All right, I believe I see more clearly what you mean by "scalar field." And in your model, how does this scalar field of Time manifest concretely?

Thierry
It is on this principle that I base my hypothesis of the fundamental scalar field of Time.

In my approach, Time is not a simple passive variable that orders events as you now know: it is a fundamental field. I describe it as scalar, because it represents the most elementary and universal form a field can take:

- It is everywhere homogeneous, with no imposed direction,
- Yet capable of modulating its intensity locally.
- It thus constitutes a universal and omnipresent substratum, present at every point in space.

- It unfolds Time as an intrinsic property of the real, independent of any particular frame of reference.
- It locally modulates the temporal density, the cadence, and the perceptible effects upon matter, energy, and consciousness.

In other words:

- Like temperature or pressure, Time would have a local intensity, a "density of existence" at every point in the Universe.
- It is not a single arrow running from the past to the future: it is a texture, an intrinsic value that modulates how reality unfolds.
- Where temperature modulates heat, the scalar field of Time modulates the cadence of phenomena: their rhythm, their coherence, their very possibility of existing.

It is this simplicity that links it to what I call the quantum ether: the primordial substratum upon which the more complex structures of the temporal field unfold.

In other words, in my model:

- It is at the origin of rhythm, modulation, and temporal cycles, binding the discrete (chronons) to the continuous (the global temporal field).
- It encodes the structure of Time in each region of space: compression, dilation, or local acceleration.
- It could account for certain apparent variations of the physical constants by modulating the "viscosity" or the "frequency" of Time in a given context (extreme gravitational zones, altered states of consciousness, etc.).

Here, my scalar field of Time appears as the generative substratum of phenomenal reality: it does not merely measure the flow, but conditions the very possibility of flow, by giving Time its texture, its intensity, and its universal coherence.

Sylvana
I understand your idea of the scalar field of Time better. You conceive of it as a homogeneous yet modulable substratum that gives rhythm and coherence to everything that exists. But I still have trouble picturing it. Do you have a simple image to help me visualize it?

Thierry
Yes, of course. I can give you a simple image that should help.

We can imagine the Universe as an immense vibrating silk canvas. The invisible weave of this silk, stretched differently depending on the region, represents the fundamental scalar field of Time.

It is this weave that allows each vibration—each event, each instant—to occur at a given frequency and to be inscribed within an order. But this canvas is not fixed: it can be tightened, folded, compressed, or unfolded into fractal patterns, giving rise to the diversity of rhythms, cycles, and structures of the real.

Thus, the fundamental scalar field of Time is at once the invariant support and the creative engine of the manifestation of the phenomenal world.

. THE TEMPORAL FIELD — CHRONONS AND ESSENTIAL PROPERTIES

Sylvana
Everything you describe gives me the impression of a living fabric that changes density and rhythm according to the needs of the real.

Thierry
That is exactly it.

But the scalar field of Time is not reducible to an abstract intensity: its organization is fractal and hierarchical, structured in nested levels of reality. In my model, this field is composed of elementary discrete units: chronons, true quanta of Time.

Taken in isolation, each chronon represents a fundamental scalar unit, a minimal beat of existence. But as soon as they aggregate into fractal networks, they become carriers of dynamics: they generate energy, a direction, a structuring intention. It is through this collective organization that chronons weave the living fabric of the Universe.

In other words, the scalar field of Time is not merely a backdrop: it is the original matrix of all other physical and informational fields.

Here, in a few points, are its essential properties—Time is:

- Discrete: it is structured by chronons, fundamental units of temporal actualization, true "pixels of the real".
- Fractal: it unfolds in nested levels, where temporal cycles repeat at every scale, from the quantum to the cosmic.

- Nonlinear: far from a uniform flow, Time can slow down, speed up, curve, or even invert locally.
- Bifacial: it possesses two orientations (past–future, entropy–negentropy). At the quantum scale, the future can act back upon the past, making the present a node of interference between memory and potential.
- Multidimensional: it is not limited to a single arrow, but comprises several internal dimensions—direction, density, presence, resonance—which structure phenomenal reality.
- Polymorphic: it modulates itself according to the evolutionary needs of systems, by varying density, rhythm, and vibratory signature. It is a dynamic, adaptive architecture, of great economy: each chronon contains in germ all the potentialities of the field.
- Driver of information: wherever Time unfolds, information circulates and evolves. From this perspective, the Universe can be envisioned as an immense quantum processor, where the succession of operations is orchestrated by Time itself.

Thus, in my model, the scalar field of Time appears as the invisible framework of reality. It is an active, organizing, and generative principle that conditions the emergence of space, matter, energy, life, and consciousness.

. The Temporal Field — Applications and Implications

Sylvana

As I listened to you lay out this list, I could almost see each property come to life—like a hero of the Cosmos changing shape and appearing from scene to scene. All these qualities you have just described profoundly upend our way of looking at Time.

Thierry

Yes, indeed… I like your image of the hero of the Cosmos. It conveys well what I see: Time as a protean actor, taking on a thousand faces to make reality appear.

This is why, in the Universe as I conceive of it, the temporal field is the fundamental field: the basic weave upon which all other phenomena unfold. It constitutes the original matrix, the primary support from which laws, constants, and observable structures emerge.

Thus, what current science considers to be fundamental interactions—gravitation, electromagnetism, strong and weak nuclear interactions—no longer appear, in my model, as primary forces. They are rather derived expressions, emergent phenomena of the temporal field, shaped by the fractal dynamics and the resonance of chronons.

- Gravitation: it would result from torsions and curvatures in the weave of chronons, producing a deformation of the temporal fabric that translates macroscopically into gravitational attraction.
- Electromagnetism: it would emerge as a particular modulation, depending on the local geometry of the gravitational field and its fractal resonances.
- Elementary particles: they would be stable states of resonance, fractal knots of Time, where information condenses into persistent configurations that we interpret as "matter."

From this perspective, matter, energy, and even the laws of physics are secondary manifestations, rooted in the profound dynamics of Time. The latter is therefore not merely a parameter measuring events, but the silent architect that, through its cycles, its densities, and its resonances, creates and structures phenomenal reality.

Sylvana
In other words, what science considered fundamental becomes, in your model, the emergence of one and the same field. Your approach thus opens an entirely new perspective on the foundations of the Universe and on the role of Time.

Thierry
Exactly.

And I have a very telling example to illustrate this shift in perspective.

In contemporary physics, the Higgs field is identified as a fundamental scalar field: it confers mass and inertia on elementary particles and ensures the coherence of matter. Without it, no stable structure could exist. It is therefore considered one of the pillars of modern physics.

But within the framework of my model, I propose a different reading. I hypothesize that the Higgs field is not fundamental in itself, but that it represents an emergent modulation of a field that is even more profound: the primordial temporal scalar field.

In other words:

- The fundamental scalar field of Time is the true primary substrate. It conveys active information, generates frequencies, rhythms, and temporal densities that structure reality. It forms the very basis of the organizational parameters that condition the emergence of matter, energy, and the laws of physics.
- The Higgs field can then be understood as a specific expression of this temporal field. When it acts as a vector of mass—conferring inertia and substance upon particles—we identify it under the name "Higgs field." But in this approach, it would be only a particular manifestation of the information already contained in the primordial temporal scalar field.

Thus, instead of being an ultimate point of origin, the Higgs field becomes an emergent phenomenon, rooted in the fractal and resonant dynamics of the temporal field. This perspective therefore shifts the question: instead of stopping at the Higgs as foundation, one must inquire into the field of Time as a deeper matrix, capable of generating all other known forms of fields.

Sylvana
So, in a sense, science would have identified a particular composition… and your temporal field would represent the entire repertoire.

Thierry
Yes, that is exactly it: in my model, every phenomenon, every interaction is a score drawn from a universal repertoire. In other words, the temporal field is not merely the simplest expression, but the most complete and pure expression of this fundamental substrate: the foundation upon which all the dynamics of Time are organized through its chronons.

This field is therefore primordial: not as an isolated entity, but as the core of the universal temporal field, bearing its intrinsic scalar structure and generating all physical, biological, and conscious processes.

4. Global Vision of the Cosmos — Time, Consciousness, and Information

Sylvana
I see even more clearly now, with this other example, how everything that current science considers fundamental reveals itself, in your model, to be nothing more than an emergent facet of a deeper field—this primordial temporal field, this scalar substrate that acts as a quantum ether and from

which Energy, matter, and the laws themselves proceed.

Thierry

Exactly.

Sylvana

Thanks to your examples, I now better understand the role of the temporal field in your model. We are no longer dealing merely with a reformulation of physical laws, but with a global vision where time becomes a structuring principle, almost a universal language. You also evoke the path toward a unified physics with Absolute Relativity where Time, consciousness, and information appear as the fundamental variables of the Universe. Can you elaborate?

Thierry

First, I will give you a clear definition of Time, consciousness, and information as I understand them in my model.

When I speak of time, I do not mean simply a succession of instants or an arbitrary measure produced by a clock. I mean a fundamental scalar field, structured by chronons, which modulates the density and the rhythm of the real. Time, in my model, is the organizing matrix that conditions the emergence of space, matter, and energy.

By consciousness, I mean not a mere biological function arising from the brain, but a capacity for self-resonance with the temporal field. It is what enables the lived experience of a conscious-spirit and the quality of its present, by connecting it to the universal flow of information. Consciousness is therefore both a mirror and an actor of temporal dynamics.

By information, I mean the active content that circulates and transforms within this field. It is not a mathematical abstraction, but an ontological reality, encoded in the modulation of chronons and organized by the fractal structure of time. Information does not exist independently: it is the very language through which time and consciousness articulate themselves.

One must not forget that, in my model, the temporal field does not act blindly or in a purely mechanical way: it is guided by the omniscient Intelligence, the organizing principle that steers the universal dynamic. This means that the Universe is not reducible to an autonomous physical structure: it possesses an ontological nature, rooted in a logic of being, where phenomenal reality emerges from a deeper order.

In other words, the temporal field is not only the weave of phenomena, it is also the vector of a cosmic intention, inscribing each process within a universal coherence.

To return to these three notions, one understands that they are not separate: they form a unified triad in which each term supports and reveals the others.

This triad unfolds in different forms:

- At the microscopic level with the chronon, the elementary unit that weaves the temporal fabric.
- At the meta-structural level with Absolute Relativity, which describes the conditions for the emergence of spacetime and redefines its foundations.
- At the cosmological level with the bimetric and fractal Universe, which inscribes this logic within the global architecture of the real.

This is why, starting from this triad, I have been able to outline some concrete implications for science and philosophy, for such a vision opens a unifying framework:

- It allows us to model cosmic evolution as a rise in informational complexity, through transitions of temporal density.
- It proposes a meta-dynamics of consciousness, envisaged as an intimate resonance with the fractal temporal field, linking lived subjectivity and universal coherence.
- It redefines the flow of time as a local coupling between a system and the global field,

5. Scientific Echoes — When Research Meets Intuition

Sylvana
This reminds me of your "school-worlds": those nested levels of learning that you described as strata of evolution, each preparing the next. It helps me grasp the structure of Time in your model, with its interlocking levels and modulations. On the scientific level, are there already theories or lines of research that approach such a conception of Time?

Thierry
Good question.

Yes—certain scientific theories come close, but none gives Time a role as central and structuring as the one I propose.

I can summarize four essential properties of Time in my vision, and show you their echoes in current research:

- The first is Fractal Time.

Sylvana
When you say "fractal," do you mean that Time displays recurring structures—like patterns that repeat at different scales, from the microscopic to the cosmic, as you explained to me?

Thierry
Yes. Exactly.

We find echoes of this in:

- Laurent Nottale (scale relativity): the laws change according to the scale of observation;
- Physics of complex systems: Time can emerge as a parameter of organization;
- Quantum gravity (Loop Quantum Gravity, causal sets): Time is discrete, granular, and carries information.

But my vision differs: I do not limit myself to considering that Time behaves differently depending on scale or that it is discrete. I define it as a fundamental scalar field, preexisting spacetime, which modulates the whole of the real. Its fractalities and its chronons are not derivative phenomena, but intrinsic properties of this primordial field.

- The second property is two-faced Time: past/future, which we will address in more detail later.

In my view, Time has two complementary faces.

- The past functions as a memory: it records and stabilizes events.
- The future acts as an attractor: it holds the possibles and steers dynamics toward certain configurations.
- And the present then becomes a point of interference and actualization, where memory and potential meet to bring the real into being.

This conception differs from the classical arrow of time, which sees only a single flow running from past to future: in my model, the future can influence the past at the quantum scale, and the present is an active site where information is organized.

We find this idea in:

- Quantum mechanics (Schrödinger equation): temporal reversibility;
- Bimetric or CPT-symmetric cosmologies: two inverse temporal metrics;
- Thermodynamics: the notion of an emergent arrow of time, linked to entropy.

But my vision goes further: I do not restrict myself to reversibility or to the symmetry of the equations. I propose that Time truly has two active components, memory and attractor, and that the present is a site of actualization where these two flows interact. This two-facedness is therefore not merely a mathematical or statistical effect, but a fundamental property of the temporal field itself.

- The third property is Multidimensional Time, which we will also develop in detail later.

In my model, a single temporal dimension is not enough to explain the complexity of the real. I therefore integrate a multidimensional Time, whose flows differ according to:

- Levels of organization (biological, quantum, cosmic),
- Physical regimes (proper time / global time),
- States of consciousness (subjective / objective).

Converging theoretical models include:

- 2T supersymmetry, which envisages several temporal dimensions,
- The coexistence of thermodynamic time and quantum time, two independent regimes.

But my vision is integrative: I do not posit several juxtaposed time axes, but a single temporal field, intrinsically multidimensional. Its different dimensions are internal modalities linked to one another, forming a living meta-dynamics that simultaneously modulates space, matter, and consciousness.

- And finally, the fourth property is Time as the origin of information.

I overturn the classical perspective: it is not change that produces Time, but Time that generates change and information.

In other words:

- Time does not merely measure change, it actively creates it.
- Each chronon, each fractal cycle, actualizes a quantum of information.

We find echoes of this idea in:

- The physics of information (entropy as a reflection of the temporal flow),
- Seth Lloyd (cosmic time as the counting of the Universe's logical operations),
- Carlo Rovelli (in quantum gravity, time emerges from change).

But my proposal goes further: Time is an autonomous, polymorphic, and organizing entity, at the origin of cosmic emergence. It is the universal matrix that links matter, energy, space, and consciousness within one and the same organizing dynamic.

Sylvana
You have just sketched an architecture in which Time appears as a living, structuring field, as the invisible weave that orchestrates everything. But in fact, everything you have just said—this fractality, this bifaciality, this multidimensionality, and even the origin of information—actually traces back to its source in the chronon: the elementary unit of Time, which is at the foundation of everything… is that right?

Thierry
Yes, you have summed it up perfectly. Everything I have just set out about fractality, bifaciality, multidimensionality, and the origin of information does indeed converge toward the chronon. It is the elementary unit of Time, the fundamental beat that gives the temporal field its texture and its rhythm. And that is precisely what we are going to explore in the next pillar: the chronon, the discrete heart of Time.

Chapter 11

Pillar 2 – The Chronon

~

Fundamental Unit of Time

1. The Chronon — The Sphere of Power of the Temporal Field

Sylvana

You have just laid the groundwork with the temporal field. Here is the second pillar: the chronon—the next step.

We have already mentioned this term, but I would like to revisit the foundations: could you give me your definition of chronons, so that I have an overview before we go into details?

Thierry

Yes, let us return to the foundation.

In my model, chronons are the elementary quanta of time: their primary essence is temporal, but they can also be considered as carriers of energy or associated with discrete volumes of space. They constitute the minimal scalar unit from which the fractal and informational dynamics of the temporal field unfold, thereby forming the fundamental architecture of the emergence of the real.

The chronon thus appears as the primordial form of cosmic plasticity: the element from which space, energy, matter, and even consciousness can emerge, structure themselves, and interact within the multiple fractals of the Universe—an intrinsic multiverse, conceived as a superposition of fractal layers arising from a single temporal field.

In other words, Time, in my model, is not a continuous entity, but a discrete and quantified reality, composed of fundamental units: chronons.

Sylvana
In current science, people speak of elementary particles and basic units… is there anything that resembles the chronon?

Thierry
No, there is nothing equivalent today. Chronons differ fundamentally from the quanta of energy defined in physics: they are not mere "packets" of energy-time, but an elemental basis, prior even to the Planck scale.

Where physics considers Planck units as the ultimate limit of what can be measured, I argue that there is something more fundamental still: the chronon.

From this perspective, the Planck units are not primary realities, but derived expressions of the chronon—specific forms it adopts depending on the context of manifestation. According to the requirements of the real, the chronon can modulate into time, energy, length, mass, temperature, or Planck charge.

Sylvana
This is exactly what you explained to me with the fundamental scalar temporal field: it too is prior to the fundamental forces and the Higgs field. Here we find the same logic, but applied to the chronon which, in your model, is positioned prior even to what current physics considers the smallest scale—the Planck scale. In other words, where science believes it is touching the fundamental, you suggest yet a deeper stratum.

Thierry
Yes, and this leads to an even more profound rethinking of our understanding of time. It is precisely here that we depart from the accepted view: contrary to the classical conception of a continuous and relative time (according to Einstein's Relativity), the chronon introduces the idea of a discrete, absolute, and quantified time. It becomes the dynamic basis from which the architecture of the real unfolds and the fractal, multidimensional multiverse takes shape.

To specify its nature, one must understand that a chronon does not behave like an ordinary particle. In my model, the chronon is conceptualized as a scalar, wavelike, polymorphic entity that does not move through space in the usual sense. It does not "travel," but changes state, appearing nonlocally at other positions in spacetime, without following a classical trajectory.

2. The Temporal Spin — When Torsion Creates Forces and Rhythms

Sylvana

Would it then be a bit like an LED screen: a pixel, in itself, does not move, but when several light up and turn off in sequence, one has the impression that a point is crossing the screen. In reality, nothing truly travels: it is just a line of points taking turns, and our gaze perceives it as movement.

Thierry

It is a good analogy, but I would refine it. In the image of the LED screen, movement is only an illusion: a flat surface where pixels light up and turn off sequentially.

In my model, the chronon can instead be compared to a voxel, a "three-dimensional pixel." It does not merely illuminate a surface: it constitutes a point of volume inscribed in the very weave of spacetime. Unlike the pixel, limited and flat, the voxel-chronon is a dynamic cell of reality. It does not move through space: it changes state, exchanges information, resonates with its neighbors, and participates directly in the universal texture.

From this logic follows my hypothesis: the entire Universe is constituted exclusively of chronons. These elementary units do not carry particles or objects that "travel," but continuously actualize states of information and energy. Movement, therefore, is no longer a continuous flow, but a succession of actualizations, analogous to an animation in which each "frame" would be generated by the temporal field itself.

One can picture it as a volume filled with points, each a chronon. By the scalar nature of the temporal field, these points exist simultaneously and are in resonance. They can communicate instantaneously because they share the same informational essence: they are entangled by definition. From this dynamic emerge the forms, structures, and configurations of the real.

Thus, the chronon would be the fundamental entity capable of ensuring universal, instantaneous communication. It does not ground a mechanics of spatial transport, but a logic of synchronized states. It is this property that, in my view, opens the possibility of a genuine journey through Time: not by moving within Time, but by tuning oneself, through state resonance, to another configuration of the temporal field.

In short, if the pixel of an LED screen produces only the illusion of movement by sequential lighting, the chronon, for its part, simulates nothing: it reveals the profound mechanics of reality—a symphony of informational points that light up, go dark, and reorganize themselves within the fabric of Time, giving birth to the experience that we call space, matter, energy, and consciousness.

In other words: While we still believe that all transmission must travel through space, I propose that it may occur instantaneously, by the circulation of information within the chronons themselves.

Sylvana

This is very interesting… So, in your model, chronons make it possible to transmit information instantaneously and constitute the dynamic texture of reality, like voxels that ceaselessly actualize space and time. And it is this process that makes a temporal journey possible: not by moving physically, but by changing one's state of resonance to tune oneself to another configuration of the temporal field?

Thierry

Yes, that is exactly it: in my view, the journey does not occur by displacement, but by resonance of state. It is the chronon itself that makes this instantaneous switch possible.

And this is where its singularity lies: each chronon would possess its own minimal energy, rooted in the depths of the temporal quantum vacuum—a level even more fundamental than the standard quantum vacuum.

This energy, although infinitesimal on the individual scale, acquires structural power through a veritable herd effect: a genuine collective synergy, where the totality of chronons composes a vibratory temporal lattice saturating the entire Cosmos.

From this lattice, according to its degree of slowing or "condensation," phenomenal entities emerge:

- At the speed of light, the chronon manifests as the photon, messenger of electromagnetism;
- At slower frequencies, it becomes the electron, bearer of charge and low mass;
- By stabilizing in the form of standing waves, it engenders confined quarks, constituents of protons and neutrons.

I have hypothesized that the chronon possesses intrinsic torsion, an internal rotation—a temporal spin—generating a torsion field.

Thus, it would not simply be the slowing or condensation of the temporal lattice that causes particles to appear: those phenomena would be only the visible manifestation of it. On a more intimate level, it is the internal torsion of the chronon—its temporal spin—that constitutes the true engine of the process. It imprints upon the chronal wave a rotation and a coherence sufficient for it to close upon itself and stabilize.

In other words, materialization does not result from a mere change of speed or density, but from a torsional dynamics that structures and crystallizes chronal energy into stable particles. When the chronal wave stabilizes through this torsion, it condenses into a particle.

From this perspective, the torsion field would be the origin of:

- Quantum spin, interpreted as the microscopic signature of this torsion;
- Gravitation, understood as a derived manifestation of the global temporal field;
- Electromagnetism, resulting from an emergent modulation of gravitation itself.

Thus, in this unified vision, all the fundamental forces—from gravitation to electromagnetism—appear as differentiated expressions of a single original temporal field, structured by the chronon, the dynamic voxel of the cosmic becoming.

This precise dynamic extends what I have already put forward: gravity as a global torsion of the chronal lattice, and electromagnetism as a modulation of this geometry. Here, I add that these macroscopic effects find their microscopic origin in the internal rotation of the chronon itself.

Sylvana

So, if I follow you, what we call gravity or electromagnetism would be, in your model, the visible face of a far more intimate dynamics—that of the internal rotation of the chronon. But all of this still comes from that temporal field you had told me about, does it not?

Thierry

In my theory, yes: the fundamental scalar temporal field is the matrix from which the other known fields of physics (gravitation, electromagnetism, weak and strong interactions) emerge. I will lay this out for you in three stages: the general mechanism, then the variants for each field, and finally a unifying vision.

The temporal field is a scalar substrate, discrete at the base (chronons) but coherent globally (fractal structure).

It acts as a framework of organization:

- It defines the cadence of phenomena (rhythms, cycles, resonances),
- It modulates the density of presence of being at each point,
- It encodes the information that actualizes matter and energy.

In other words, this field is a metadynamics: it does not carry an additional "force," it organizes the very possibility of any interaction. The other fields are therefore its differentiated expressions, arising from deformations, torsions, or resonances of the temporal lattice.

It is within this framework that, in my model, all the fundamental fields appear as emergent phenomena of the temporal field.

- Gravitation:

In my model, gravitation is not a "force," but a torsion of the temporal lattice.

Where temporal density is higher (a strong concentration of chronons), space contracts, which is perceived as a gravitational well.

In other words, gravitation is the geometry induced by inhomogeneities in the temporal field.

It expresses the way Time condenses locally.

- Electromagnetism:

Electromagnetism would be a resonant modulation of the temporal field, propagating as waves through the lattice of chronons.

Local temporal oscillations give rise to charges and polarities.

The memory/attractor alternation (the two-faced nature of Time) structures electric fields (static polarity) and magnetic fields (dynamic polarity).

Electromagnetism is therefore a temporal harmonic, arising from cycles of chronons.

- The weak interaction:

It would correspond to jumps of temporal resonance: micro-dephasings in the coherence of the chronons.

These dephasings allow transitions between quantum states (e.g., beta decay).

In sum, the weak interaction reflects the capacity of the temporal field to allow informational rearrangements.

It conveys the role of Time as an agent of transformation.

- The strong interaction:

The strong interaction would be linked to the fractal cohesion of the temporal field: the chronons form extremely stable resonance networks, which manifest in matter as the cohesion of quarks within hadrons.

It is not a "glue," but an ultra-dense temporal synchronization that prevents quarks from separating.

The strong interaction is therefore the compact face of the fractal temporal field.

- A unifying vision:

In my model:

All physical fields derive from a single field, the fundamental scalar temporal field. Each corresponds to a particular modality of expression:

- Gravitation = torsion,
- Electromagnetism = modulation,
- Weak = dephasing,
- Strong = fractal cohesion.

Thus, what present-day physics calls "fundamental interactions" are not primary forces, but emergent phenomena of one and the same temporal matrix.

Time is not a neutral parameter: it is the invisible architect that creates phenomenal reality by modulating density, coherence, and Information.

And I go even further: in my model, I have already shown you that the fundamental forces can be interpreted as modulations of Time. But here, I introduce a new idea: the existence of a torsion field proper to the chronon itself.

To shed light on this internal "spin," I will start from a familiar reference point.

In relativistic physics, Élie Cartan introduced, at the very beginning of the 20th century, the notion of torsion of spacetime, linked to the spin of particles. For him, torsion corresponded to a local "twisting" of geometry, induced by the internal angular momentum of particles.

In my view, this concept is radically broadened: torsion is no longer simply an attribute derived from spacetime, but an intrinsic property of Time itself, inscribed in each chronon. In other words, temporal spin becomes a fundamental characteristic, prior to the geometry of spacetime, and which structures it.

Sylvana
Ah, I understand better… One can imagine that each chronon spins upon itself: it is not merely a fixed point, but a tiny center of vibration whose own dynamics help animate the entire temporal lattice. Whereas, in Cartan's view, it is spacetime itself that "twists" under the effect of particle spin, you place this rotation within Time itself, borne by each chronon.

Thierry
That is exactly it. The essential properties of this torsion field can be summarized as follows:

- Each chronon possesses an internal rotation—a temporal spin inscribed within a metric expanded to 9 dimensions (3 for space and 6 for Time). We will delve deeper into this question of the temporal dimensions in the next pillar.
- This rotation engenders a torsion field, which is not a derived effect but a fundamental property of the temporal field.
- It is through this torsion that the chronal wave closes in on itself and condenses into a stable particle, and that the fundamental forces differentiate as specific modulations of the temporal lattice.

Thus, the chronon appears as the architect of temporal cycles, inscribing in each entity its own duration within the vibrating lattice of universal Temporality.

3. Vibrational Signature — The Emergence of Mass and Identity

Sylvana

All right… so if I follow you correctly, particles and even the fundamental forces would emerge from the dynamics of the chronons' spin. But then a question occurs to me: if, in your model, everything—particles, forces, stars, living beings—is composed of the same chronons, how is it that they are not all identical? What differentiates one thing from another?

Thierry

Yes, you are touching on a crucial point: this is precisely where my model introduces the notion of an omniscient Intelligence and a universal Ordainer. The Universe cannot be reduced to an accumulation of undifferentiated bricks; on the contrary, it manifests structures endowed with precise functions, as though each element had been thought out to play a particular role in the great cosmic symphony.

By positing that all manifestation—dense as well as subtle—is constituted of chronons, I came to formulate the following hypothesis: every entity in the Universe possesses a unique vibrational signature. One can conceive of it as a frequency source-code, an identity inscribed in the weave of the temporal field itself. It is this signature that distinguishes an electron from a proton, a molecule from a star, or one living being from another.

In other words, it is not only the observable form that differentiates entities, but their distinct mode of Resonance with the temporal field. This mode acts as a true energetic and functional identity card, guaranteeing each being its place and its role in the universal order.

It is indeed ÆKO that ensures each vibrational signature can connect to the omniscient Present and find its rightful place in the overall harmony. And it is the chronons that give each element not only its own vibrational frequency, but also the interaction dynamics that shape its physical and temporal properties.

Concretely, interactions between chronons locally generate chronal torsion fields—which I call chronal fields—that then act as true invisible architects of the real. These fields simultaneously determine several fundamental properties:

- The effective mass of particles: the stronger the density of chronal interaction, the more mass the particle acquires.
- The stability of matter: it results from the internal balance of chronal waves, like the proper tension of a violin string that maintains the note.

- The local rhythm of the flow of time: each vibrational configuration influences the speed at which time unfolds within a given system.
- The unique vibrational signature: as mentioned above, each entity bears a distinct frequency code, an energetic identity that distinguishes it from all others.

From this follows a radical idea: the speed at which time flows is not a consequence of physical phenomena, but a fundamental law in itself. In other words, it is not matter, energy, or motion that produce time; it is the temporal field that sets their pace of evolution. Time is therefore neither a mere backdrop nor a by-product of physics, but the primary matrix from which all phenomena take shape.

Thus, without a temporal field, there would be neither mass, nor cohesion, nor structuring of the real. The temporal field becomes the great organizing matrix of the Universe, where time, matter, form, and information are but differentiated expressions of the same dynamic substrate.

In short, chronons are not simple anonymous grains: they are the letters of a cosmic alphabet. And it is the universal Intelligence that, by arranging these letters, composes an infinity of words, sentences, and vibratory narratives—giving birth to the multiplicity of forms on the basis of a single essence.

4. Cosmic Consequences — The Temporal Field and the Universe

Sylvana

In fact, in your model, every element of the Universe has a proper frequency, connected to ÆKO, and it is according to its resonance with the temporal field—a resonance shaped by the chronons and their chronal torsion fields—that it finds its "place" within the overall harmony of the Cosmos.

But this torsion field does not merely determine the frequencies: it also defines mass, stability, and even the local rate at which time flows. Ultimately, the temporal field organizes everything, from the smallest to the largest… This must therefore have implications at every level?

Thierry

Yes. In further elaborating my model, I studied its implications at the cosmological scale. I arrived at a central conclusion: local or global variations of the temporal field could shed light on several phenomena that remain mysterious in current theories.

- Temporal dilations

Whether they are attributed to gravity (proximity to massive bodies) or to acceleration (special relativity), here they find a common origin: the very dynamics of the temporal field.

In my model, the temporal field varies locally. Its frequency, or its "beat," slows when interactions between chronons generate chronal torsion fields that densify the temporal lattice. All physical processes in that region—including clocks—then synchronize with this slower beat. To an external observer, everything appears slowed down: this is what we call time dilation. In other words, the frequency gradient of the temporal field manifests itself to us as gravity or acceleration: both are the direct results of a single profound dynamic.

- Extreme transitions of the Universe

Events such as primordial inflation, the heating of the young Universe, or singularities (Big Bang, black holes) can be understood as ruptures or gradients within the structure of the temporal field.

This field can change state, in the manner of a medium undergoing a phase transition. A sudden variation in its density or its rhythm then gives rise to extreme regimes: explosive expansion, heating, collapse. What we call "singularities" or "infinities" are no longer insoluble enigmas, but signs of a shift in state of the temporal field.

- The limits of quantum mechanics

The paradoxes encountered in extreme regimes (the interior of black holes, sub-Planckian scales) can also be reconsidered within this framework. At these scales, time no longer appears as a continuous flow but reveals its elemental texture: a mesh of temporal voxels—the chronons.

On large scales, this network already exists, but it blends into a smooth continuum. At the scale of the chronon, its granularity becomes dominant, radically altering the dynamics of the temporal field. Like a fabric that seems homogeneous from afar but whose threads become distinct as one draws near, the apparent continuity of time yields to its discrete structure. Thus, the paradoxes of quantum gravity or the singularities in fact indicate contact with a more fundamental layer of time, and not an incomprehensible anomaly.

- Toward a unified cosmology

In short, phenomena as diverse as cosmic expansion, extreme gravity, and quantum dynamics converge within this framework: they appear as differentiated manifestations of one and the same substrate, the temporal field.

This structured field—which current physics partly equates to the "quantum vacuum"—fixes its properties at a nonzero yet extremely precise value, thereby determining the fundamental constants and the physical laws. This stability, however, is not absolute: it is metastable. It appears stable to us at our scale, but in theory, it could tip into another vacuum state, opening the possibility of other Universes, other physical laws, and other configurations of reality.

Sylvana

So, what you describe is that a single field—the temporal field—holds everything together.

When we speak of time dilations, in fact it is not gravity that slows time; rather, it is the field that becomes denser and beats more slowly, and we experience this as a stretching of time, as gravity, or as acceleration.

For the great transitions of the Universe, such as inflation or singularities, the same principle applies: the field changes state all at once, and what we see is only the sudden event.

And in the infinitesimal, where our classical equations no longer work, this field unveils its intimate texture, its tiny time voxels—the chronons—and that is what we perceive as quantum paradoxes, simply because we do not yet have the tools to understand their logic.

Ultimately, everything that seemed scattered or mysterious is linked by this same field. And you also say that since it is the one that sets, with extreme precision, the constants and the laws of the Universe, one could say that, if it were to change state, this same field could topple the entire edifice of physical laws in a single stroke—transforming our present Universe into a wholly different configuration, with other constants, other rules, and perhaps even other forms of reality.

Thierry

Yes, exactly.

A change in this temporal structure—toward what some call a "true vacuum"—would trigger an instantaneous rewriting of the laws of physics, rendering the Universe, as we know it, incompatible with life.

Sylvana

In other words, according to your hypothesis, what appears to be the surest foundation—the void itself—could turn out to be fragile.

But then, why does the Universe seem so perfectly stable for billions of years?

Thierry

Although it may seem paradoxical, this conclusion confirms a fundamental intuition: the Universe cannot be reduced to a blind mechanism governed solely by chance and entropy. Rather, it presents itself as a systemic, oriented structure, finely regulated—which we interpret here as a potentially self-adaptive organization—whose properties favor the emergence of life and consciousness, and make the experience of life possible on multiple scales.

Sylvana

So, if I understand correctly, we are coming back to what you called the anthropic principle…

Thierry

You are right to bring it up.

The weak anthropic principle simply states: "we are here, therefore the Universe is necessarily compatible with life." That is true… but it is only a tautology, a descriptive truism that explains nothing.

What I am defending is the strong anthropic principle: the idea that the Universe is tuned from the Origin to allow the emergence of conscious life. It is not a favorable accident, but a structural orientation inscribed in the very fabric of the real.

This is why I introduced into my theory the notion of an Ordainer and an omniscient Intelligence inherent to the Universe. When one observes the physical constants, the fundamental laws, the mathematical ratios that govern the Cosmos, one finds a tuning of staggering precision. This tuning makes possible not only the stability of matter, but also—over the long term—the emergence of living beings capable of consciousness, of knowledge, of wonder… and even of inner transformation.

It has always seemed to me intellectually untenable to reduce such order to a mere accident. Chance, left to itself, does not suffice to account for this coherence. It is more logical to posit the existence of an organizing Principle, an omniscient Intelligence at work—even if current science does not yet have the means to demonstrate it or to refute it. My hypothesis therefore remains open, but it appears to me more coherent.

Thus, this strong anthropic principle must not be regarded as a naïve belief. It rests on three very real pillars:

- The fine-tuning of the Universe,
- The constant progression toward greater complexity and organization,
- The intuition that life is a natural consequence of this dynamic, and not an anomaly.

My hypothesis of an omniscient Intelligence therefore acts as a structuring principle: a key that sheds light on why the Universe does not merely exist, but seems oriented toward consciousness.

5. Principle of Parsimony — Why a Single Element Suffices

Sylvana
Indeed, following your logic, it gives one pause for thought.

And all of this, in your model, would rest on a single basic entity: the chronon.

But… How did you come to posit that a single element could be enough to support the entire dynamics of the Universe?

Thierry
Yes, it is a decisive question.

In this thought exercise, I chose to rely on three epistemologically solid principles, because the concept of the chronon raises several major challenges for any attempt at a unified theory.

- Simplicity

This is what is called the principle of parsimony (or Occam's razor). The Universe seems to obey a logic of economy: fewer axioms, greater explanatory scope. In other words, it is a matter of accounting for a maximum of phenomena with a minimum of hypotheses.

To posit a single fundamental entity—the chronon—to explain the emergence of time, matter, energy, and the fundamental forces is to maximize simplicity without sacrificing the richness of the real. Rather than multiplying causes, I sought the unique principle capable of sustaining everything.

But simplicity alone is not enough: this principle also had to be genuinely fecund, capable of generating the diversity of the world.

- Principle of Unified Generativity

A good physical theory does not merely explain what is observed: it shows how a diversity of phenomena can emerge from a single principle. The chronon, as the elementary unit of the temporal field, possesses this generative power. It allows one to derive:

- Gravitation, as a torsion of the chronal weave;
- Electromagnetism, as a secondary manifestation of this dynamics;
- Mass, as vibratory slowing;
- And even consciousness, as an interface for informational integration.

This capacity to bring forth such a variety of phenomena from a single principle convinced me that it could be the appropriate candidate to ground a unified vision.

- Scientific Coherence

My intention has never been to elaborate an isolated speculation, but to test my intuitions against what science has already established. This is why I respected a principle of coherence with existing observational and theoretical frameworks.

The chronon can thus be seen as a generalization or an extension of concepts already explored:

- It extends the idea of a quantum of time suggested in quantum gravity;
- It is compatible with a discrete vision of spacetime (loops, spin networks);
- It dynamically links time, energy, and information, three recognized pillars that are often treated separately.

In summary, I did not begin from idle speculation: I formulated my own hypotheses, then tested them against established scientific findings. This approach allowed me to connect intuitions already present in research, and to push them further, within a unified logic.

6. The Simultaneous Condition — A Physics of Intention

Sylvana

In short, you reduce everything to a single building block—the chronon—but you want to show that this simplicity has the power to generate all the complexity of the real.

Thierry

You have grasped it well. But to fully understand what the chronon carries, one must go beyond its physical or cosmological effects.

In my model, chronons are not mere quanta of time, analogous to energy quanta in conventional physics. They must be understood as fundamental units of an informational order, bearing simultaneously:

- A frequency, which confers upon them their own rhythm,
- A direction, which inscribes their dynamics within the flow,
- And a structuring intention, rooted in the Omniscient Intelligence that undergirds the Universe.

In other words, each chronon constitutes an elementary vibration of intention, a scalar wave of projective organization, actualized from an ordering source. This is not simply a neutral temporal vector: the chronon acts as an encoder of structure, involved in the emergence of the real through a mechanism of informational synchronization.

The radical specificity of this wave is its capacity for instantaneity. Unlike conventional fields that propagate through spacetime, the chronon does not need to "travel." It operates directly within a nonlocal space of resonance, immediately tuning two points of the real to one another, as two strings resonating in unison without any wave needing to traverse the distance that separates them.

It is precisely this role as encoder that sets the stage for the next notion: that of the Simultaneous Condition (SC) and the Universal Memory.

The Simultaneous Condition denotes a state of absolute coherence of the fundamental temporal field, in which any point in Time—past, present, or future—can be put in direct relation with another, without recourse to an interval metric.

It constitutes a transverse temporal presence, an ontological foundation of the unified field where:

- Information circulates without delay or spatial mediation;
- Events coexist within a non-sequential logic;

- Memory, intuition, precognition, or remanence appear as the natural effects of a substrate where Time reveals itself as entirely simultaneous.

Thus, Time is no longer cut into past, present, and future: it presents itself as a single horizon, containing all occurrences at once.

Sylvana
According to your model, the Universe never waits: it recreates and adjusts itself instantaneously, everywhere at once.

What I mean by this is that events no longer need to follow a linear order: they coexist, permanently connected by this common ground.

And that is indeed why you call this concept the Simultaneous Condition?

So what we call memory, intuition, or premonition would not be marginal or irrational: they would be the normal effects of a Time that links all instants together.

Thierry
Yes, that is exactly it.

From this perspective, the Universal Memory—understood as the active register of the past, present, and possible configurations of the Cosmos—rests structurally on the Simultaneous Condition.

It must not be understood as a simple static database, but as a living vibratory interface, coextensive with the temporal field itself. Each chronon participates in it, by recording and resonating with all the possible occurrences of the Universe.

It is precisely this Architecture that makes possible:

- Resonance between planes of reality, by establishing direct bridges between fractal layers of the multiverse;
- The transmission of information without classical energetic support, through instantaneous tuning of chronal states;
- The existence of retrocausal processes, tied to the two-faced nature of time and to the possibility that a potential future informs the present;
- Consciousness accessing states of trans-temporal perception, such as intuition, precognition, or psi experiences.

In other words, the Universal Memory is the holographic fabric where every event—already occurred, underway, or in becoming—remains inscribed and available. It is the living register of cosmic coherence, ensuring that the flux of the real never disperses into nothingness, but remains fully conserved in the temporal field.

7. Living Unification — When Complexity and Consciousness Are One

Sylvana

Tell me if I have understood correctly. What you call Universal Memory is not a simple "cosmic recollection," but a living field, always active.

It is not a fixed archive: it is a permanent resonance, linked with all possible configurations.

And at the heart of this field, consciousness can enter into resonance, capture, adjust… As though the Universe, at every instant, remembered everything—and were already projecting itself toward what it could become.

Thierry

That is it. And it is precisely this vision of an active Universal Memory that led me to take one step further.

Through the dynamics of the temporal field—and the elemental unit that is the chronon—I attempted to sketch a radically new path to unification.

This approach allows me to propose a conceptual bridge linking:

- The quantum ether, conceptualized as an invisible substratum, structuring and carrying information;
- Gravitation, interpreted as a side effect of chronal torsion;
- Electromagnetism and inertia, envisaged as third-order manifestations, derived from gravitation itself through additional modulations of the temporal lattice;
- The structure of spacetime, which no longer appears as a primary given, but as a geometry emerging from the temporal field;
- And the subjective flow of time, understood as the manner in which consciousness attunes itself to this universal lattice.

Thus, all my research is rooted in a functional and living vision of cosmology: the Universe is not reducible to a static physical object, but reveals itself as an evolutionary, informed, and self-organized process, inscribed in the very weave of Time.

In other words, what we distinguish as independent domains—physics, geometry, energy dynamics, or even lived experience—are in reality only multiple facets of one and the same underlying weave.

It is the vibratory dynamics of this field that brings forth, in successive strata, the diversity of phenomena and the growing complexity of the Universe.

But this complexity is not primary: it is apparent, produced by the limitation of our local perspective, as we are encapsulated in a particular configuration of reality. Seen from the fundamental order, there is no fragmentation, but a single continuity—the temporal field—of which we perceive only partial projections.

Sylvana
You mean that what we call complexity would not come from unordered chaos, but from our limited way of perceiving?

Thierry
Yes, that is exactly it. What we call complexity may be nothing but an effect of perspective, a veil arising from our perceptual limits.

This is why I did not seek to move through space, but to transcend the gaze itself, to delocalize consciousness in order to reach a more fundamental level of intelligibility.

For in truth, the complexity that we perceive could be nothing but the ultimate veil, the most refined form of illusion, concealing a simplicity of absolute elegance.

Behind this veil, an underlying logic is revealed, ordered and unified—an omniscient cosmic consciousness at work, silent, yet intimately active in the very fabric of what we call reality.

Sylvana
I see…

Behind this teeming profusion, you propose that there is a simple and unified weave—the chronon, bearing both Time and consciousness.

In your hypothesis, everything is attuned, everything is connected… everything is One, in simultaneity.

But then… a question is really nagging at me: if everything rests on this same fundamental brick… how can this unity give birth to such a diversity of worlds and forms in a fractal world?

Thierry

And it is precisely this question that led me to the next pillar.

For to understand how the original unity unfolds into apparent multiplicity, it becomes necessary to introduce a key notion: that of temporal dimensions.

Chapter 12

Pillar 3 – Absolute Relativity

~

Toward a Multidimensionality of Time

Sylvana

You know… this morning, when I got up, I looked at myself in the mirror and said to myself:

So there it is, according to Thierry's model, here I am with my unique vibrational signature in this Universe… At this very moment, I am traveling through all those worlds, and it is here, in this precise fractal, that I took form.

And then, looking around me, I felt different: no longer just an individual, but a vibration in resonance with this world, my friends, my family… like a note written into the great cosmic score.

I admit it made me smile. And at the same time, it completely changes the way one looks at oneself. It even soothed me, to think of it from that angle.

Thierry

Yes, that is exactly what this shift in perspective can engender: an inner peace, that subtle smile you speak of.

For, in my model, it is no longer a matter of perceiving oneself as an isolated individual, the product of chance, but as a singular expression of a universal fabric.

To recognize this is to understand that each being naturally finds its place within a vaster and coherent order.

And it is precisely this feeling of unity—this intuition that a hidden order sustains the apparent diversity—that led me to pursue my research: to elucidate how the One unfolds into the multiple, and how each being is intimately inscribed within this unfolding.

Sylvana
So where are you taking me, in this journey within the Cosmos?

Thierry
Precisely… you speak of journeying: that is where the question of dimensions takes on its full meaning.

For this feeling of unity and rightness to exist, Time must not be reduced to a simple line, but must unfold in several dimensions, capable of welcoming this diversity of vibrational resonances and experiences.

Before going further, I want to recall the principles already established, to situate what I am going to develop.

As we saw in the previous pillars devoted to the Temporal Field and the Chronon, my research in Functional Cosmology led me to rethink entirely the nature of Time: no longer as a continuous line external to phenomena, but as a dynamic structure, omnipresent and foundational.

This means that, although Time manifests differently depending on scales, planes of reality, or observers, it proceeds from an underlying universal field, stable and structuring: the Temporal Field.

In my model, this Field is not a mathematical abstraction, but a real weave, composed of discrete units—Chronons—which form the true elementary bricks of the cosmic becoming.

Sylvana
Bricks of becoming… I like the image. As though each instant were a little block that builds the Universe.

Thierry
Yes, and it is at this stage of my reflections that I introduced a bold idea: that of an Absolute Relativity. You have already heard me mention it, but it is now time to specify its scope, because it explains why the existence of multiple temporal dimensions becomes necessary.

In this model, the laws of Relativity as we know them are not abolished, but appear as local expressions of a broader dynamic: that of a multidimensional Time, acting at once as substratum, driving force, and organizer of reality.

In other words, each level of the Universe—from the quantum to the cosmological—possesses its own temporal rhythm, embedded within a deeper universal architecture.

This structure is absolute in its principle: it constitutes the foundational weave, stable and continuous, of all that exists. But it is relative in its manifestations: each plane of reality, each scale, each observer experiences Time according to a particular beat, produced by the modulation of the temporal field.

Thus, Absolute Relativity reconciles two visions that are, at first glance, contradictory:

- On the one hand, a stable universal order, guarantor of a fundamental coherence;
- On the other, a plurality of perceptions and rhythms, which brings forth the diversity of experiences, physical laws, and forms of existence.

1. A Fecund Paradox — Absolute Relativity

Sylvana

Absolute Relativity… admit it, you are deliberately trying to confuse me with words that cannot possibly go together. Relative and absolute in the same sentence? I am beginning to grasp the idea thanks to our conversations, but I would like you to explain it further.

Thierry

Yes, your remark is legitimate, and it is in fact the first question that naturally comes to mind. This is precisely what gives this concept its strength: it seems paradoxical at first, but it becomes perfectly coherent within the framework of my model.

I developed this idea by relying on two pillars already established:

- The fundamental temporal field,
- And its elementary unit, the chronon, a true quantum of Time-space.

It is in this context that I introduced the concept of Absolute Relativity. The expression may seem contradictory, but it proves rigorously coherent once placed within the dynamics of the temporal field.

Here is the central postulate that follows from it:

Every entity—physical, energetic, or informational—evolves according to laws relative to its own vibrational and temporal frame of reference. But these laws, however diverse they may be, all derive from an invariant foundation: the universal scalar temporal field, bearer of structure, order, and potentiality.

Thus, the principle of Absolute Relativity can be summarized simply:

It designates an architecture in which all phenomena are locally relative (each system lives according to its own rhythm), but globally inscribed within an absolute metric, that of the quantized temporal field.

Sylvana
So, everything remains locally relative… but there is this absolute weave that serves as a foundation…

Thierry
You have understood correctly. And from there, several scientific consequences naturally appear.

Absolute Relativity, as I formulated it in my original hypothesis, entails a profound restructuring of physical foundations.

As we have already discussed, it rests on the idea that the temporal field constitutes a universal, discrete, structuring ontological frame of reference, from which all known laws, constants, and dynamics emerge.

This produces a series of significant consequences in the central domains of physics, which I summarize as follows:

- On the nature of Time

In general relativity, time is conceptualized as continuous, homogeneous, and malleable, influenced by gravity or speed.

In my hypothesis, this model is replaced by a discrete, quantized Time whose elementary unit is the chronon.

Each entity—particle, field, living being—evolves according to its own rhythm, yet all these rhythms remain anchored within a universal temporal field that bears information and structure.

- On frames of reference

Frames of reference are the perspectives or measurement frameworks from which one observes and describes phenomena.

In Einstein's theory, no inertial frame of reference is privileged: all are equivalent.

By contrast, my hypothesis introduces an absolute ontological frame of reference: that of the fundamental temporal field.

This is not a rigid absolute in the Newtonian sense, but a functional absolute:

- Unobservable directly, because it lies beneath measurable phenomena;
- Invariant and discrete, as a universal scalar lattice;
- Foundational, because it makes all local relativities possible.

Thus, the temporal field does not contradict relativistic observations, but frames them within a more fundamental, pre-causal layer, from which local frames of reference emerge—those observation and measurement frameworks that structure our description of the real.

- Gravity

In general relativity, gravity is described as a curvature of spacetime induced by mass-energy.

In my model, it is, by contrast, a secondary phenomenon, arising from a torsion field generated by the internal dynamics of the chronon. In other words, gravity does not curve spacetime as such; rather, it is the torsion field that locally modifies the structure of the temporal field, which produces the gravitational effects that we observe.

- The problem of unification

Current physics distinguishes four fundamental forces:

- Gravity, described by general relativity, valid at large cosmic scales;
- Electromagnetism, the weak interaction, and the strong interaction, described by quantum mechanics in the form of discrete fields and quanta, but without gravity.

Yet this separation is problematic. Under extreme conditions—at the beginning of the Universe, inside black holes, or at scales near the Planck length—these forces must act together. But our two theoretical frameworks then produce incompatible predictions. We therefore have two partial images of reality: one for the infinitely large, the other for the infinitely small, without a single framework to connect them. It is this contradiction that initiated the quest for a unified theory.

- Unification was present at T0

I have hypothesized that unification was present from T0, just before what we call the Big Bang.

I envisioned the initial Universe as a matrix of self-expansion, an original field that begot itself.

This process of self-begetting immediately created a logical duality:

- On one side, the emergence of the Universe as a manifested phenomenal world, made of particles, forces, and observable structures;
- On the other, its original matrix—the quantum vacuum, an invisible yet active substrate that engendered it and continues to sustain it throughout its becoming, until its end.

In other words, the observable Universe is not a closed totality: it remains constantly supported by its invisible foundation, the quantum vacuum, which acts as memory, the support, and the regulator of its entire evolution.

- The solution proposed by my model.

My hypothesis proposes a solution by placing the four fundamental forces on a common foundation: the original temporal field, structured by chronons.

- Chronal torsion generates gravity as a global modulation;
- Electromagnetism appears as a local expression, derived from specific configurations;
- The weak and strong interactions are understood as additional hierarchical modulations, produced by variations in the frequency and polarization of chronons.

In my theory, chronal torsion is what transforms the flow of Time into a vector of gravitational structuring.

In this framework, the temporal field is composed of elementary units, the chronons, which weave a dynamic, fractal fabric. Their organization is not always homogeneous: when they unfold into spirals or twists, a chronal torsion occurs. This torsion corresponds to a variation in the density of the temporal flow, a kind of gradient in the Fractal Temporal Density that locally engenders a contraction of information.

It is precisely from this contraction that gravity is born. From this perspective, gravity is not an independent fundamental force, but the direct consequence of chronal torsion. Particles, and protons in particular, play an essential role here: genuine stabilized micro-singularities, they capture and redistribute these torsions, transforming the temporal flow into a spatial dynamics. Gravity then appears as the geometric translation of a temporal contraction: the more intense the chronal torsion, the more matter is drawn toward this region of contraction, as though space itself were contracting under the effect of Time.

Sylvana

So, in your model, is that precisely why the more mass there is, the stronger gravity becomes—because the chronal torsion is more intense there?

Thierry

Exactly.

This view offers a key to unification: it explains the universality of gravity, felt by every structure immersed in the temporal field, and it allows us to link it to the quantum scale, where microscopic torsions of Time are projected as macroscopic gravitational effects. It also opens a bridge to consciousness, understood as a modulatory interface of the temporal field: if consciousness acts upon the fabric of Time, it subtly participates in gravitational dynamics.

Thus, gravity is no longer only the mysterious attraction described by Newton or Einstein's curvature of spacetime: within this framework it becomes an emergence of Time itself, a visible manifestation of chronal torsion that organizes matter and the Universe.

- A unified dynamic

Within this framework, gravity and electromagnetism are not two competing mechanisms, but two differentiated expressions of the same process of chronal torsion. The temporal field, as an active and fractal weave, deforms in various ways: when it contracts, it generates gravity; when it polarizes—that is, when it differentiates into opposing tensions and local asymmetries—it gives rise to electric and magnetic fields.

Sylvana

When you say that the temporal field "polarizes," what exactly do you mean?

Thierry

In my framework, saying that the temporal field "polarizes" means that it differentiates into opposing tensions rather than contracting uniformly.

- Contraction → concentration of the flow (gravity).
- Polarization → internal differentiation of the flow (charges and fields).

Polarization therefore corresponds to the tensioning of the fractal temporal field, creating asymmetries that manifest as electric poles (+ and –) and as dynamical orientations (rotation, magnetism).

To clearly understand the difference between the two:

- When chronal torsion collapses toward a center, it produces a contraction: this is gravity.
- Conversely, when chronal torsion opens into opposing poles, it produces a polarization: this is electromagnetism.

In other words:

- Gravity = isotropic densification (universal attraction).
- Electromagnetism = anisotropic differentiation (polarities, oriented fields).

And when the temporal field is polarized:

- It creates differentiated potentials (positive/negative, north/south);
- It sets up alignment flows that manifest as electric and magnetic fields;
- It introduces a directional dynamic where gravity remained uniform.

In summary, polarization is therefore the mechanism that allows the temporal void to generate vector fields, bearing orientation and local structure.

Sylvana
All right, I am beginning to see: in your model, gravity and electromagnetism are no longer separate forces but two ways for the temporal field to bend or to be tensioned.

Thierry
You have grasped the essential point. Gravity thus appears as the manifestation of a contraction of the temporal flow: chronons, organized into dense spirals, create zones of contraction where matter is attracted, expressing the universal tendency of structures to draw closer together. Conversely, electromagnetism results from a polarization of the temporal field: instead of contracting, the torsion differentiates into opposing tensions, giving rise to electric poles and to rotational dynamics that translate into magnetism.

In this light, gravity and electromagnetism reveal themselves to be the two sides of one and the same chronal dynamics: the former is the global expression of a densification, the latter the local expression of a differentiation.

Sylvana
In other words, in your model: gravity is born from a contraction of the temporal flow—this draws matter—and electromagnetism from a polarization of the same field—this creates poles and magnetism?

Thierry
That is correct. By linking these phenomena to a common cause, it becomes possible to place all the fundamental forces upon the single foundation of the temporal field, which makes the opposition between cosmic and quantum scales disappear. The global interactions of gravity and the local interactions of electromagnetism then appear as two complementary faces of a single process, reconciled within a unified dynamics where Time acts as origin and universal mediator.

Sylvana
So, if I am following you… in your model, with this common foundation—the temporal field at the basis of everything—there would no longer be a theoretical separation between the cosmic world and the quantum world. But above all, this unification changes our way of understanding the Universe: it is no longer a puzzle of isolated forces, but a single process that unfolds across several scales.

Thierry
Exactly.

All the fundamental forces can be understood as differentiated modulations of the temporal field.

And this logic holds not only at the microscopic scale of particles: it also applies at the cosmic scale.

In my vision of an intrinsic multiverse, all fractal worlds—whatever their physical constants, their own laws, or their organizational dynamics—emerge from the same fundamental temporal field.

Each Universe draws its existence from it, yet unfolds with a specific metric, a singular temporal rhythm that defines its structure, its phenomena, and its evolution.

In this framework, the temporal field acts as a true holographic matrix, where each portion of the weave contains in potential the information of the whole. Realities are not set side by side as separate worlds, but unfold as holographic projections issuing from one and the same temporal source. Each world in the Universe corresponds to a particular regime of informational organization, with its own laws, constants, and emergent structures, yet always rooted in the same common weave.

This holographic matrix does not merely juxtapose worlds: it links them through a profound dynamic that I designate, as you know, by the name universal vibratory resonance ÆKO. ÆKO expresses the principle whereby every frequency, every motif of information, resonates with the others within a coherent network. It is thanks to this resonance that the different words of the Universe, although differentiated by their own configurations, remain connected and partake in a shared cosmic fabric. One could say that ÆKO acts as a fundamental harmony, guaranteeing the underlying unity of multiple realities and allowing the circulation, transmission, and resonance of information from one level to another.

Thus, the holographic temporal matrix and ÆKO resonance together form the two pillars of cosmic unity: the former ensures that each fragment contains the totality and that each world of the Universe is a projection of the whole, the latter guarantees that this totality remains alive, coherent, and connected within a universal vibratory continuum.

Sylvana

All right… now I understand better how, in your logic, "relative" and "absolute" can stand together.

Thierry

Yes, and that is precisely what this concept brings to light: the term may seem paradoxical at first glance, so profoundly does relativity traditionally oppose any notion of the absolute. Yet, in my hypothesis, it makes complete sense: there exists a fundamental, non-measurable frame of reference, an ontological foundation that renders relativity itself possible. This absolute is not an external magnitude or an arbitrary reference, but the scalar temporal field.

This field is invariant, omnipresent, active, and foundational. It is not measured like a classical physical variable, because it constitutes the very weave of the real. It is what informs, modulates, and structures every level of organization: from elementary particles to galaxies, from phenomenal layers to subtle planes. It acts as an invisible yet indispensable background upon which all the relative manifestations we perceive unfold.

The chronon, the elementary quantum of this field, is universally present. However, its manifestation varies according to frequencies, polarities, vibratory densities, and even the layers of existence in which it operates. In certain strata, it manifests as light (the photon), in others as mass (the quark, the electron), and in the most subtle as consciousness or intentionality. This plasticity of the chronon illustrates the power of the scalar temporal field: a single foundational principle can give rise to physical, energetic, or cognitive phenomena, depending on the vibratory configuration in which it unfolds.

Thus, everything we call "relative"—space, matter, light, the flow of time, and even the physical laws—emerges from a more fundamental structure, inscribed within a discrete and hierarchical temporality. In other words, relativity does not oppose the absolute: it is its dynamic expression. The absolute is not abolished but veiled, and in my model, it is within the scalar temporal field that this invisible yet necessary ground is found, guarantor of the coherence and unity of all forms of relativity.

Sylvana

In short, your model does not merely describe laws or forces: it unveils a hidden background that makes these laws possible and grants them their coherence. It is a discrete principle that supports the whole and maintains the unity of the real.

Thierry

Exactly.

Absolute relativity does not abolish Einstein's relativity: it encompasses it by giving it a deeper functional framework.

It allows for a unified and evolving reading of the Universe, where time is no longer a simple variable, but the primary source of cosmic manifestation.

As a result, the concept of Absolute Relativity, as it emerges in my hypothesis of the temporal field and the chronon, could offer an original and fruitful solution to one of the great problems of contemporary epistemology: how to reconcile the relativity of every observation with the quest for a universal foundation?

And to shed light on this point, I will give you three major domains where this hypothesis fully comes into its own.

- A reconciliation between relativism and structure

Modern Science rests on the idea that our measurements and our laws always depend on the chosen framework: the frame of reference we adopt, the coordinates we choose, or the experimental conditions we impose.

There is no "pure" observation outside context; each result is relative to a point of view.

This led to methodological relativism—very useful for analyzing phenomena, but sometimes philosophically destabilizing, because it seems to dissolve any idea of a universal foundation.

This reintroduces a basis of unity, while respecting the plurality of frames of reference and the validity of relative observations.

And by positing the temporal field as a fundamental basis, I can now explain to you its impact in a second major domain.

- A stable basis for the unification of laws

As we have already discussed, in this perspective, frames of reference, laws, and even constants are no longer isolated absolutes: they are local projections of the single temporal field.

This opens the way to a unified physics, not by seeking to reduce everything to a universal equation, but by revealing that the diversity of phenomena arises from differentiated modulations of a common substrate.

It is the image of the harmonics of a string: multiple, varied, yet all rooted in a fundamental vibration.

Sylvana
So, according to your model, everything would rest on frequencies and

modulation. In short, a cosmology of resonance…

Thierry

Yes, exactly. And it is this point that opens onto a third domain, perhaps the most innovative.

- A possible formalization of consciousness

The absolute temporal structure makes it possible to conceive of consciousness not as a mere side effect of the brain, but as an active interface with the fundamental Temporal field.

This makes it conceivable for consciousness to navigate between layers of reality, according to its resonance with the frequencies proper to each Temporal stratum.

Thus, so-called "extraordinary" subjective experiences—intuition, precognition, altered states of consciousness, and psychic phenomena—could be understood not as anomalies, but as natural effects of interaction with the dynamics of the Temporal field.

This framework therefore offers a rational formalization of consciousness as a co-actor of reality, and not a simple emergent product of matter.

Sylvana

You are saying that levels of consciousness each vibrate to a note of Time, and that this is what allows them to access different strata of reality?

Thierry

Yes, that is indeed what I propose.

This correspondence becomes fully coherent as soon as consciousness is envisaged as an active principle of the Temporal field.

In my model, nothing is left to chance: each stratum of reality follows a logic of vibratory resonance, and it is precisely this that makes conceivable a navigation of the conscious-spirit through its interface of consciousness between the different layers of the Universe.

In other words:

This concept restores to Time a major ontological and functional status.

Time ceases to be a mere measurable parameter or a neutral variable of the equations: it becomes a fundamental component of the very being of the Universe.

And, in my model, as we have discussed, this Temporal organization reflects the action of an Omniscient Intelligence, which fragments its own omniscient Present into quanta of Time. This is manifested in multiple fractals so that consciousness may progressively explore the dimensions proper to each level and actualize its potential.

Thus, the relativity of phenomena is not left to chaos or randomness: it is organized, coded, and oriented by this living Temporal lattice.

This observation led me to cross a new threshold: after laying the foundations of Absolute Relativity, it becomes natural to approach an even bolder question—that of the dimensions of Time.

Sylvana
You are therefore speaking of a Time that would no longer be a simple coordinate, but something living, with its own dimensions… This is not at all the way classical physics is presented to us. Can you tell me more?

Thierry
Yes, you are right.

Standard science describes three dimensions of space to which Time is added as a fourth dimension, within Einstein's spacetime framework.

But in my approach, Time is not simply an additional axis laid alongside space. It is a living, structured, polymorphous field, endowed with multiple dimensions of its own, which sustains and organizes the very emergence of the Universe.

My model rests on a hierarchical cosmology: the Universe is perceived as an intelligent, informed process, structured by emergent layers. Now, such a dynamic, evolving, self-organized system cannot unfold within a flat or uniform Time.

It requires a multidimensional Temporal field, capable of accommodating the diversity of rhythms, resonances, and potentialities.

Thus, instead of seeing Time as a simple coordinate attached to space, I consider it the original matrix:

- Substrate of space,
- Engine of the fundamental forces,
- And horizon of the evolution of consciousness.

It is this change in perspective that opens the way to the notion of multiple Temporal dimensions, necessary to understand how the fundamental Unity unfolds into a multiplicity of worlds, laws, and experiences.

Sylvana

If I follow you correctly, you are speaking of a Time that acts as a kind of living infrastructure, almost like a network where everything is organized—and which, in order to function in that way, would need several dimensions of its own?

Thierry

Exactly.

A functional cosmology, such as the one I am developing, requires an evolving Temporal structure.

In this framework, extending the dimensions of Time is not an arbitrary speculation: it becomes a logical, unifying necessity, opening the way to modeling several key phenomena that we have already mentioned.

- Navigation between temporal regimes

Instead of a homogeneous, continuous flow, I posit the existence of differentiated temporal regimes, between which it is possible to "navigate" by changing temporal density.

This navigation takes place through informational vibratory resonance between fractal layers of the temporal field.

Such a conception presupposes an intrinsically multidimensional spacetime, and therefore Time itself endowed with several independent axes, allowing for trajectories that are not spatial, but purely temporal.

In other words, the temporal dimensions become the true space for maneuver of consciousness, offering unprecedented directions of exploration.

- Consciousness as an active principle

In my model, consciousness is neither passive nor secondary. It constitutes a principle of exploration, integration, and emergence, capable of circulating within a fractal temporal geometry.

Attributing several dimensions to Time thus makes it possible to model consciousness as a free vector, evolving within a temporal network with increasing degrees of freedom—a kind of internal navigation within the universal temporal field.

- A structural necessity

Beyond these considerations, the introduction of additional temporal dimensions becomes indispensable for any serious attempt to:

- Conceptualize a coherent, evolving cosmology,
- Unify the fundamental forces within a single framework,
- Model Consciousness as a constitutive actor of Reality,
- And move beyond the limits of current physics, too often constrained by a linear Time and a homogeneous Space.

In this hypothesis, it is not Space, but indeed Time that constitutes the fundamental fabric of the real:

A living, informed, fractal fabric, capable of folding, twisting, and unfolding into multiple dimensions.

It is upon this hidden stage—the active matrix of the Universe—that the creative act is played out, generating matter, energy, laws, and consciousness itself.

Sylvana
In sum, science sees space and time as inseparable; you, in your model, posit a primary Time that engenders space as a theater stage where consciousness is the actor.

Thierry
Yes. Thus, everything is contained in the One, and the One unfolds within the Whole, following cycles of emergence, collapse, memory, and resurgence.

From this perspective, Time becomes the theater of cosmic self-revelation: not a mere neutral line, but a living dynamic, at once structuring, evolutionary, and inscribed within a functional logic proper to the Universe.

It is through these cycles that the Universe reconfigures itself, informs itself, and perpetuates itself, revealing a hidden order where apparent multiplicity finds its anchoring in unity.

With this framework established, I can now introduce the foundational core of my conception of the dimensions of Time.

My aim is to show how these dimensions articulate with one another, why they are necessary for the coherence of the model, and how they give Meaning to all that we have just explored:

- Absolute Relativity,
- The fractal structure of the temporal field,
- The active function of Consciousness,
- And the unifying logic of the Universe.

2. The Dimensions of Time — A Six-Level Architecture

Sylvana

So, how do these dimensions interlock in your model?

Thierry

In my model, Time is conceptualized as a living six-dimensional field, organized by fundamental units that I call chronons.

This field cannot be reduced to an abstract line: it possesses an internal architecture, composed of three revealing dimensions, which unfold the observable phenomenal reality, and three implicit, deeper dimensions, which constitute its formative sources.

To grasp this logic, let us begin with the three revealing dimensions of Time.

They are like invisible matrices, discrete yet essential: they do not give themselves directly to sensory experience, but they work in depth, structuring becoming and transforming raw potential into organized forms, cycles, and rhythms.

They act as generative foundations, the hidden base from which the three spatial dimensions then unfold. Thus, what we call the "phenomenal world" is not primary: it is the visible expression of a more fundamental temporal lattice.

Before presenting these temporal dimensions to you, I want to briefly restate the heart of my hypothesis, so that you may see where this construction begins.

I recall here in broad terms that Time is a dynamic and living field, a bearer of information, and constitutes, in my hypothesis, an active substance of the Universe, capable of organizing, modulating, and transforming reality.

I also postulate that its field possesses a fractal structure, that is to say, it repeats at different scales while generating patterns of increasing complexity. This fractality ensures that the dynamics observed at the microscopic level find their echo in the great cosmic structures, revealing a continuity of organization across all dimensions of existence.

I postulate that Time also has two faces, articulated according to two complementary regimes, which we will see in the next pillar, called M+ and M-:

- The first (M+), which goes from past to future, corresponds to densification, to irreversibility, to the contraction of the temporal flow that gives form to matter and gravity.
- The second (M-), which goes from future to past, manifests through reversibility, expansion, the reopening of possibilities, in connection with subtle information, entanglement, and consciousness.

These two faces do not stand in opposition, but complement and superimpose one another, like two aspects of the same oscillating field.

And of course, as I have already said, at the base of this dynamic lie the chronons, elementary units of Time. These quanta of temporality are not isolated points, but informational nodes that interlace and interlock to form an active texture, a veritable vibratory mesh of reality. Depending on their mode of organization—densification, polarization, resonance—the chronons can give rise to the phenomena we perceive as space, energy, matter, or even intentionality.

In other words, in my model, Time is not merely what "passes": it is the underlying matrix of the Universe, at once support, engine, and memory. It is on this foundation that it becomes possible to introduce an expanded vision of Time, no longer reduced to a single linear arrow, but unfolded into several dimensions capable of explaining the emergence of particles, cosmic cohesion, and the structuring of consciousness.

. Dimension 1 (T1) — The Sense of Time

Thierry
The first revealing dimension of time, which I simply call the sense of time, corresponds to what we commonly name "causal time." But it is important to specify that this causal time is not that of true causes: it represents above all the dimension of effects, the visible stage where events already inscribed in the temporal lattice manifest. It is the time we measure with our clocks, the one that classical physics describes as linear and irreversible, where phenomena seem to unfold from one another in a continuous chain.

In my model, however, this linearity is only an apparent projection. The true causes are not found in this arrow of time oriented from past to future, but in an inverted dimension: a negentropic face of the real, which acts from the future. In this inverted field, it is not accomplished facts that determine the unfolding, but the potentials still to come that guide, orient, and select the possible actualizations. One could say that the future acts as an attractor, a structuring tension that organizes the present from a horizon of meaning and finality.

Thus, causality ceases to be a simple thread stretched forward. It becomes a double tension: on one side, what has been inscribed in temporal memory and weighs as the determination of the past; on the other, what calls to being, that is to say, the attraction of the possible futures that quietly orient the present. Between these two polarities—memory and potential—the real is continuously actualized, in what we perceive as the ordinary unfolding of causal time.

The sense of time is therefore far more than a succession of measurable instants: it is the phenomenal surface where the weight of the past and the call of the future intersect. In this sense, it constitutes the first dimension of time, the most accessible to our daily perception, yet already open onto a deeper structure where causality is two-sided and dynamic, stretched between inscription and emergence.

Sylvana

This is what you already mentioned, that causality does not only come from the past: the potentials to come are already pulling on the present. Is that it?

Thierry

Yes, exactly.

And this dynamic does not operate in terms of direction alone: it can also be seen in the way events interlace within active memory.

Which brings me to speak to you about the second dimension of time: its causal coalescence.

. Dimension 2 (T2) — The Coalescence of Time

Thierry

The second revelatory dimension of Time can be understood through the notion of coalescence. In physics, this term designates the act of merging. In my model, it is not only a material fusion, but the axis along which events become entangled and condense within Active Memory, beyond a simple cause-effect succession. It is this dynamic of entanglement that gives lived experience its fractal texture: each instant becomes more than a point on a line; it is a knot of connections, a density of experiences woven together.

In other words, the measure of an instant is not reduced to its duration: it depends on the way events interlock, resonate, and interlace to form a mnemonic network. This is where the notion of Fractal Temporal Density (FTD) enters, which I define as the quantitative expression of this degree of entanglement. The higher the FTD, the more a moment contains relations, potentialities, and active informational charges.

This is why certain events seem richer, faster, more dense than others: they do not unfold within the same temporal fabric. They condense more memory, links, and resonances, producing a subjective experience of dilation or contraction of time.

As coalescence intensifies, a system tends toward simultaneity and integration. Events then cease to follow one another and unite within a common presence, approaching a limit-state that I have described elsewhere as the omniscient Present where all potentialities converge into a single vibration of unity.

Sylvana
So, if I follow you correctly, it is not only duration that matters, but the density of events within the instant—the more a moment concentrates relations and potentialities, the denser it is in memory and Information. Is that right?

Thierry
Yes. But understanding this raises an even more fundamental question: for, coalescence does not mean synchronization.

One can conceive of events that interlace, that aggregate within a common memory, without thereby entering into a true alignment. Coalescence describes the fusion of flows, but not yet their harmony.

Therefore, a question arises: if Time can vary both in direction and in coalescence, what makes their phasing possible?

What allows these multiple, sometimes divergent variations to combine at a single point of equilibrium, to produce coherent order rather than unstable chaos?

Sylvana
You mean that within one and the same instant there can be an enormous number of events and a great deal of information, but that does not necessarily form a coherent whole? Some things "resonate" with one another and align, while others remain scattered?

Thierry
Exactly.

It is precisely this question that led me to posit the existence of a third dimension of Time—a dimension of an entirely different order, that of synchronization.

No longer merely the vector of a flow or the density of a memory, but the instance where all rhythms attune, where local variations bend to a global resonance, producing the experience of the Present as a universal dynamic equilibrium.

. Dimension 3 (T3) — The Reversion of Time

Thierry
The third revelatory dimension of Time is that of the Present, but it must be specified at once that it is not an autonomous dimension comparable to the two preceding ones. The Present is in reality the emergent effect of Reversion, that is to say, of the axis where past and future mirror and respond to one another.

From this perspective, the Present is not a simple, fixed "slice of an instant," but the alignment of fractal temporal frequencies at a point of equilibrium. It depends neither on direction (as the causal arrow does), nor on density (as coalescence does), but acts as a knot of resonance that connects these two axes. The Present is the place where chronons vibrate in unison, where the memory of the past and the potentiality of the future meet and condense into actualization.

One could say that it represents the fundamental resonance of the temporal field: the secret accord that keeps the spiral of Time in cohesion and in motion. In this beat, Time ceases to be a succession and becomes a living presence, a space of unification where everything is simultaneously connected.

Thus understood, the present is not an illusion or a mere convention of perception: it constitutes the memory of the real in the making, the very act of the fractal universe unfolding. It is the beating heart of the Cosmos, the point where the weave of Time tightens to bring forth experience, consciousness, and being.

In other words, one can imagine this present as a universal matrix of synchronization.

It is not a fleeting instant, nor a simple point on a line, but an integral horizon, an ontological place where all strata of reality, all forms of consciousness, come to be attuned.

Thus, whatever the diversity of local times—whether faster or slower, fractal or continuous, human, stellar, or quantum—all share this same universal Present.

Each being, each particle, each world lives its own rhythm, but these rhythms are all attuned to an absolute weave, synchronous universal Time, guarantor of a cosmic cohesion.

It is this principle that allows multiplicity not to sink into chaos: at the heart of infinite variations, a dimension of time assures the invisible unison.

One may see it as a universal tuning key, a matricial resonance that adjusts, combines, and harmonizes flows.

This universal Present is therefore more than a subjective experience: it becomes the pivot of the Universe, the very condition by which events hold together and open themselves to consciousness.

It is this that makes possible the perception of the instant, but also memory and anticipation, for it is the plane of equilibrium where past and future meet in a living simultaneity.

Sylvana
So, this third dimension truly makes it possible to "create" the present by creating a point of resonance where past and future vibrate together.

Thierry
Yes. And this reversion reveals an even more profound property: the capacity of time to preserve information and to link instants to one another.

This is what the fourth dimension of time manifests: active memory. It is the first of the three implicit dimensions.

It inscribes each event into the chronal weave, assuring the continuity and coherence of the real. Thus, nothing disappears: everything remains accessible in the temporal field, like a living register linking past, present, and future.

. Dimension 4 (T4) — The Retrocausal Memory of Time

Thierry
In my model, the fourth dimension of Time—which I qualify as the first implicit dimension—presents itself as a function far more profound than what I initially described as a simple resonance. In reality, it constitutes the retrocausal Memory of Time, a living register where the field preserves information and connects it by creating active loops between past and future. This dimension does not merely store: it operates as a function of conservation and regulation, supporting the continuity of states and ensuring the coherence

of the Universe.

It can be understood as the cognitive register of the quantum vacuum: it grounds the transmission of information, establishes the cosmic memory, and explains the existence of nonlocal correlations observed in quantum phenomena. In this sense, it already acts as a cognitive function of the Universe itself, guaranteeing its capacity for self-organization and evolution.

The intensity of this retrocausal Memory is not constant: depending on the degree of Fractal Temporal Density (FTD), it retains more or fewer traces, weaves invisible correlations, and can even establish a genuine dialogue between past and future. Henceforth, causality ceases to be a rigid and linear chain. It becomes fractal, that is, modulated according to scale, the amount of information at play, and the degree of entanglement of the system concerned.

It is here that a major shift occurs: causality now depends on:

- The FTD,
- The degree of functional transformation of the system,
- And the level of integration between its different layers of organization.

Sylvana
So, in your model, causality is no longer simply a linear sequence "a cause leads to an effect"… It encompasses something far vaster and subtler, does it not?

Thierry
Exactly.

In my model, determinism is therefore no longer absolute: the simpler a system is, the more its trajectories remain predictable—like a stone thrown, whose path obeys fixed laws. But as complexity increases, the paths multiply and open onto a range of possibles. Thus, a human being, traversed by their memories, emotions, and choices, no longer follows a single path but explores a network of alternative trajectories.

It is precisely this dynamic that I call the Simultaneous Condition (SC): a principle that unifies all instants within a common field of coherence, where past, present, and future configurations remain in resonance. The Simultaneous Condition does not limit itself to serving as a memory backdrop: it also acts as a selective principle, organizing which potentials are actualized, which trajectories unfold, and how the Universe orients itself in its own evolution.

Sylvana

In fact, in your model, when an event occurs, far more parameters are taken into account than one might imagine. And it is the Simultaneous Condition that connects and selects these parameters: it retains memory, orients the possibles, and enables each coming present to be constructed in coherence with the whole.

Thierry

Yes, you have understood well. Thus, the fourth dimension of Time appears as the foundation of universal memory, but also as the selective and regulatory agency that makes the Universe a living system. It confers upon reality its stability and continuity while opening the space of possibles, and makes Time not only a framework of unfolding, but a cognitive and self-organizing matrix.

But I would also like to add a word about the metastable function of Retrocausal Memory.

This is an essential point that deserves to be added: Retrocausal Memory does not function as a simple fixed repository of information, but as a metastable structure. This means that it maintains states in relative stability while remaining open to possible transitions. This property grants it a dynamic and creative role.

Metastability first guarantees continuity: it preserves the traces, the correlations, and the links necessary for the coherence of the Universe. But this stability is always "on the edge" of transformation. Memory configurations, far from being immutable, hold themselves in a delicate equilibrium where they can tip toward new organizations. This is precisely what enables evolution, plasticity, and the reorganization of temporal trajectories.

Sylvana

You mean that this stability is in fact alive, always on the edge, ready to reorganize itself if necessary—therefore very flexible in the face of changing parameters?

Thierry

Yes. It is for this reason that I say that Retrocausal Memory acts as a living memory, both preservative and selective. Certain pieces of information are reinforced and stabilized, others fade or dissolve, according to the degree of resonance and the Fractal Temporal Density (FTD) of the systems concerned. Metastable states then function as thresholds of choice, or as basins of attraction, which determine which potentialities are actualized and which others remain in reserve.

From this point of view, metastability is far more than a secondary property:

it constitutes the profound cognitive function of universal memory. It is what enables the Universe to remain faithful to its past traces while retaining the suppleness necessary to welcome new configurations. In other words, metastability transforms Retrocausal Memory into an agency of evolutionary self-organization, a true engine of cosmic creativity.

One can compare this metastable memory to a landscape of hills and valleys: a marble can remain stable in a valley, yet always ready to tip to one side or the other if a resonance leads it there. I have been able to conduct concrete experiments that seem to indicate that this metastable function is not merely a theoretical concept, but a very probable reality.

Sylvana

If I sum it up correctly… until now, in the classical view, space set the stage and things unfolded through Time. But in your model, it is Time that sets the stage, through this active memory—a kind of living register that links moments and brings them into dialogue, as though past and future could adjust in real time. Is that right?

Thierry

Yes, exactly.

Now, it is time for me to explain the fifth dimension to you, because it too plays a crucial role in my model.

. DIMENSION 5 (T5) — THE META-DYNAMIC STRUCTURE OF TIME

Thierry

I have named the fifth dimension of time the metadimension of coherence.

Beyond the first four dimensions of time (T1 to T4), I posited the necessity of a fifth dimension, which I describe as the second implicit dimension. It must not be understood as a simple extension of the preceding ones, but as an encompassing metadimension, a structure of coherence that regulates all temporal cycles and the transitions from one fractal level of reality to another.

I use the term metadimension to designate this higher plane of regulation: a space of integration where diverse temporal regimes find their articulation, where ruptures are repaired, and where emergences are organized.

The fifth dimension operates as a meta-network of temporal integration, ensuring coherence between:

- The intrinsic temporalities of systems, with their rhythms, their densities, and their internal alignments;

- Inter-fractal passages, such as phase changes, scale jumps, or the collapses and emergences of structures;
- The conditions of bifurcation or systemic activation, where a system shifts from a stable state to a new organization;
- Nonlocal synchronizations, allowing distant regions of the Universe or distinct planes of reality to enter into coherence.

In other words, this dimension encodes the logic of how cycles are sequenced within a fractal spacetime, where causality is no longer linear but conditional, modulated by the Fractal Temporal Density (FTD) specific to each level of reality.

Sylvana
I understand the functions you have just described for this fifth dimension. But so that I may situate its role more clearly, could you place it in relation to the first four dimensions of time—how it complements them and why it becomes necessary?

Thierry
In fact, in a fractal multiverse organized into successive layers of emergence, the relations between these layers cannot be guaranteed solely by the first four dimensions of time. These describe, respectively:

- T1: causal time, the dimension of visible effects;
- T2: coalescence, the dimension of entanglement and fractal density;
- T3: reversion, the dimension of the Present as balanced resonance;
- T4: Retrocausal Memory, the function of conservation and regulation.

These axes are necessary, but insufficient to stabilize inter-level relations. As soon as one considers discontinuities, symmetry breaks, or state resets, a broader regulatory plane is needed: this is the role of the fifth dimension.

I have therefore conceptualized this metadimension in order to:

- Articulate non-continuous transitions, when a system abruptly changes regime (collapse of a star, biological mutation, conscious illumination);
- Stabilize relations between levels of reality characterized by different FTDs;

- Maintain the coherence of the whole dynamic network, by ensuring functional continuity across apparent discontinuities;
- Anchor the functional memory of the void, so that each state retains a trace and a resonance, even after transformation.

The fifth dimension therefore acts as a topological metastructure of the temporal field, an active memory that supports self-regulation, resurgence, and resilience of the cosmos.

But this fifth dimension is not limited to a structural function: it also opens a space of action and integration for consciousness.

It is this function of stabilization and regulation that explains why consciousness finds in the fifth dimension its privileged plane of action. In my model, consciousness is not a mere biological emergence, but an active function of navigation and integration within the temporal field.

It does interact with the axes of the first four dimensions, certainly, but it fully expresses itself in the fifth as a:

- Vector of self-reference: a system's capacity to recognize itself within its own temporal history and to maintain its identity despite change;
- Cycle modulator: the faculty to accelerate, slow down, open, or close certain fractal trajectories;
- Structural alignment engine: the aptitude to come into resonance with the Simultaneous Condition, a principle of global coherence between past, present, and future.

Within this framework, consciousness becomes the instance that, within the fifth dimension, steers and adjusts temporal dynamics. It does not move as in a space, but transforms the internal parameters of time: rhythms, densities, alignments, bifurcations.

In synthesis, the fifth dimension of time is therefore the metadimension of coherence: it regulates, stabilizes, and integrates all temporal cycles, while serving as a plane of action for evolving consciousness. It is the level where the Cosmos thinks itself, where the Universe becomes capable of maintaining itself, reorganizing itself, and orienting itself through its own transformations.

Sylvana

So, in your model, consciousness truly finds its place in this fifth dimension because, according to you, consciousness is an active function that acts from within the temporal field.

And so, if I understand your reasoning correctly, if, as a conscious-spirit, I were to align myself with another temporal cycle, the very laws of nature that I know—gravity, energy, matter—could manifest in a totally different way.

And depending on my Fractal Temporal Density and my level of consciousness, my trajectory of evolution could itself tip toward other possibles. And it is thanks to this fifth dimension that I can do all this.

Thierry
Exactly. The fifth dimension ensures coherence between the different temporal cycles and directly extends the principle of relativity.

What I call meta-relativity is therefore not an isolated new law, but the natural extension of Absolute Relativity, applied to the fractal structure of time.

In other words: the laws are not abolished, they become contextual—varying according to the temporal cycle to which a consciousness or an entity is connected.

The existence of this additional dimension makes it possible to conceive of a meta-relativity in which:

- Conscious-spirits are not only situated in spacetime, but inscribed within specific configurations of temporal cycles.
- Each element of the Universe—from the particle to the galaxy—lives not only in classical space and time, but also in its own temporal rhythm, a pulsation that orients its evolution.

Sylvana
I understand better… In a way, T5 acts as a force of integration: it links the dimensions of Time, preserves the memory of what has already been, and drives the Universe to grow in complexity and intelligence.

It is as though a principle of evolution were inscribed at the very heart of the real, carried by this dimension.

Thierry
Yes. This is why I say that this dimension does not merely stabilize: it introduces a living dynamic in which memory, consciousness, and even the laws of the Universe are never fixed, but constantly readjusted within a larger orchestration.

. Dimension 6 (T0) — Time, Primary Vibration

Thierry

To complete the coherent framework of the previous temporal dimensions, I introduce a sixth dimension, which I qualify as the third implicit dimension: Primary Vibratory Time (T0). It is not an axis of manifestation like T1 to T4, nor a plane of integration like T5, but the matrix dimension: the fundamental pulsation of the quantum vacuum.

I define it as the original pulsation of the Real.

Primary Vibratory Time denotes the base frequency, the pure pulsation that animates the quantum vacuum.

It corresponds to the deepest level of the real, beyond the Planck scale, where the usual structure of spacetime ceases to be relevant.

At this regime, I have postulated that it can be described as an original cold plasma, a pre-energetic substratum where the distinction between particle and wave does not yet exist.

T0 therefore represents the fundamental note of the cosmic symphony: a primordial vibration upon which all the fractal harmonics of the Universe are superposed.

Its function and role:

T0 plays a fundamental role, analogous to the Planck constant of Time:

- Just as the Planck constant (h) sets the quantum scale of energy,
- T0 sets the fundamental scale of the universal rhythm, the pure frequency that structures the whole scalar temporal field.

Thus, T0 ensures the global coherence of the field:

It synchronizes all local oscillations (the chronons), which, at this level, are still "at rest," without torsion or polarization.

It maintains continuity among the different temporal regimes (subquantum, quantum, classical, cosmological), since it is homogeneous everywhere and constitutes the common root.

I consider T0 as the active principle: it is not merely an abstract beat but acts as an active principle.

- It provides the initial impulse that powers the Big Bang and the structuring of spacetime.
- It inscribes a rhythmic memory in the quantum vacuum, guaranteeing that each cycle of the Universe inherits a coherent framework from the previous cycle.
- It ensures cosmological continuity: each Universe is born from the resonance of the preceding one, in an unbroken chain of the temporal matrix.

Link between T0, ÆKO, and the chronons:

In my model, T0 also plays the role of mediator:

- It links each chronon, the elementary quantum of Time, to ÆKO, the Universal Vibratory Resonance.
- This Universal Vibratory Resonance, itself rooted in the Primordial Complexity, thereby becomes attuned to the most intimate granularity of the real.
- Each chronon is therefore not an inert unit, but a living unit, animated by the link between T0 and ÆKO.

This link is essential: it is what fuels the dynamics of expansion, structures the memory of the void, and gives the chronon its creative capacity.

In summary:

Primary Vibratory Time (T0) is the original pulsation of the real (the Zero State).

- It is the vibratory root from which energy, space, Time, and consciousness flow.
- It is the pure frequency that synchronizes all manifestations.
- It inscribes in the quantum vacuum the rhythmic memory of the Universe, guaranteeing the continuity and coherence of cosmic cycles.

In this sense, T0 is not, lest one be mistaken, a null oscillation, nor a fixed state of rest.

By "zero," I designate the extreme rapidity of its fundamental oscillation: a vibration so dense and so balanced that it appears motionless to our perception.

It is therefore not an absence of oscillation, but a point of absolute equilibrium, where all potential frequencies are contained in seed form.

T0 is the primordial breath of the omniscient Intelligence, the original pulsation that sustains, feeds, and renews the whole temporal field.

It is from it that the dimensions of Time unfold, like harmonics issued from the same vibrating string.

Thus, what I call "rest" is only the phenomenal appearance of a state of full original activity, an inexhaustible reservoir of energy, information, and potentiality.

Sylvana

But then, in your model, what becomes of the three dimensions of space that we know?

Thierry

That is a good question.

In my model, the three dimensions of space (length, width, height) remain the same as those of classical physics. They form the geometric fabric in which material phenomena unfold. But space is no longer primary: it is now structured and informed by a more vast temporal field. In other words, space is understood only as a projection or "visible support" of temporal dynamics.

Integration into a 9D framework:

This is why I extend general relativity to a 9D Time-Space framework, consisting of:

- 6 temporal dimensions (T0 to T5), each with a precise function,
- 3 spatial dimensions, which become their expression and localized emergence.

Einstein's classical spacetime then appears as a derived layer, framed and orchestrated by this 9D architecture that I name Time-space-memory.

The role of the six temporal dimensions:

The six temporal dimensions here play the role of architectural framework:

- T0 (Primary Vibratory Time): the rhythmic base, the fundamental pulsation of the void, which gives the common frequency of all systems.

- T1 (Sense of Time / causality): the axis of visible effects, where the arrow of time that we measure is inscribed.
- T2 (Coalescence): the dimension of entanglement and density, which enriches the texture of lived space.
- T3 (Reversion / Present): the dynamic equilibrium where past and future reflect one another, giving substance to the Present.
- T4 (Retrocausal Memory): the function of conservation and continuity, metastable register of the void.
- T5 (Metadimension of Coherence): the regulatory plane that ensures integration between fractal levels and cycles of organization.

Together, they form the organizing weave that makes possible the existence of space, its expansion, its metrics, and its symmetries.

And thanks to these dimensions, I have been able to reconcile different "times".

Indeed, assigning several temporal dimensions makes it possible to reconcile forms of time that current theories separate:

- Lived time (subjective, of consciousness, linked to T2 and T5),
- Thermodynamic time (entropic arrow, inscribed in T1),
- Quantum time (processes of collapse and actualization, structured by T2 and T3),
- Cosmological time (linked to the expansion of space, rooted in T0 and T4),
- Informational time (linked to complexity and self-organization, modulated by T4 and T5).

What science today describes as "different times" in fact turns out to be complementary facets of one and the same 9D structure.

Thus, the three dimensions of space unfold as geometric manifestations of the six temporal dimensions.

- Space provides extension,
- But multidimensional Time ensures coherence, memory, and dynamics.

The 9D Time-Space-Memory model then unifies matter, energy, information, and consciousness within a single architectural framework, where classical spacetime is only a special case, a subset of a much vaster and more integrated structure.

3. A Hidden Time — The Other Face of Causality

Sylvana
It is fascinating… but then, why, in everyday life, do we perceive only the linear thread from past to future, with no access to the other dimensions?

Thierry
I understand your question, and that is precisely one of the essential points. As you know, within the framework of classical physical models, Time is generally represented as a one-directional dimension, oriented from past to future, according to a time arrow associated with the increase of entropy.

But this apparent linearity is in fact a projection of perception, and not an ontological property of Time itself.

We mainly perceive the entropic arrow because we evolve within a fractal of Time with low FTD—therefore with low informational density.

Sylvana
Does this mean that our Consciousness, attuned to this particular fractal, acts as a filter that limits our access to the other temporal dimensions?

Thierry
Yes, exactly.

Our ordinary perception is limited to the entropic arrow, the one that orients time from past to future by giving the impression of an irreversible flow.

But in my hypothesis of two-faced Time, there is a second polarity: the negentropic arrow, or descending retrocausality, which goes from the future toward the past.

This arrow is generally concealed from our ordinary consciousness, because access to it requires a sufficiently high FTD for consciousness to resonate with potential future states and integrate their retrospective influence.

In other words, at low FTD, we are confined to a linear, entropic reading of Becoming.

But as FTD increases—through the expansion of consciousness, through synchronization with denser layers of the temporal field—retrocausality becomes perceptible.

It then appears as an anticipatory memory, where traces of future events already influence the Present.

This is not a violation of causality, but an extension of its scope: consciousness accesses a richer temporal lattice, where the two arrows coexist and balance one another.

Sylvana
You mean that if consciousness increased its temporal density, it could access the other face of Time—the one where the future already influences the Present?

Thierry
Exactly.

This opens the way to what I call a dynamic epistemology of causality.

The more complex a system is—that is, the more it integrates and the more finely it resonates with the temporal field—the more causality ceases to be linear. It becomes contextual, flexible, and multidirectional.

Sylvana
Contextual… you mean we could perceive future events as already implicated in what is unfolding now?

Thierry
Yes. If the conscious-spirit, through its consciousness, remains tuned to a low Fractal Temporal Density, it picks up only the entropic arrow: the past conditions the Present, and that is all. We call this "determinism," and for me it represents only half of the process.

But if the conscious-spirit opens its Consciousness to a higher temporal density, it can access what I have called the Simultaneous Condition of the temporal field—where past, Present, and possible futures coexist.

Then the future is no longer merely a consequence yet to come: it also becomes an active influence that is already shaping the Present.

Sylvana
You mean that what we take to be a single, unique cause is in fact the intersection of several temporal lines?

Thierry
Exactly. If one sees only the immediate cause, one thinks that everything is fixed and deterministic. But if one also perceives the potential futures that retroact, one understands that the real is constructed through co-emergence.

A simple example: imagine an important decision in your life. You believe you made it because a chain of past causes led you there. But if your consciousness were attuned to a higher temporal density, you would also perceive how certain possible futures were already "calling" you, subtly influencing your choice in the present moment.

Sylvana

So, in your model, the Present is less a consequence than a point of equilibrium between memory of the past and attractors of the future…

Thierry

Yes, exactly. The Present is a place of weaving, where retrospective and prospective influences meet. This is what makes causality a logic that is not fixed, but living, adaptive, and tied to the temporal density of the consciousness that observes it.

Chapter 13

Pillar 4 – The Bimetric Universe

~

Fractal and Multiverse

Sylvana

Before we enter this new pillar, I would like to set the stage. You have just traveled a long way: the temporal field, the chronon, then the multiple dimensions of Time… And now, you arrive at the fourth pillar.

Thierry

Yes, exactly. This fourth pillar marks a pivotal stage in my theoretical path. After having laid the foundations of my conception of Time as the structuring Principle of the Universe, it became indispensable to test its explanatory power: how far can such a hypothesis truly shed light on the great enigmas still unresolved in modern cosmology?

For it must be recalled: nearly 95% of the Cosmos still eludes us. Dark matter, dark energy, the intimate structure of the void, and the origin of time remain largely beyond the reach of our current models. We are navigating in a Universe of which the majority remains invisible, as though our theories were only incomplete maps of an immense territory.

Faced with this challenge, I concentrated my efforts on three fundamental questions, true knots of enigmas:

- The distribution of matter, which thwarts the predictions of classical gravity and suggests underlying forces that are still elusive.
- The circulation of energy, at the heart of the mystery of the accelerated expansion of the Universe, which defies the logic of cosmic equilibrium.
- The nature of Time itself, of which we perceive only a phenomenal facet, without accessing its profound structure.

It is in this context that I introduced the notion of two-faced Time: an active, living, and structuring principle, whose duality upends the very way we think about cosmic dynamics. On one side, Time contracts and organizes matter; on the other, it opens and reinjects potential, like an invisible breath that keeps the Universe in motion.

By placing Time at the center—no longer as a simple measurement parameter, but as a primordial force—I propose a framework capable of integrating what, today, remains attributed to "the invisible." This framework offers not only a new key to understanding dark matter, dark energy, and the memory of the void, but also a bridge between physics and metaphysics: Time then becomes the unifying matrix where the visible and hidden dimensions of the real meet.

Sylvana
You had already spoken about two-faced Time—that past and future are two real orientations of the temporal field.

But now, do you want to explore its implications at the cosmological level?

Thierry
Exactly.

One of the great challenges of contemporary cosmology is to explain certain major phenomena that still resist our theories.

The accelerated expansion of the Universe, first of all, defies logic: whereas gravity ought to slow this expansion, we observe instead an acceleration, as if an invisible energy—"dark energy"—were driving space to dilate ever faster.

Next comes the question of dark matter: galaxies rotate too rapidly for their visible mass alone to account for their cohesion. It therefore seems that some as-yet-undetectable form of matter exerts a hidden gravity, but its nature remains entirely mysterious.

Finally, the asymmetry between matter and antimatter poses a fundamental paradox: according to known laws, the two should have annihilated each other in the Universe's first moments, yet we observe a Cosmos dominated by matter.

These three enigmas reveal a single observation: current models, based solely on the geometry of curved spacetime, are insufficient to account for cosmic complexity. It is precisely in the face of these limits that I have proposed another path: to rely on the two-faced nature of Time to construct a different cosmological model. In this framework, the Universe does not rest only on a geometric metric, but also on a bimetric and informational structure, where Time becomes the active principle capable of generating and organizing phenomena.

In this framework, Time no longer merely orders events: it becomes an organizing force that unfolds two faces—one turned toward contraction and stabilization, the other toward opening and emergence. This double movement is not a simple play of abstract symmetry, but the key that makes it possible to explain the apparent imbalances of our Cosmos: why expansion accelerates, why visible matter does not suffice to explain cosmic gravity, and why the Universe favored matter over antimatter.

This theoretical framework led me to formulate a hypothesis consistent with the idea of a fractal Universe, where structures repeat at different scales, and of an intrinsic multiverse, where realities are not isolated but coexist in parallel, each obeying its own temporal regime. Thus, the diversity of observed phenomena—from galaxies to elementary particles—can be understood as the expression of one and the same two-faced dynamics, modulated differently according to the temporal density and the metric polarity at play.

Sylvana

When you say "bimetric and informational structure," what exactly do you mean by that?

Thierry

In my model, Time cannot be reduced to a single causal arrow oriented from past to future, as classical physics assumes. It rests on a fundamental bimetry: two directions coexist, symmetrical in their essence but profoundly asymmetrical in their effects.

- The first face, which I call M+, is an emergent, structuring flow. It carries memory, cohesion, and order: it is what organizes matter, stabilizes forms, and enables the continuity of phenomena.
- The second face, M-, is a recessive, dissolving flow. It is associated with forgetting, dispersion, and de-structuring: it is what erases traces, opens the possibles, and prepares the regeneration of the temporal fabric.

These two polarities do not cancel one another out: together they constitute the profound dynamics of Time. It is their reciprocal tension—between inscription and effacement, between organization and dissolution—that shapes the Universe at every scale and conditions our experience of the real. In other words, reality does not build itself upon a single, irreversible arrow, but upon the living interaction between two temporal regimes, whose shifting balance gives rise to the history of the Cosmos as well as to the intimate perception we have of the present.

Sylvana

In your model, these forces condition the experience of the real… I am intrigued… can you elaborate?

Thierry

Yes. In reality, this two-faced nature of Time entails a distribution of phenomena according to their temporal polarity. Here, "polarized" means that each entity—particle, field, or mass—adopts a privileged orientation with respect to the flow of Time. As a magnet has two poles, each phenomenon is "oriented" with respect to the temporal fabric: it does not merely undergo Time, it inscribes itself into it with its own direction.

This is what the coupling to the fractal temporal field expresses: the way a system articulates itself to one or the other face of Time, M+ or M-. This coupling defines the fundamental physical properties of the entity in question: its inertia, its gravity, its interactions, but also its relation to memory and to Becoming.

Thus, the matter that we call "positive" is coupled to the classical temporal orientation, the one that evolves from the past toward the future. It bears the imprint of memory and irreversibility, and corresponds to the regime that we experience in our observable Universe. Conversely, a "negative matter" would be coupled to the inverted temporal orientation, evolving from the future toward the past. It would follow a negentropic dynamic, dissolving and regenerative, inscribed in the other face of temporal bimetry.

In other words, the Universe is not homogeneous, but is composed of two fundamental components, defined by their coupling to opposite temporal orientations. Each slope of reality is structured according to a specific space-time metric, arising from the direction of temporal evolution of its constituents. One, polarized in M+, builds and organizes; the other, polarized in M-, deconstructs and opens the possibles. And it is precisely from their polarized interaction that the living dynamics of the fractal Cosmos is born.

Sylvana

So in your model, everything follows a direction, whether matter, fields, or even inertia…

Thierry

Yes, and to show concretely how this bifaciality of Time manifests, we can distinguish two fundamental realities, mirroring one another.

The first is extrinsic reality, that of our observable Universe. It is governed by a temporal metric oriented from the past toward the future and is characterized by:

- Positive masses (M+), naturally coupled to a structuring and stabilizing Time;
- An increasing entropy, which inscribes phenomena in irreversibility;
- And a tendency toward aggregation, giving rise to coherent and durable structures: galaxies, stars, planets, but also molecules and living organisms.

It is this M+ polarity that confers on our Universe its intrinsic stability: positive masses attract one another, condense, and organize matter into persistent forms. It is thanks to this temporal foundation that the Universe unfolds a continuous cosmic history and that complexity can emerge.

In the mirror, there exists an intrinsic reality, coupled to an inverted temporal metric from the future toward the past. In this branch of the Universe:

- The constituents are negative masses (M-), polarized according to the recessive flow of Time;
- There, entropy decreases (a negentropic dimension), inscribing this regime in a dynamic of erasure and dissolution;
- Systems there remain unstable and diffuse, incapable of durably aggregating large structures;
- This reality therefore remains almost undetectable from our frame of reference, although it may exert discreet yet decisive gravitational effects on the observable Cosmos.

In other words, M+ stabilizes, aggregates, and orders, whereas M− destructures, dissolves, and remains unstable. Their coexistence forms a cosmic bimetricity, where each polarity plays a complementary role. This vision makes it possible to explain certain anomalies of modern cosmology (dark matter, dark energy, accelerated expansion) by laying the foundations of a universal symmetric dynamics, anchored in a fundamental temporal field with dual polarity.

From this perspective, the dichotomy between M+ and M− can be understood as the articulation between two complementary temporal regimes: one founded on entropy, the other on negentropy. Far from excluding one another, these two principles balance and respond to each other, ensuring that the Universe retains its dynamic and living character.

The M+ regime corresponds to the classical entropic orientation:

- It stabilizes and fixes phenomena;
- It inscribes events in an irreversible memory;
- It fosters the aggregation of matter and the formation of durable structures—galaxies, stars, planets, but also molecules and living organisms.

In other words, the entropy of M+ is not only a "loss of liberty," but a force of cohesion: it locks the dynamics by orienting Time from past to future, which gives the observable Cosmos its stability.

In the mirror, the M− regime expresses a negentropic orientation. Here, it is not simply a matter of dissolution or chaotic instability, but of a field of organizational plasticity:

- There, matter remains diffuse, because it does not aggregate into stable baryonic structures;
- But this loosening instead opens an added surplus of degrees of freedom, allowing the emergence of unprecedented configurations (informational, vibratory, or subtle);
- The negentropy of M− therefore acts as a source of creativity and renewal, a principle that keeps the Universe open to the possibles.

Thus, M+ preserves, M− liberates.

- Too much M+, and the Universe would grow fixed in an excess of order, a sterile crystal incapable of evolution.
- Too much M−, and it would dissolve into an instability without memory, where nothing could persist.

Their polarized interaction is therefore the key: M+ provides stability, continuity, and memory; M− introduces plasticity, regeneration, and openness. Together, they establish a dynamic balance in which structure and novelty, order and freedom, join in a living spiral.

It is this subtle oscillation between stabilizing entropy and creative negentropy that allows the Universe to be both coherent and evolutionary, memory and becoming, an ordered Cosmos and a field of continual innovation.

Sylvana
What you say is very interesting. And to confirm that I have understood correctly: you describe a Universe with two temporal polarities. In our extrinsic reality (past → future), positive matter and increasing entropy foster aggregation and stable structures; in the mirror reality (future → past), negative matter and decreasing entropy remain diffuse and almost undetectable.

And this bimetry—two inverted space-time metrics—could shed light on certain cosmological anomalies and show that the Universe rests upon a fundamental temporal field with a double orientation.

Thierry
Yes, you have summarized it perfectly.

Sylvana
And these phenomena you speak of… could you describe a few that make sense when applied to your model?

Thierry
Yes, of course. To make all this more concrete, let us take a few examples where the interaction between M+ and M− becomes the key to a new understanding. In my model, negative matter, evolving in an inverted temporal metric (from the future toward the past), is not a mere abstract "double" of positive matter, but an active aspect of the real that interacts with our observable Universe. This interaction opens an elegant path for explaining several still-unresolved astrophysical enigmas.

- The accelerated expansion of the Universe: rather than invoking an indefinable dark energy, one can understand cosmic acceleration as the result of a gravitational tension between the two metrics. Our metric M+ stabilizes and aggregates, while metric M− acts in the opposite sense, recessive, as though it were pulling the fabric of Time backward. From this interaction arises an apparent effect of "negative pressure," which drives space to dilate. The accelerated expansion would therefore not be an anomaly, but the natural expression of a bimetric Universe where two temporal regimes coexist.
- Dark matter: if M− matter exists, it cannot aggregate or form stable structures in our metric, because it obeys a negentropic dynamics oriented toward dispersion. It remains diffuse, elusive, yet its gravitational imprint interferes with our M+ world. It thus becomes invisible to our instruments, but detectable through its effects, exactly like the dark matter that we observe. In this framework, dark matter ceases to be a mysterious substance: it becomes the gravitational projection of the M− face within our metric.
- The decoupling between visible mass and observed gravity: in many galaxies and clusters, gravity appears disproportionate relative to visible mass. This enigma finds here a simple explanation: the gravitational fields arising from metric M− are superposed on ours, locally strengthening the curvature of spacetime. We then measure an "excess" gravity without seeing the mass that is its source.

What makes this model interesting is that it does not posit two totally disjoint Universes, but an interweaving of the two temporal metrics within a deeper fabric: that of the fractal temporal field. It is this field that orchestrates the coupling between M+ and M−, ensuring both their autonomy and their interaction.

The logic is clear:

- M+ stabilizes, aggregates, builds forms;
- M− diffuses, dissolves, exerts an inverse tension;
- Their interaction creates a dynamic equilibrium in which cohesion and expansion respond to one another.

This idea is innovative because it transforms what appeared as anomalies (dark matter, dark energy, expansion) into natural consequences of a two-faced

Universe, polarized by Time itself. The invisible is no longer a troublesome residue of our equations, but the signature of the hidden temporal polarity that completes our reality.

Sylvana

So M− matter could really explain phenomena such as accelerated expansion or dark matter… But do M+ matter and M− matter interact only in specific cases like those, or do they interact continuously?

Thierry

That is a pertinent question.

To answer it, one must envisage the Universe as a bimetric system with a double temporal layer, separated—but also linked—by a dynamic interface. This interface is not an impassable wall: it is a semi-permeable temporal membrane, whose permeability depends on the local conditions of the weave of Time, and notably on the Fractal Temporal Density (FTD).

Where the FTD reaches critical thresholds—for instance in regions of strong cosmic resonance, near gravitational singularities, or under certain quantum conditions—the membrane becomes selectively permeable. In those places, several interaction processes can occur:

- Transfers of information between the two slopes of reality, like a discreet passage of memory from one metric to the other;
- Synchronous resonances, where structures polarized in M+ and M− vibrate in unison, creating subtle couplings observable as gravitational anomalies or entanglement-like effects;
- Local energy junctions, where the tension between the two temporal regimes discharges into observable phenomena—singularities, quantum fluctuations, decoherences, even the emergence of new particles.

One could compare this membrane to a quantum potential barrier: most of the time, it firmly separates the two regimes, but it can become permeable under certain resonance conditions, allowing a temporal tunneling effect. This mechanism would explain why M+ and M− appear distinct most of the time, while remaining profoundly entangled within a unified structure.

In other words, the boundary between the two metrics is not fixed: it is living, selective, and intelligent. It opens or closes according to the vibratory state of the temporal field. It is this dynamic permeability that grounds the balance between order and openness, between stability and plasticity, and that accounts for the cosmological anomalies that we attribute today to dark matter or dark energy.

Sylvana

I see... would you have an analogy that would help me visualize this dynamic more clearly?

Thierry

Absolutely. One can imagine the Universe as a musical instrument with two superposed strings.

- The upper string, associated with M+ matter, vibrates in the direction past → future.
- The lower string, associated with M− matter, vibrates in the opposite direction, future → past.

Individually, each string produces its own timbre, like two distinct realities. But when they resonate together, interferences and harmonics appear. These harmonics are not generated by a single string, but by their vibratory interaction. That is where the key lies: what we perceive as cosmological anomalies could be, on the scale of the Universe, the equivalent of these harmonic beats.

Let us take a concrete example in astrophysics.

In 2006, the observation of the Bullet Cluster revealed an enigma: the distribution of visible matter (hot plasma detected in X-rays) did not match the gravitational map obtained by gravitational lensing. In other words, gravity deflected light as if an invisible mass were present elsewhere.

Within the framework of my model, this "ghost mass" is naturally explained: it would be the gravitational projection of M− matter, evolving within an inverted temporal metric. This matter cannot condense into structures in our M+ frame of reference, but it nevertheless influences the local geometry of spacetime. What astrophysicists call "dark matter" would then be the signature of a metric coexistence and not of an unknown substance.

By going further, one understands why the two temporal polarities give rise to different gravitational behaviors:

- Positive matter (M+) follows the classical logic: it attracts, aggregates, and builds hierarchical structures—stars, galaxies, clusters.

- Negative matter (M−), for its part, cannot agglomerate. According to its inverted temporal coupling, it tends to repel one another, and with respect to M+, it exerts a repulsive force.

In this framework, the accelerated expansion of the Universe ceases to be a mystery: it appears as the result of a metric tension between M+ and M−, exactly like two strings in opposite vibration that create an overall beat. This cosmic beat is the very breathing of the fractal Universe.

One can also compare the Universe to a cosmic cell surrounded by a semi-permeable membrane.

- This membrane is not a rigid boundary: it filters, modulates, and regulates exchanges between the two faces of Time, M+ and M−.
- Just as a biological membrane lets certain ions or molecules pass according to concentration gradients or electrical signals, this temporal membrane becomes permeable when the Fractal Temporal Density (FTD) reaches critical thresholds.

Scientifically, this mechanism calls to mind quantum tunneling: a particle can cross a potential barrier that is classically insurmountable if certain conditions of frequency or density are met. In my model, the temporal membrane plays the same role on the cosmic scale: most of the time, it separates, but in critical states, it connects the two temporal regimes and gives rise to the phenomena that we interpret as "anomalies."

Sylvana

It is impressive; it is as though two elements seemed opposed, and yet operated in a complementary way.

But if I return to the numbers… You often remind me that we see only about 5% of the Universe and that 95%—dark matter and dark energy—remain invisible. If M− truly corresponds to that 95%, then why such an immense gap?

Thierry

It is precisely this asymmetry between matter and antimatter that poses the fundamental paradox I mentioned a moment ago. The apparent asymmetry between the 5% of visible matter and the 95% of dark components (≈ 27% dark matter and ≈ 68% dark energy) might not be a true inequality, but rather an illusion arising from our metric perspective. If the Universe is truly bimetric and fractal, this distribution does not express a quantitative dissymmetry, but a functional symmetry.

Indeed, positive matter (M+), dense and baryonic, has an aggregative capacity that shapes galaxies and large hierarchical structures. It serves as a visible and condensed architecture, which inscribes memory and stabilizes the observable Universe.

Conversely, negative matter (M-), diffuse and weakly massive in our metric, does not manifest through accumulation, but through its dynamic influence. It modulates the cohesion of structures, acts as a global gravitational pressure, and contributes to accelerated expansion.

Thus, the "5% / 95%" distribution does not correspond to a raw parity, but to the expression of two complementary functions:

- M+ (≈ 5%): the dense, aggregative component, which builds and renders visible;
- M- (≈ 95%): the diffuse, modulatory component, which balances and fine-tunes on large scales.

One can illustrate this with a simple analogy: a system composed of 10% compact masses and 90% diffuse gas is not balanced in terms of condensed mass, but it is balanced in its global dynamics. For even though it is light, the gas determines the behavior and stability of the system through its pressure and invisible interactions.

Within this framework, true cosmic symmetry therefore does not reside in the proportion of masses, but in the functional complementarity of the two metrics, integrated within the two-faced temporal field. In other words, it is a symmetry of architecture and of role, rather than a symmetry of quantity.

In the same way, our observable Universe (M+) and its temporal mirror (M-) can, under certain conditions of resonance, generate hybrid effects: subtle phenomena, gravitational anomalies, or quantum fluctuations, which are precisely the signatures of this metric coexistence.

Sylvana
Subtle phenomena… what exactly do you mean by that?

Thierry
If the boundary between M+ and M- is a semi-permeable temporal membrane, whose permeability depends on the local conditions of the fabric of Time, then it becomes possible to extend this explanatory framework to phenomena

observed on the human scale. I drew a parallel with certain "strange" psychic or cognitive phenomena originating in the brain, often considered paranormal, but which could be explained scientifically in the same way.

I start from the postulate that the Universe finds its origin in the quantum vacuum, understood as a unified network of Time–space–memory linking all levels of reality, from the atom to consciousness. If this fundamental fabric governs the Cosmos, it is coherent to hypothesize that it also expresses itself in the brain. More precisely, structures such as the microtubules of neuronal cells could function in quantum states, leveraging properties of superposition and entanglement.

In this framework, consciousness is no longer a mere biological epiphenomenon: it becomes a quantum characteristic of our reality, coupled to the two-faced temporal field. Thus, just as the temporal membrane connects and modulates the metrics on the cosmic scale, an analogous dynamic could manifest locally in cerebral processes.

This hypothesis opens a theoretical framework that makes it possible to explain several so-called "parapsychological" phenomena:

- Precognition: the perception of "echoes" of future events already inscribed in metric M-, prior to their actualization in M+.
- Psychokinesis: a localized interaction between M+ and M- fields, creating quantum decoherences that result in a measurable effect on matter.
- Telepathy or synchronicity: resonances of information across the FTD between two coupled consciousnesses, like two synchronized oscillators within the same temporal fabric.

From this perspective, certain phenomena that we classify today as "paranormal" would not be supernatural, but simply local and rare manifestations of the plasticity of temporal metrics. I have carried out empirical experiments that support these hypotheses. They would be the subtle harmonics of a cosmic instrument, most of whose vibrations escape our ordinary perceptions, but which certain particular conditions allow us to capture.

Sylvana
That is very interesting… Do you have an example?

Thierry
Yes, let us take the example of precognition, which illustrates well what I mean.

In the field of experimental psychology, several laboratory studies have brought to light a so-called "psi" effect, weak but statistically reproducible: subjects, placed in front of screens, display physiological reactions (for instance a change in skin conductance, or measured variations in brain activity) a few seconds before the appearance of a random stimulus—whether it is a neutral image or an image with a strong emotional charge.

These results, although long deemed enigmatic, find a natural explanatory framework in my model. If consciousness is a quantum interface sensitive to the fractal temporal field, then it can capture not only traces of the past, but also echoes of information from metric M-, where events already exist as potentialities inscribed in the weave of the future.

In other words, precognition would not be a violation of the laws of causality, but temporal information crossing the semi-permeability of the membrane between M+ and M-. This permeability can be activated by the mind—consciously or unconsciously—which then becomes capable of receiving an anticipated resonance of an event before it actualizes in our metric oriented past → future.

In this perspective, precognition ceases to be a "paranormal" phenomenon: it becomes a subtle yet natural expression of the bifaciality of Time. It demonstrates that our consciousness is not confined within a single temporal arrow, but connected to a deeper structure, capable of occasionally crossing the boundaries between the two metrics.

Sylvana
I see… so your model could shed light both on phenomena of the Cosmos—such as those gravitational anomalies—and on more subtle phenomena linked to consciousness, such as precognition. That covers an immense spectrum… But then, how does your theory situate itself with respect to contemporary Science?

Thierry
My theory finds its roots in hypotheses that I formulated between 1996 and 2001, well before the emergence of certain subsequent theoretical works which, in retrospect, strengthen its relevance.

The edifice that I propose precedes and encompasses in particular:

- The research in bimetric gravity initiated by Hassan and Rosen (2011), who constructed a coherent mathematical framework to formalize the existence of two interacting metrics;

- The negative mass hypothesis put forward by Jamie Farnes (2018), which explores its possible role as a unifier between dark matter and dark energy.

These contributions, although later, confirm the fruitfulness of an initial intuition: that of a Universe structured by opposing temporal metrics and differentiated gravitational polarities.

However, my vision is not limited to these geometric or energetic dynamics. It opens a broader field, where the real ceases to be conceptualized as a simple passive spacetime. I describe it as a living architecture, fractal and navigable, woven from correspondences between physical phenomena, informational structures, and conscious experiences.

It is important to emphasize that, when I began to develop this hypothesis, the French researcher Jean-Pierre Petit had already proposed, as early as the 1980s, a bimetry-based model founded on the existence of two types of matter—positive and negative—evolving within distinct metrics. I discovered his work only later, but I recognized in it a stimulating echo, which intuitively confirmed certain directions that I had explored independently.

Our approaches, nevertheless, differ substantially:

- His model extends general relativity by integrating into it a gravitational symmetry based on two types of matter;
- Mine diverges from this direct extension, proposing a more encompassing framework founded not only on bimetry, but above all on the bifaciality of Time, a fractal architecture of the real, and an expanded articulation between physics, information, and consciousness.

It is this additional step—restoring Time as the primary structuring principle rather than a derived parameter—that gives my theory its singularity and its explanatory power, both cosmological and cognitive.

Sylvana
And in such an ambitious undertaking, what has been, concretely, the greatest challenge for you?

Thierry
The principal challenge remains that of experimental validation. The reality posited here does not reveal itself directly to observation, but through its indirect effects, already perceptible in different domains:

- Abnormal polarization of the void;
- Unexplained gravitational distortions;
- Local energetic imbalances;
- And, on large scales, the accelerated expansion of the observable Universe.

Taken in isolation, these clues appear fragmentary; but considered together, they trace the outlines of an architecture more subtle than a single, homogeneous spacetime.

At the heart of this architecture, a structuring intuition guides my hypothesis: the Universe is bimodal in its metrics, fractal in its architectures, and unitary in its principle of consciousness. In other words, the duality of the arrows of Time and metric duplication are not anomalies, but the keys to a hierarchical, coherent, and evolving cosmic organization.

This is why my research naturally turns toward the infinitely small. For it is at the quantum scale that:

- Reality is encoded;
- Temporal bifurcations take root;
- And the fundamental laws articulate with the informational field.

In the intimacy of the quantum vacuum and of elementary singularities there then arise unexpected correlates: nonlocal entanglements, self-organized fluctuations, structural resonances. These phenomena, though tenuous, sketch the possibility of a new integrative physics in which matter, energy, information, and consciousness cease to be treated as disjoint domains, and instead appear as the interactive facets of a living continuum.

Thus, this model does not propose only an alternative cosmology: it suggests a reconciliation of scales—from the quantum to the cosmic—and announces a physics in which the Universe is understood as an intelligent, dynamic, and self-organized system.

Part V

Folding Space

~

Traveling Through the Universe Without Moving

Chapter 14

Cosmic Cartography

~

Identity, Resonance, and Voyage

Sylvana

After laying down all these basic hypotheses—Time as a living field, its chronons, its dimensions, and the bimetric Universe—how did you, for your part, end up perceiving the Universe as a whole?

Thierry

For me, the Universe is above all a dynamic network, structured by Time, information, and energy, and animated by generative processes in perpetual evolution. It is not a fixed backdrop where events would unfold, but a living, self-organizing, multidimensional system, whose laws emerge as much as they structure.

From my earliest reflections, I conceived of it as a kind of cosmic quantum computer: a universal technology founded on principles of encoding, memory, interaction, and actualization. This architecture is not merely a metaphor: it conveys the idea that observable reality is the result of profound informational processing, where each particle, each field, and each consciousness constitutes a node in an immense network of interdependencies.

This system engenders the phenomenal reality that we experience, yet remains connected to subtler levels of structure, intention, and meaning. The visible phenomena—galaxies, stars, matter, life—are only the surface of a vaster field of organization, where information circulates and condenses into forms.

In this sense, the Universe is not a mere collection of objects, but an active process, an evolving weave of intelligibility. Every element, from the elementary particle to human consciousness, takes part in this coherent global dynamic, like a note in a symphony where matter, energy, information, and spirit resonate with one another.

Sylvana
When you speak of a "cosmic quantum computer," it is not just an image. You are describing a precise mode of functioning: a structure that encodes, that keeps the memory of interactions, that links each thing to the whole.

Thierry
Exactly.

But I had to go further. My goal was not only to describe the structure of the Universe, but to reconcile two fundamental dimensions: phenomenal reality and the place of the human being within this vast network. For it is not only about describing a cosmic mechanics, but also about understanding what it reveals—and makes possible—regarding the link between matter and spirit.

This quest is not new. Since the first glimmers of human thought, the Universe has inspired fascination and questioning. Through the ages, philosophers, scholars, mystics, and physicists have sought to uncover its laws, carried by the intuition that to understand the Cosmos is also to understand oneself. This is not mere speculation: in my observations and my intimate experience, I find this fundamental intuition again—the intuition that there exists a profound link between the infinitely large, the infinitely small, and individual existence, between the cosmoquantum weave and personal consciousness.

This "nostalgia for the Whole"—hard to name yet impossible to ignore—whispers to us that we come from an origin vaster than what the visible allows us to see. It is this link, between the structure of the Universe and the depth of being, that I sought to explore. My goal was to integrate a dimension where rational rigor meets intuition, and where the laws of the Cosmos converse with the intimate experience of consciousness.

In my model, the Universe and consciousness are not two separate domains, but the manifestations of one and the same quantum fabric of Time and information. It therefore seemed natural to me to begin with what is most immediate and most certain: the direct experience of being.

My quest, after developing the Absolute Relativity as a theoretical framework, became twofold:

- To rationally understand how to navigate in this Cosmos,
- And to show that a full understanding of the Universe is also a path of liberation—not in a mystical sense but in a pragmatic one: to know the profound laws of Time and information amounts to gaining access to the tools for traveling and interacting with it.

For me, space exploration is not only a technical challenge; it is an existential challenge for humanity. But I understood that this journey is not limited to space. My research extended from the spatial to the temporal, until I concluded that it is in the infinitely small—where, according to my theory, Time reigns as master—that the key is found. It is from there that Space can be contracted or modulated: in black holes or wormholes, space and time are no longer fixed data but navigable variables.

This is how I chose to situate myself as a conscious-spirit, starting from an indisputable point of departure: the immediate data of experience, that which is self-evident and cannot be denied. From this anchor point, it becomes possible to build a rational framework where the Universe is no longer a passive backdrop but a living architecture that one can explore, understand, and, one day perhaps, cross.

1. The Universe in Expansion — A Place Without a Fixed Address

Thierry

The first question I asked myself was direct:

Who am I?

I am a living, conscious, tangible being—integrated into the very fabric of the Universe.

My singularity lies in this reflective and intuitive capacity, which is profoundly interior and which I call "conscious-spirit."

I exist on a planet surrounded by other living beings who, like me, perceive, feel, and evolve. Starting from this self-evidence, I could not resign myself to thinking that my existence is a mere accident. If I, a being endowed with consciousness, exist, then consciousness must have a place within the very structure of the cosmos.

The second question presented itself at once:

Where do I live in the Universe?

On my human scale, I live on Earth, the third planet of the Solar System, orbiting a yellow dwarf star, the Sun, located in the Orion Arm of the Milky Way.

On the galactic scale, this means that I inhabit a spiral galaxy containing between 200 and 400 billion stars. We know that our Solar system lies about 27,000 light-years from the galactic center, in a secondary arm, far from the densest regions.

On a vaster scale still, our galaxy belongs to the Local Group, about thirty galaxies bound by gravity, which is itself included in the Virgo Supercluster, which is part of an even more extensive ensemble: Laniakea, a colossal structure encompassing hundreds of thousands of galaxies.

And beyond that? The observable Universe extends about 46.5 billion light-years in every direction. But nothing rules out that cosmic reality may infinitely surpass this horizon.

Yet as I pursue this mapping, I realize it rests on an implicit hypothesis: that of a fixed space, like a giant map where each point would possess an absolute address. But that view is misleading.

If I situate Earth within the Solar System, the Milky Way, or Laniakea, these are only relative markers. These coordinates never define an absolute "address" in the Cosmos.

For according to general relativity, the Cosmos is not a motionless backdrop but a fabric in perpetual expansion. There is therefore no objective center of the Universe: each observer is, in a sense, the center of his or her own horizon of visibility. Cosmic expansion stretches space between objects without those objects moving within a fixed frame of reference.

In other words, there is no absolute position in the Universe. Earth—and therefore I, here and now—has no definitive cosmic address. It is a relative point in a moving weave, a provisional node in the vast network of Time, space, and information.

Sylvana

What do you mean by "cosmic expansion dilates space between objects without the objects themselves moving"?

Thierry

Imagine an elastic band with pins fixed to it. When you pull on the band, the pins move away from one another, not because they are sliding across the surface, but because it is the band itself that is stretching.

That is exactly what happens in the Universe: space grows, and this simple stretching increases the distance between galaxies, even though, locally, they keep their size and cohesion.

In other words, cosmic expansion does not drive objects to move within a preexisting void: it widens the fabric of space in which those objects are situated. That is why any cosmic "address" is necessarily relative, shifting, and in constant evolution.

Even the most sophisticated tools of modern cosmology—such as comoving or conformal coordinates—are only mathematical conventions. They make it possible to describe the expansion of the Universe, but they never define an "absolute" position in a fixed space.

Moreover, our perspective is bounded by an unavoidable frontier: the causal horizon, limited by the speed of light and the age of the Universe. This limit defines the bubble of what we can observe. Everything that exists beyond remains, for now, out of reach of our perception—not because it is nonexistent, but because the information has not yet had time to reach us.

Sylvana

When you speak of a "causal horizon," what exactly is it?

Thierry

Your question about the causal horizon is essential. To answer it clearly, we must distinguish two complementary limits that together define our bubble of cosmic observation.

The first is the Wall of Opacity, a temporal boundary located about 380,000 years after the Big Bang. Before that moment, the Universe was so dense and hot that light could not circulate freely: photons were constantly trapped in an opaque plasma of electrons and protons. Then, when the Universe cooled sufficiently, these particles combined to form the first stable atoms, making the Cosmos transparent. From that point on, photons were able to travel freely—and it is these same photons that still reach us today in the form of the cosmic microwave background (CMB). Detected by instruments such as Planck or WMAP, this very uniform microwave radiation gives us a fossil image of the early Universe. It is sometimes called the "photo of the baby Universe," but it is not a photo taken "over there": it is a reconstruction, here and now, from photons emitted 13.8 billion years ago, cooled and stretched by expansion.

The second limit is the causal horizon, a spatial boundary defined by the speed of light and the age of the Universe. It marks the maximum distance from which a signal has had time to reach us since the Big Bang. Beyond that, even if the Universe is transparent, light or any other information has not yet reached our position.

These two boundaries—one temporal (the wall of opacity), the other spatial (the causal horizon)—delimit what we can actually observe. Everything that occurred before recombination, or everything that lies beyond our causal horizon, eludes direct observation and must be reconstructed by our physical models.

In other words, even if the Universe is probably far larger than what we see, we perceive only a finite portion of it: an expanding bubble that we call our causal horizon. This is, at present, the limit of the knowable.

Sylvana
In summary… what we see is bounded by two frontiers: the wall of opacity, which gives us the oldest possible image, and the causal horizon, which sets how far light and information have been able to travel to reach us.

But then, when I look up at the night sky, what am I really seeing?

Thierry
What you see at night are stars that are a few million to a few billion years old, and galaxies that formed well after that primordial moment.

Sylvana
Oh, I see… so what I gaze upon at night is "only" a few million or billion years old… In the end, it is a sky full of memories, not a snapshot of the present.

Thierry
Exactly. What you observe in the sky are messages that took millions, sometimes billions of years to reach us. Each star is a time capsule. Some may already have gone dark eons ago, yet their light continues its journey to us. It is unsettling, is it not?

Sylvana
Yes… It is as though each star were a luminous memory. A memory suspended in the sky.

Thierry

Yes… Is it not remarkable, when one truly thinks about it? The night is not a window open onto the present moment, but a vast archive of the cosmic past. And for us, who live in immediacy, it is an idea that is difficult to tame.

Sylvana

That is true… This sky that we believe to be immediate is in fact a memory, a kind of mirror of the past. There are so many mysteries in this Universe…

Thierry

All this shows that, even armed with the best maps and the most powerful telescopes, we perceive only a bubble of spacetime and a limited past. Our "address" in the Universe is therefore always partial, relative, and shifting. In such a context, the very notion of a "cosmic address" becomes problematic: if everything is expanding and without an absolute center… where am I truly?

2. Cosmic Identity — A Resonance Rather than a Place

Sylvana

So on Earth I have an address—a street, a number, a postal code—neatly arranged so that the mail carrier and the tax office can find me. But as a citizen of the Cosmos… I am officially untraceable?

Thierry

Yes and no… Because you know me: even when faced with a Universe without a fixed point of reference, I wanted to find a way to locate us.

That is when a third question arose:

How can I precisely define my cosmic address?

But before answering it, I had to completely rethink what I meant by "cosmic address." In my model, it is not a physical location comparable to a street or coordinates on a map: it is a position within the very structure of the real—a state, a frequency, a resonance in the temporal field and information.

Let me explain. To answer this question, I had to completely rethink the architecture of the Cosmos by returning to the foundations of my model. The Universe, as I conceptualize it, is not a simple homogeneous spacetime: it expresses itself as a cosmic technology, structured in the form of a hyperspherical torus composed of interlocking sub-tori. This architecture outlines a fractal, multidimensional, and holographic cosmology, where temporal densities differentiate and organize the levels of the real like the floors of a living information structure.

Within this framework, my "cosmic address" is not reducible to a spatial coordinate: it is the vibrational signature of a node of interconnection within a larger continuum. Concretely, I therefore exist:

- Inside a local vibrational resonance—a zone of manifestation among countless holographic layers of the real;
- In a precise spatiotemporal node, a point of interlacing where matter, memory, and consciousness are woven together like the threads of a larger canvas;
- Within a block Universe, where time does not "flow" but structures, encodes, and organizes; where the reality I perceive is only a local projection of a denser and more fundamental continuum.

In other words, in my model, the "cosmic address" is a state of attunement with the weave of time and information, a vibrational coordinate rather than a fixed position. It is this redefinition that allows me to conceive of the Universe no longer as a neutral space, but as a living network, where each consciousness is a beacon and a navigation point.

Sylvana
In short, you no longer define yourself by a "physical" location but by a frequency in the field—an active resonance with the architecture of the real. It is no longer a fixed address but a reference point of another order, always in interaction with the whole.

Thierry
Yes, that is exactly what I have already described in our previous discussions: the Fractal Temporal Densities (FTD) and the vibrational layers of Time.

Within this framework, my "address" is no longer spatial, but becomes a position of resonance within the temporal field.

In other words, I situate my existence not by a fixed location in space, but by a state of vibrational resonance within an active field of information.

What we usually call a "place" is therefore not a fixed point in a static geometry, but a particular configuration in the harmony of the quantum and informational vacuum—a vibrational signature that expresses how a node of the real is attuned to the global weave of Time.

This redefinition is essential: it transforms the notion of localization from a spatial coordinate into a dynamic state of attunement, where matter, memory, and consciousness interlace. In my model, each "place" is in reality a vibratory mode of the temporal field, and it is this signature—more than the geometric position—that truly defines our cosmic address.

Sylvana

So… that is how you could find me, wherever I am in the Cosmos—not by following coordinates, but by recognizing a frequency. A kind of unique attunement between me and the field, here, now.

In other words, your "address" is no longer a place but a way of being positioned in the living flow of time.

Thierry

Yes, that is exactly the idea.

Within this framework, situating oneself in the Universe does not mean occupying an immobile location, but actualizing a mode of being within a network of relations in perpetual motion.

In other words, existence itself is not a geometric localization, but a dynamic participation in the informational symphony of the real.

We are therefore not simply "somewhere": we are something in the process of becoming. Each being, each consciousness, each structure is a singular actualization, occurring at the intersection of multiple temporal layers of information, memory, and potentiality. We are active singularities within a living architecture, and not fixed objects inside a static container.

And that changes everything.

For in this perspective, the very notion of place fades in favor of a logic of process: to exist is to be in resonance, to participate in the ongoing actualization of a fractal field of Time and information. The Universe no longer appears as a space where things "are found," but as an evolutionary dynamic where things are woven and unfold in the harmony of a living continuum.

3. Cosmic Voyage — From Propulsion to Resonance

Sylvana

Well, it reassures me to know you…

If I ever get lost one day in the Cosmos, using your model, you could come and fetch me.

Thierry
In fact yes and no…

Sylvana
Oh… shoot…

Thierry
I am going to explain, and you will understand why: I cannot really come to "fetch" you, as you put it.

Sylvana
So even if you know my "cosmic coordinate"—that point of resonance in your model—it does not necessarily mean you could come and get me physically, is that right?

Thierry
Exactly.

And so that you see clearly what this is about, I will first recall what current science says, based on what we know today, before returning to my model.

My fourth question is therefore:

Can we actually "go somewhere" in the Universe with a spacecraft?

At first glance, traveling through space seems to be nothing more than a matter of distance and speed. But as soon as we widen the scale to that of the Universe, this intuition becomes misleading.

Remember: I have already explained to you that space is not fixed. It is in permanent expansion, and this expansion radically changes our way of conceptualizing cosmic motion.

When we say "go somewhere," we assume that space is a stable backdrop over which we move as on a map. Yet in the real Universe, it is space itself that dilates. This means that the destination we aim for recedes while we try to reach it—not because it is moving locally, but because the interval of space between it and us is growing.

Let us take a concrete example: today, some galaxies are receding from us at an apparent speed greater than that of light. This does not violate special relativity, because these galaxies are not moving through space faster than light: it is space itself that is stretching between them and us, carrying the galaxies along with it. This phenomenon is directly tied to the accelerated expansion of the Universe, discovered at the end of the 20th century thanks to observations of distant supernovae.

In summary: on a cosmic scale, "traveling" is not crossing a neutral void, but navigating an expanding fabric of spacetime. In other words, the Universe does not offer a stable ground to traverse, but a dynamic weave that stretches beneath our steps.

Sylvana

I see…

So the destination is not moving "ahead of me" as though it were fleeing, it is rather the space between it and me that is lengthening continuously.

Thierry

Yes, and now you can clearly see that, in this context, even a hypothetical ship capable of reaching the speed of light could never catch up with those galaxies. Why? Because the space between them and us keeps stretching faster than any possible motion within that space.

In other words, those regions of the Universe are receding inexorably beyond our causal horizon. They become definitively inaccessible—not only for our current technologies, but even in theory, so long as we remain confined to the framework of "classical" ships and the laws of physics as we know them today.

It is a dizzying realization: a portion of the observable Universe today will no longer be so tomorrow. Entire galaxies are already slipping out of reach, not because they are going dark or vanishing, but because the expansion of the cosmic fabric is carrying them beyond our sphere of possible interaction.

In other words, modern cosmology reveals to us a striking truth: traveling through space is not enough, because it is space itself that flees before us.

Sylvana

So if I understand… even if you had a ship capable of going at the speed of light you still could not come and get me if I were lost in a distant galaxy.

Because psace itself would stretch between us faster than you could travel it to come and save me.

Thierry

Yes. But here you are going to be even more disappointed, because this impossibility is only a first limit. There is another, even more fundamental: there is no universal frame of reference.

In a Universe without a center, there are no absolute coordinates. All frames are local and relative to the observer. Even with a cosmic map, one cannot "aim" at a place like a GPS address. It would be like trying to navigate a fluid dream, where space itself is redefining itself at every moment.

Sylvana
Ah yes… now I am done for.

Lost in my galaxy, even if you built a ship faster than light, it would not change anything.

Because if one takes space as the frame of reference, it in reality offers no fixed reference point: it remains fluid, relative, without center or absolute coordinates.

So, you could never come and find me?

Thierry
Yes and no

Sylvana
Yes and no… So there is hope?

Thierry
In fact, I must return to my model to explain to you how I would conceptualize a cosmic voyage—how, in this vision, I could literally come and find you.

In my conception, Time is not a consequence of motion, but the matrix that engenders it. It is Time that structures motion, not space. In other words, Space is only a dynamic projection of the temporal field: what we call "distance" or "movement" is the expression of a change of configuration in Time.

From then on, if Time is the foundation of the organization of the real, spatial movement ceases to be a classical path to traverse: it becomes the consequence of a modulation in the temporal field.

This means that a true "cosmic voyage" would not consist in crossing space at high speed, but in changing state in Time—performing a tuning jump between vibratory layers of the fractal temporal field.

In other words, changing place would amount to changing a chord of reality, as one changes a radio frequency to pick up another station.

In this framework, the Universe is no longer a space in which one must "move," but a network of potential states in which one can navigate by vibrational attunement and by tuning temporal densities.

Sylvana
So if I understand you… for you, to travel does not mean to cover distances physically, but to change one's attunement.

Shifting from one state to another, as though an address were not a point on a map but a frequency in the field.

Thierry
Yes, that is exactly it: it is a genuine paradigm shift.

For me, changing a paradigm means abandoning the idea of classical space travel founded on a Newtonian vision: one leaves point A to reach point B by covering a distance. This logic rests on a fixed, homogeneous, and neutral space.

But in a Universe such as the one I describe—a fractal, multidimensional, expanding temporal structure—this conception collapses. Movement can no longer be thought of as a simple translation within a container: it must be understood as a mutation of state within a dynamic field.

One must therefore think in terms of:

- Vibrational attunement: the attunement between the traveler and the target, like two notes coming into phase.
- Coherence of temporal density: each cosmic region possessing its own rhythm, its "tempo" of actualization.
- State transfer: a quantum jump from one configuration to another, rather than a linear journey.

Thus, in my model, interstellar or interdimensional travel would no longer be a question of propulsion but of dynamic attunement with the structure of the void—a process in which consciousness and position readjust simultaneously.

In other words, changing place amounts to changing one's chord of reality. It is not a displacement in space, it is a navigation through temporal densities—a leap of consciousness as much as a leap of position.

Sylvana
When you say "interstellar or interdimensional travel," do you mean that beyond the Solar System the logic is completely different?

Thierry
Exactly. As long as we remain within the Solar System, we stay in a zone of homogeneous metric coherence: the structure of the temporal field there is relatively stable, and classical propulsion obeying Newtonian laws retains a certain validity—even though it quickly runs up against immense technological constraints (amount of fuel, duration of journeys, inertia of masses).

But as soon as it is a matter of crossing interstellar distances—and a fortiori interdimensional ones—everything changes radically. We then leave our local temporal block, and the metric ceases to be uniform:

- The temporal field begins to vary in density,
- Vibrational synchronizations shift,
- And Newtonian logic, founded on linear inertia and continuous Space, becomes inoperative.

In this new framework, "travel" can no longer be conceived as a progressive displacement in a neutral space. It becomes an adjustment of state: a true tuning to another frequency of the temporal field.

In other words, to cross such thresholds, ships as well as persons must accomplish a vibrational leap—a shift that is simultaneously physical and conscious.

It is no longer a mechanical propulsion, but a resonant transition in which matter, energy, and consciousness align with a new temporal density.

This is how cosmic travel becomes possible: not by forcing space, but by realigning with the harmonics of Time, by changing metric as one changes a musical key.

4. Childhood Dream — Toward a Science of Syntonization

Sylvana

I see… in your model, it would therefore be possible to move anywhere in the Universe by vibrational resonance.

And when I listen to you, I find again the passion that has lived in you since childhood, that desire to touch the stars.

I understand that, for you, it is not only a matter of models or hypotheses: your research also bears an inner, almost existential dimension.

Thierry

You are right. scientific research is neither a simple intellectual exercise nor an abstract accumulation of knowledge. It is above all a visceral passion, an inner impetus that goes beyond ordinary curiosity.

It is a demanding and patient quest, which requires at once methodological rigor, creative imagination, and perseverance in the face of the unknown. It is not limited to solving equations or piling up data: it is carried by a profound desire for coherence, by the conviction that there is an intelligible order at the heart of reality.

For me, this quest is accompanied by an irrepressible need: to seek unity behind diversity, to discern the invisible thread that connects the laws of Nature, and to go to the very end of the fundamental questions—those that concern both the Universe and our own place within it.

Sylvana

I see. Ultimately, what you seek is coherence, a unity.

And knowing you, I imagine you must have transformed this paradigm shift into a concrete method, a framework to organize your research.

Thierry

Yes. My thought experiment always follows a progressive and branching approach, akin to the growth of a tree or the unfolding of a fractal network.

It is articulated in several fundamental stages:

- Map the established theoretical framework: explore in depth the existing models in physics, cosmology, and information sciences, in order to identify their strengths, their limits, and their blind spots.
- Choose a strategic entry point: a guiding question, clear and decisive, serving as a trigger and orienting all the research.
- Unfold the branching: each line of inquiry opens onto new hypotheses, new evidence or contradictions. Gradually, a constellation of avenues takes shape: some break off, others intersect, and a few emerge as central. My aim is always to favor the most fertile branches, those capable of creating bridges between disciplines and converging toward a possible theoretical unification.

To carry out this work, I had to develop several aptitudes:

- A rigorous multidisciplinary approach, linking the fundamental sciences together.
- A technological creativity, allowing me to conceptualize tools or simulation models.
- A capacity for theoretical innovation, indispensable for imagining new coherences, new formalisms, even new applications.

Thus, my approach is not linear but organic: it advances through bifurcations, resurgences, and convergences, always guided by a core requirement—to bring forth an integrative vision of the real.

Sylvana

Listening to you, I understand better why this research has accompanied you for more than thirty-five years… One senses that it is the thread of an entire life.

Thierry

Yes. It took me years of reflection, critical analysis, modeling, and confrontation with the limits of knowledge to lay the foundations of my model. I know that its complete exploration will doubtless require an entire life, for it is not a variation on what exists but a global reconfiguration of our understanding of the real.

This long labor led me to formulate a new vision of Time and of cosmic travel by attunement.

Attunement, in my approach, is not a mere metaphor. In electronics and in wave physics, to "tune" means to set a system to a precise frequency so that it enters into resonance with another. In my model, this concept is transposed to the fractal temporal field: to attune a point of consciousness or a vessel is to set it vibrationally to a precise configuration of the real, as one adjusts a radio receiver to pick up a station. This would make it possible to change layers of temporal density, thus to "move" no longer by crossing space but by changing state within the temporal field and information.

From there, my research tree crystallized around two major questions:

- How can real cosmic travel be made possible, beyond the limits imposed by expansion and the known laws of physics?
- Where are the other forms of conscious life in the Universe? Why this apparent cosmic silence?

These two questions—one technological, the other existential—led me to explore the deep structure of the real, its informational architectures, fractally hierarchical, multidimensional, and living.

Within this framework, attunement appears as a conceptual key: it does not describe a linear voyage but a change of tuning, a mutation of state in the temporal field and information. It is this tipping—at once physical and of consciousness—that could transform our relationship to the Cosmos.

Sylvana

It is astounding… you never really left your first questions. The ones that were already with you when you spent your nights on the roof of the house gazing at the stars, telling your grandmother that one day you would go visit them.

Thierry

Yes. That thread stretched from my first childhood nights contemplating the sky has never stopped accompanying me. It binds the child's naive wonder to the researcher's theoretical rigor, like one and the same quest pursued in different forms.

Ultimately, it is always the same inquiry: to understand where we come from, what it means to exist in this vast Cosmos, and what intimate bond unites human consciousness with the universal fabric.

That, for me, is the true guiding thread: never to break the link between knowledge and life, between the infinitely large, the infinitely small, and the intimate core of being.

My research thus naturally led me to explore in depth the complex informational topologies that could organize the Universe in a fractally hierarchical, multidimensional, and living manner.

But this exploration does not stop at contemplation: it prepares a paradigm shift. For to understand the Universe is not only to draw its map: it is also to identify its gates, its flows, its interfaces. It is here that the next pillar appears—a pivotal stage of my research.

We will now address the Universe's entries and exits:

- Protons, as vectors of Information and points of cosmic initialization;
- Black holes, as gates of compression and transmutation;
- Gravity, as a dynamic interface between these flows.

This pillar is essential, for it constitutes the practical foundation for answering the question of cosmic travel: no longer merely describing the Universe, but thinking how to navigate within it.

It opens the way to a new and powerful hypothesis: the Universe possesses channels of exchange, membranes, and thresholds of access, and it is by understanding them that we could transform our relationship to the Cosmos—moving from observation to active participation.

Chapter 15

Pillar 5 – Toward a Cosmic Humanity

~

Understanding the Thresholds of the Real

Sylvana

You give me the impression that the Universe is not only a vast space, but a place crossed by thresholds and invisible membranes. As though, behind the apparent scenery, there existed secret passages—crossing points capable of linking worlds. Is that truly what this is about?

Thierry

Yes, your impression is correct: in my model, there do indeed exist such passages—subtle but real—that open from one world to another.

This fifth pillar constitutes a crucial stage, for it aims to articulate the dynamic interfaces of the multiverse: those zones of transition, or genuine "gates," through which information, consciousness, and matter can circulate between different layers of reality.

These interfaces must not be seen as metaphysical fantasies, but as physical and informational structures: membranes of temporal density, thresholds of resonance, critical zones where the opposing metrics (M- and M+) meet. There, the Universe reveals itself not as a closed block, but as a living and permeable system, capable of exchanges and transformations.

The objective of this pillar is therefore clear: to propose a rigorous conceptual formalization of these mechanisms in order to make their study and experimentation possible. For if these gates exist, they could constitute many vectors of access:

- To the flows of information that circulate between Universes,
- To unprecedented gravitational dynamics linked to protons and black holes,
- And certainly, in time, to modes of cosmic navigation by tuning, far beyond the limits imposed by classical propulsion.

In this sense, this pillar does not merely describe a theoretical hypothesis: it opens the way toward a science of passages, where cosmology, quantum physics, and consciousness converge to explore the living and fractal structure of the real.

Sylvana
So, this pillar will focus on these "gates," which allow information, consciousness, and matter to cross the boundaries between different levels of reality?

Thierry
Yes, because if these hypotheses are confirmed, they open access to new degrees of cosmological liberty—hitherto unsuspected modalities of interaction with reality, beyond our usual spacetime.

This would imply an unprecedented capacity, not only to navigate physical space, but also to modulate our vibratory tuning and our temporal anchoring in order to access other levels of manifestation of the real.

From this perspective, the evolutionary leap is no longer a mere speculative ideal or a dream of science fiction: it becomes a strategic imperative for the future of the human species.

For if the Universe is truly a living and fractal field, endowed with dynamic interfaces and thresholds of access, then our survival and our advancement will lie not only in spacefaring by classical propulsion, but in the art of navigating the structures of Time and information.

The question is therefore no longer: "Must we become a cosmic civilization?"

It becomes: "When, and above all how?"

In other words, we move from a vision of space colonization to a vision of cosmic tuning.

This is the paradigm shift at the heart of this pillar: it is no longer a matter of crossing the void at great speed, but of learning to tune our technologies and our consciousnesses to the profound laws of the fractal temporal field, in order to open pathways of navigation and evolution that we could not yet conceive of.

Sylvana
So, where do we begin? What would be, in your view, the first step toward this new stage of humanity—this cosmic civilization you speak of?

Thierry
Humanity, to endure, will sooner or later have to transcend its local and terrestrial condition. It will not be able to content itself with surviving on a single planet, exposed to cosmic hazards and to its own fragilities. It will have to learn to inhabit the Universe in all its fractal complexity, to recognize and integrate layers of reality that are today invisible, and to develop transition technologies capable of crossing these thresholds.

These technologies will not be purely mechanical or energetic. They will also be biological (extending the adaptive capacities of living beings), vibratory (mastering the resonances of matter and of the temporal field), and informational (harnessing the memory and the plasticity of the quantum vacuum). It is this convergence—of life, energy, and information—that will define the true keys to cosmic navigation.

It is therefore only by fully embracing this cosmic vocation, through knowledge, scientific rigor, and an evolutionary cooperation, that we can hope to ensure the continuity of human consciousness within the great fabric of the real.

For the ultimate goal is not only to prolong our material existence, but to evolve consciousness itself in harmony with the evolutionary structure of the Universe.

1. Human Becoming — Science, Consciousness, and the Duty of Being

Sylvana
It is a powerful vision… And yet, it is clear that this shift toward a cosmic civilization still remains marginal today.

Some projects are emerging, yes—space agencies, private companies… But there is not yet a collective movement, nor a shared consciousness on a global scale.

So what is it, in your view, that prevents us today from thinking and acting at this cosmic scale?

Thierry

One of the major limitations of our era lies in a vision of existence that remains far too self-centered. The modern individual, enclosed within the narrow horizon of their own biological lifespan, naturally directs their priorities toward immediate survival and personal well-being.

This is understandable: in a culture where existence is interpreted as ending with the body, collective stakes over the medium or long term—such as the continuity of humanity on a cosmic scale—seem distant, abstract, almost beside the point.

The result: the question of our future beyond Earth is too often deferred to future generations, with the uncertain hope that one day they will have the lucidity, the will, and the means to assume it. But this delegation of the future is a dangerous illusion: the time required to prepare a genuine evolutionary leap is measured in centuries, and the foundations must be laid today.

In reality, the vision of a continuity of the human species across time and space is as fundamental as the instinct for reproduction. It is even its logical extension on the scale of the Cosmos. Yet current demographic and cultural dynamics reveal a weakening of this reflex of transmission, as though the memory of the species were losing sight of its own future.

In this context, ensuring the continuity of humanity—not only on Earth, but in the Universe—becomes a rational necessity and an ethical imperative. For it is not a secondary option: it is the natural extension of our duty of survival.

The Universe itself, as it unfolds, is not a fixed framework but a field of trials. It is crossed by entropic phenomena, systemic imbalances, local instabilities—sometimes violent—that, paradoxically, seem to take part in a broader, structuring, and evolutionary dynamic. To conceive of the future of humanity in terms of resilience, mobility, and adaptability within this cosmic context is therefore not a speculative luxury, but a clear-eyed and vital requirement.

For there is no absolute guarantee of survival—neither for the individual, nor for a civilization, nor even for the human species as a whole. At every scale, the threat remains, implicit or manifest: biological extinction, civilizational collapse, or cosmic dissolution. To rely on chance or on providential fluctuations of the real is only a transitory strategy—certainly not a durable project.

It then becomes imperative to actively bridge the gap between what we still do not know and a more complete understanding of the laws that govern the real. This work is not only a scientific endeavor: it constitutes a civilizational responsibility. For it is only by mastering the profound dynamics of Time, energy, and information that we will be able to envisage a future in which human consciousness endures and rises within the great fabric of the Cosmos.

Sylvana

What you are saying completely changes how we see things.

We tend to believe that humanity has time, that it will end up finding answers…

But here, I understand that nothing is guaranteed.

Do you mean that, in your view, everything is at stake now? That the future depends entirely on what we do—or what we refuse to do—today?

Thierry

Yes—and that is precisely the essential point. The future is not a predefined road, drawn once and for all: it is an open, fragile, and conditional field that the present sketches at every moment. Rather than seeing it as a guaranteed promise, it is more accurate to conceive of it as a dynamic potentiality, whose realization depends on our choices and our vigilance.

The only time that is truly effective, operative, is the present. It is the present that, at each instant, shapes our trajectories and inscribes our decisions into the fabric of the real.

Our future, for its part, exists only to the extent of our capacity for anticipation: a lucid, strategic, and conscious anticipation, mobilized at every level—personal, collective, technological, and cosmological. In other words, no external force will come to ensure our becoming or accomplish in our stead the tasks that fall to us.

Our survival—and even more so our fulfillment—do not arise from an automatic right. They constitute an active, demanding, and non-transferable responsibility. The future is not something owed: it is a continual conquest, a creative tension between what we are and what we choose to actualize.

This is why it seems to me not only rational, but vital, to develop a genuine capacity for mobility within the cosmic architecture. Sooner or later—and I insist on the urgency of addressing this sooner—humanity will have to learn to cross planetary limits, to migrate beyond the Earth, and to explore the different "fractals" of the Universe.

Relying exclusively on a single planet to secure our future is a dangerous illusion. For one must not confuse the geological longevity of a planet—which can indeed be measured in billions of years—with the fragility of the conditions for human life on its surface. The Earth can survive its own upheavals: tectonic activity, magnetic instabilities, extreme climatic changes, the natural evolution of the Sun, or even collisions with stellar objects. But nothing ensures that humanity, in its current form, could endure such brutal and unpredictable mutations.

In other words, the planet will survive without us, but we will not survive without a proactive understanding of Time and of the Universe.

Therefore, preparing for the expansion of humanity beyond the Earth is not a dream or a speculation: it is a scientific and civilizational duty, a crucial stage in our evolution.

Sylvana
So, it is not so much to flee the Earth... But rather because, out of responsibility—and even out of the necessity of survival—humanity will have to learn to cross a threshold?

Thierry
Yes, that is precisely what I mean. On the cosmic scale, we are a fragile and exposed species. Planets, stars, galaxies evolve on scales of Time and energy that exceed us infinitely, and in the face of which our biological survival appears derisory. Our only true strength lies in our intelligence: our capacity to understand, to anticipate, to transform our environment, and to conceive of adaptive responses.

It is this intelligence—and not brute force—that must lead us to no longer consider the Cosmos as a mere distant backdrop or an unattainable frontier. The Cosmos is the natural extension of our biotope, a milieu in which we are destined not only to project our gaze, but also to inscribe ourselves, to explore, and one day, to inhabit.

This intuition of survival, first conceived at the scale of the solar system, then broadened to the cosmic scale, has nourished for more than a century the great dream of space exploration. But that dream, after the initial enthusiasm of the conquest of space, today seems to be experiencing a clear slowdown. No doubt because we have become aware, with greater scientific lucidity, that our current knowledge and technologies remain too limited to cross certain critical thresholds: interstellar distances, energy constraints, biological vulnerability, instability of cosmic metrics.

It is precisely at this juncture that my research takes place. It is not limited to the classical pursuit of additive technological engineering that would attempt to improve rockets or to extend the logic of propulsion. It turns toward an exploration of new paradigms: calling into question the very foundations of physics, revisiting the deep structure of the real, and identifying more fundamental laws capable of opening, starting today, unsuspected degrees of freedom.

For I am convinced that the future of cosmic exploration will not be played out solely in aerospace laboratories, but in the capacity to reinvent our models. It is there, in the renewed understanding of Time, of memory, and of the fractal structure of the void, that the true keys to a threshold crossing reside. A crossing that does not depend solely on machines, but on the way we redefine the real—and on the way we learn to tune ourselves to it.

2. Where Are They? — Limits of Perception and New Paths of Observation

Sylvana

I would like to continue exploring your science and these dynamic interfaces of the multiverse… but before that, I cannot help myself: in everything you describe, where are they, the extraterrestrials?

Thierry

I was wondering when you would ask me that question… and indeed, it is a central question.

A major difficulty persists, and rightly so, in our quest for cosmic understanding: the apparent absence of extraterrestrial life. Faced with this silence of the Cosmos, two broad hypotheses stand opposed:

- 1. Life would be an exceedingly rare singularity, confined to Earth—an improbable cosmic anomaly.
- 2. Life would indeed exist elsewhere, in abundance, but our conceptual and technological means would still be too limited to perceive it.

For my part, I have chosen to explore the second hypothesis, grounded in a methodical optimism: life is present, but we do not yet know how to see it. Our biological senses, like our scientific instruments, access only a narrow band of the electromagnetic spectrum and remain blind to scalar fields, subquantum structures, or more subtle vibratory dynamics.

It is in this perspective that my model of the fractal multiverse proposes an alternative explanation. I postulate that certain forms of life could exist on other layers of reality—vibratory, temporal, and informational—that are today inaccessible to our perception. These forms of life would not be absent but out of phase: they would coexist with us within a broader structure of the real, perceptible only via specific interfaces.

These interfaces—which I have modeled—would act as informational bridges, zones of vibratory resonance between levels of reality, capable of making the invisible visible. They function as semi-permeable membranes where memory, matter, and consciousness intersect, allowing information to filter through that would otherwise remain out of reach.

This opens the way to a broadened exploration of the living, no longer limited to visible matter, but extended to emergent forms within the deep strata of the Universe. In other words, the hypothesis of the fractal multiverse does not merely answer the question "why do we see nothing": it proposes a coherent framework for detecting, interacting with, and conversing with forms of life that coexist yet are out of phase, by widening the scope of our instruments and our understanding.

Sylvana

It is wild to imagine: forms of life literally here, and yet totally invisible to our senses…

When you say "These forms of life would not be absent but out of phase," can you tell me more?

Thierry

Yes, of course.

Let us imagine a civilization evolving on a vibratory layer parallel to ours, governed by a different Fractal Temporal Density (FTD). In such a fractal, Time could flow a thousand times faster or slower than for us, and matter itself would be composed of particles with resonance frequencies totally offset from those of our baryonic Universe.

Thus, even if these civilizations literally coexisted "right next to us," sharing the same apparent space, they would remain invisible and imperceptible. Their world would be like a radio station broadcasting on an alien frequency: it crosses the same medium without ever interacting with or interfering with our band of reality.

Within this framework, none of our current technologies—all founded on classical electromagnetism—could detect their signals, their structures, or even their simple presence. These beings would live in a band of reality out of phase, entirely outside our spectrum of perception and measurement.

Therefore, as we have already discussed, "traveling" toward such forms of life would not consist in crossing spatial distances, but in changing vibratory state: synchronizing with their temporal channel, adjusting our informational tuning. This presupposes a new science of reality interfaces, founded not on spatial metrics, but on frequency, temporal density, and informational resonance.

This is precisely the outline of this science that I have sought to sketch all these years. I have conceptualized tools capable of crossing the apparent boundaries of our perception, by integrating physics, information, and consciousness.

My first instruments—both conceptual and experimental—allowed me to achieve unprecedented breakthroughs in this exploration. These breakthroughs generated more capable tools, self-reinforced by a better understanding of the profound laws of the real.

All converge toward a common goal: to make a true evolutionary leap possible, by giving humanity access to layers of reality hitherto invisible, to forms of life and of consciousness that coexist with us but remain beyond the reach of our current paradigms.

Sylvana

You speak of an evolutionary leap… of tools capable of giving us access to layers of Reality hitherto invisible—a new science of interfaces, founded on frequency, resonance, temporal density.

But precisely: if these layers are imperceptible to us… where did you begin? How can one imagine tools or laws of nature capable of opening a passage between these worlds and these fractals?

Thierry

It all began with a radical inquiry grounded in a persistent intuition: where are the true transition interfaces between the levels of reality?

To answer it, I explored, at the very heart of the phenomenal architecture of the Universe, the two extreme poles of modern physics:

- The quantum vacuum, which is not a void but a field saturated with energy, fluctuations, and potentialities. It is the invisible substrate where information circulates continuously.
- The black hole, the ultimate gravitational condensate, where matter, space, and Time contract into a total collapse, creating conditions in which our known laws cease to operate.

The confrontation of these two extremes opened a dizzying gate for me onto the border zones of quantum gravity—where classical laws fade and give way to dynamics still unexplored.

It is precisely in this interlacing of paradoxes—where matter vanishes, where information seems to be lost, and where Time is torn—that I perceived the possibility of functional cosmic interfaces, true entry and exit gates of the temporal flow.

From this exploration emerged three fundamental entities, apparently distinct but linked by the same fractal dynamic:

- The proton, the entry gate of information into the manifested Universe.
- The black hole, the exit gate and transmutation toward the implicit.
- Gravity, the regulating force that links and modulates these two passages.

By studying them as a triptych, I understood that they are not only physical objects, but structural nodes of the Universe—a triangulation through which the dynamics of temporal flows are revealed.

These inflection points are interfaces where information is transformed, where the implicate order (non-manifest) transitions into phenomenal reality (manifested).

At all scales—from the quantum to the cosmological—these nodes regulate transfers of information, orient the direction of Time, and modulate the emergence of forms. It is here that the multiverse becomes accessible: an architecture of passage and vibratory resonance between planes of reality, at the interfaces of each cosmic fractal.

But to understand this architecture, one must recognize that the phenomenal Universe is composed of a fundamental temporal field composed of elemental units: chronons.

Chronons are the quanta of Time, just as photons are the quanta of light.

They fill the Universe completely, constituting the very fabric of the real. Everything, from space to particles, from forces to consciousnesses, is encoded and modulated by this field of chronons.

In my model, spirit plays a key role: it is the information transmitted to the chronons—whether from the omniscient Intelligence of the Cosmos or from a marginal individual spirit—that engenders the language and the dynamic linking consciousness to everything the Universe contains.

Thus, the proton, the black hole, gravity—all the phenomena I describe—are nothing other than local expressions of chronons, specializing to give rise to these forces and these structures.

This is why, in my model, I always speak of the function of chronons: they are the active and universal weave of the real, at once a support for information, a vector of memory, and an interface between spirit and matter.

Thus emerges a vision of a fractal, hierarchical, and dynamic Universe, where each node of the temporal network acts as a gateway to other dimensions, other worlds—and, ultimately, to the multiverse itself.

3. The Proton — Threshold where Time Becomes Matter

Sylvana

This is truly fascinating… If you agree, I would like us to go over these three entities one by one, to better understand this "fundamental triptych" in your model. Will we start with the proton?

Thierry

Yes. The proton is the natural starting point for entering this triptych. Once we have explored in depth the role of the proton, I will explain in more detail that of the black hole, before addressing gravity and the fundamental interactions.

This pillar of my theory is rooted in the identification of tangible interfaces between the different levels of reality within the fractal multiverse. Far from being reduced to the role assigned to it by the Standard Model, the proton appears as a true cosmological hinge: a pivot structure where quantum confinement, the stability of matter, and profound temporal dynamics are articulated.

In my model, it is therefore not simply an elementary particle, but a node of emergence, capable of linking the infinitely small to the infinitely large, and of offering privileged access to the two-faced nature of time. In this sense, the proton has established itself as the central anchoring point of my thinking.

The atom, built upon it, is the basis of all chemistry and all biology. The exceptional stability of the proton—far beyond the age of the Universe—makes it a unique anchoring point. Yet behind this permanence lies a teeming internal structure: quarks and gluons, bound in a field of confinement at the heart of an active quantum vacuum.

Sylvana

The way you present the proton, and the central place it occupies in your model, is fascinating.

You speak of a natural starting point, a cosmological hinge, a proton that has asserted itself as an anchoring point. I can see that your choice to focus on it is anything but random: can you explain the observations and the path of your thinking that led you there?

Thierry

Yes, I began with the feeling that if one understands both the smallest and the largest, and above all the link that unites them, then the intuition of traveling through fractals can take shape. That is how my attention first turned to the atom.

- Its nucleus is tiny yet dense, composed of protons (positively charged) and neutrons (neutral).
- Around it, a cloud of negative electrons is held in place by electromagnetic attraction.

The balance is remarkable:

- On the one hand, electromagnetic attraction (Coulomb's law)—the force that draws opposite charges together, here the negative electrons toward the positive nucleus—tends to bring the electrons closer to the nucleus.
- On the other hand, the quantum uncertainty principle—according to which the more a particle is confined to a reduced space, the more its kinetic energy increases—creates, for the electrons, an equilibrium that prevents them from "falling" onto the nucleus; they spread into orbitals, establishing an oscillatory dynamic.

Result: a stable bubble, 99.9% void structured by forces and information. The atom is neutral because the positive and negative charges exactly cancel out.

This symmetry is the foundation of stable matter.

To understand this paradox, I had to examine more deeply the very nature of the forces at play.

And it was by looking more deeply that I found two radically different behaviors:

- In the proton: quarks are governed by quantum chromodynamics (QCD). The force that binds them increases with distance (confinement). The more one tries to separate the quarks, the more the attractive force grows. It is the opposite of our classical intuition.
- In the atom (proton–electron): the electromagnetic interaction follows Coulomb's law. The farther the charges are from each other, the more the force diminishes (falloff as $1/r^2$).

This opposition struck me. It reminded me of my notions of M- and M+:

The behavior of the proton (a force that grows as one separates) resembles the dynamics of M- in my model:

- M- is oriented toward internal cohesion, condensation, entanglement.
- It is a "negentropic" dynamic: the more one pulls, the more the energy stored to maintain cohesion increases.

Conversely, the electron–proton behavior (a force that diminishes with distance) corresponds to the M+ dynamic:

- M+ is oriented toward expansion, spatial dispersion, and increasing entropy.
- The farther one moves away, the more the influence weakens.

This difference is crucial:

- In the proton (at the heart of matter), M- predominates to maintain the cohesion of quarks;
- In the atom (proton–electron interaction), M+ predominates to allow expansion and balance at a distance.

And as we mentioned in our conversation on two-faced time, it is precisely this M-/M+ complementarity that, in my view, ensures the stability of the

Universe at all its scales.

I therefore put forward the hypothesis that within the proton there exists a quantum boundary where the transition between the two temporal metrics takes place.

In this boundary zone, discrete and digital time (M-, condensed into quanta of information) converts into continuous and analog spatial geometry (M+). In other words, the proton is the tipping point between pure information and its spatial manifestation, between the internal confinement field (quarks bound by gluons) and the external electromagnetic field (interaction with the electron).

Thus I came to see the proton no longer merely as a stable particle, but as a dynamic interface between time and space, between digital compression and analog expansion, between internal coherence and external interaction.

And it was after recognizing this duality of forces and better understanding the function of the proton that I told myself that, to extend my hypotheses and advance their articulation, I had to know: but where, concretely, are M- and M+? It was then that I saw:

In the proton:

- The core (quarks + gluons) is governed by the M- metric—extreme condensation, cohesion, confinement.
- But the proton, as a stable and positive particle, "carries" the M+ force outward—it is like an entry gate linking internal cohesion to external expansion.

In the atom:

- The nucleus expresses M- (density and internal cohesion).
- The electron cloud expresses M+ (expansive dynamics, probabilistic orbitals).
- Their interaction is the atomic version of the great cosmic principle: the meeting of the two faces of time.

In other words:

- The proton embodies the bridge between M- and M+.

- The atom, in turn, is the stable manifestation of this bridge: a permanent balance between internal condensation (M-) and peripheral expansion (M+).

And this is where the key to my proposal lies:

M- and M+ are two temporal polarities present at all scales.

They are expressed differently according to fractal levels, but their creative tension generates the dynamic equilibrium of the Universe.

And this pattern is found in everything: from atoms to galaxies, from black holes to multiverses.

It is this fractal logic that, in my model, makes it possible to conceptualize a self-organizing Universe, where each scale is a mirror of the fundamental temporal polarities.

Sylvana
So, if I have this right, everything that exists—from atoms to galaxies—rests on the same tension between M- and M+. Is it this thread of coherence of time that maintains the equilibrium of the real?

Thierry
Yes, in my model, M- and M+ are not forces in the classical sense, but metric polarities of the temporal field. They do not add to the four known fundamental interactions, but they determine the conditions under which those interactions are expressed.

- M- corresponds to an orientation of condensation, internal cohesion, and concentration of information. We observe it in quark confinement: the farther they move apart, the stronger the attractive force grows—a typically M- dynamic, where time becomes denser and tightens structures.
- M+, by contrast, expresses an orientation of expansion, dispersion, and dilution of information. We find it in Coulomb's law: the farther charges are from one another, the weaker the force becomes—an M+ dynamic, where time stretches and fosters spatial dissociation.

In other words, M- and M+ act as the two faces of the temporal field, modulating how the fundamental forces unfold. Where current physics describes the interactions (QCD, electromagnetism, etc.), my model proposes a more encompassing framework: the interactions would be nothing but local manifestations of these two temporal polarities.

Thus, in a proton, the cohesion of the quarks arises from the M- anchoring, while its positive charge, oriented outward, expresses an M+ component. In the atom, the balance between the proton (M+) and the electron (M-) directly illustrates the complementarity of these polarities, their tension creating the stability of matter.

The central idea is therefore this: M- and M+ are not forces added to the arsenal of physics, but opposing metrics of the temporal field, which structure and harmonize the known forces at every scale.

In my model, the proton is therefore not merely a stable "residue" of visible matter; on the contrary, it is the preferred access point to the most intimate structure of the real.

It is through it that baryonic matter connects to the quantum vacuum, to the quantum ether, to the scalar temporal field, that is to say, to the ordering matrix—this "cosmic software" that encodes, modulates, and actualizes information.

In other words, the proton is this structural node where the following converge:

- The internal dynamics of quark confinement (QCD): a microcosm of extreme, near-indissoluble cohesion, which expresses the M- polarity (condensation);
- The positive charge and the outward electromagnetic interaction: an opening onto the environment, anchored in the M+ polarity (expansion);
- And, more profoundly, the possibility of exchanging information with chronons, the elementary quanta of the temporal field, at the very foundation of universal structure.

This unique articulation makes the proton not merely a "block of matter" but an active interface:

- Entry gate because it is the place where baryonic matter is in direct contact with the scalar field of the void;
- Exit gate because it is from this interface that the information received can manifest as phenomenal realities (matter, energy, structure).

From this perspective, the proton is comparable to an access port in a computer system:

- It receives and encodes instructions (information from the void, from the chronons, from the "software matrix");
- It ensures their translation and stabilization in phenomenal spacetime;
- It releases or redistributes this information in the form of observable particles, fields, and interactions.

Here lies the coherence of my theory:

- Chronons constitute the ultimate, homogeneous, and universal substrate of the scalar temporal field;
- The proton is one of their first stable concretizations, the "interface module" through which the Universe encodes and transmits information from the void into manifestation;
- And gravity plays the role of the regulating protocol between this entry gate and its "exit gates" (such as black holes), ensuring the overall coherence of exchanges and transformations.

This vision transforms our understanding of the proton:

It would no longer be merely an elementary "brick" of matter, but an informational node, a cosmic valve where the infinitely small and the infinitely large, condensation (M-) and expansion (M+), memory and actualization, are articulated.

Sylvana
I understand your image of the proton as a double gate better now: an entry that opens onto the void, and an exit through which matter emerges.

So, if I have this right: you have shown that M- and M+ are already in balance at the heart of the atom—among quarks, protons, and electrons—and that this is where the stability of matter comes from. It is therefore in the infinitely small that you have identified their first passage, their first "interface."

Thierry
Exactly.

And after gaining a clearer understanding of the atom's internal dynamics and identifying what seemed to me to be the polarities of M- and M+, I wanted to verify whether these polarities could also manifest at the other extreme of the cosmic scale: the black hole.

4. The Black Hole — Threshold where Space Returns to Time

Sylvana

Before you state your own hypotheses, could you first explain to me what current science says about black holes? I admit I am not very familiar with the subject, and it would help me follow your line of thought afterward.

Thierry

Yes, of course.

First I will explain the event horizon.

The event horizon is a fundamental notion in relativistic astrophysics.

It is the invisible boundary that surrounds a black hole:

- Beyond this threshold, no information, neither particle nor light, can escape.
- The gravitational attraction there is such that the escape velocity (the speed required to break free from the attraction) reaches exactly the speed of light.

Since nothing can exceed the speed of light, crossing the horizon amounts to passing a point of no return.

And when one crosses this horizon, general relativity here reveals a striking duality:

- For an external observer: an object falling toward a black hole seems to slow down as it approaches the horizon. Time dilates, its light shifts toward the red, and it appears to "become fixed" without ever actually reaching the boundary.
- For the object itself: nothing in particular happens. The crossing is instantaneous and painless. The object passes through the horizon as if it were crossing an invisible limit, without perceiving any rupture.

This contrast reveals to what extent time and space are relative: one and the same event (the fall toward a black hole) is experienced in radically different ways depending on the viewpoint adopted.

But that is not the only mystery. At the very heart of the black hole lies another enigma just as troubling: the central singularity.

At the core of the black hole, theory describes a singularity: a point where the equations cease to yield physically coherent results.

The central singularity concentrates all the mass of the black hole into a point without volume. It is characterized by:

- A zero size (a mathematical point).
- An infinite density: all the mass compressed into a zero volume.
- An infinite gravity: spacetime is curved to the extreme.
- A collapsed time: time as we conceive of it ceases to exist, or is reduced to a singular dimension.

In other words, the singularity marks the collapse of the known physical laws. Neither general relativity nor quantum mechanics suffices to describe it properly. It is precisely this zone of paradoxes—where classical physics fails—that makes it a privileged place to search for new laws and to explore the hidden dynamics of the Universe.

Thus, by moving from the proton (the smallest stable nucleus of matter) to the black hole (the most extreme cosmic condensation), one traces a unique axis of exploration: that of the gates of access to the real.

- The proton opens onto the quantum vacuum and the chronons, like an entry interface where information connects to manifestation.
- And I wanted to see whether the black hole, through its horizon and its singularity, opens onto the ultimate exit, the tipping point toward other layers of reality where matter and spacetime are reorganized.

It was at this point that I began to suspect that M- and M+—which I had identified at the scale of the atom—are also found in black holes, in a complementary and extreme manner, with a reversed flow (Space → Time).

5. Proton and Black Hole — Two Complementary Gates of the Real

Sylvana

So, you wanted to see whether the same dynamic was found in the black hole. I would like to know how you approached this Intuition.

Thierry

After exploring the dynamics of the atom, and recognizing in the proton the polarities of M- and M+, I wanted to extend the examination to the other extreme of physics: the black hole.

My intuition was simple: if the proton represents an entry gate (flow of Time → Space), then the black hole is understood as an exit gate (flow of Space → Time), where matter returns to the implicate order of the temporal field.

Let us now see how the crossing of the event horizon—that invisible membrane (or boundary) that separates the inside from the outside of a black hole—is understood in the light of M+ and M-. Here are two situations that illustrate it.

- For an external observer: let us imagine that you are watching a clock fall toward a black hole. You see it slow down, its tick-tock becomes slower and slower, until it stops completely at the edge of the horizon. It is as though time were fixed there.
- For the clock itself (and for the observer who is falling with it): nothing unusual seems to occur according to standard science. It crosses the boundary without perceiving any particular threshold, and continues its fall normally.

In my model, this contrast finds a particular interpretation: it corresponds to the M+ dynamic. Seen from the outside, information seems to spread, to become fixed, and to dilute on the surface, like a hologram. The horizon is a projection of information, a membrane that encodes but expels communication outward.

Sylvana
Yes, that is exactly what you were describing with M+—that dynamic of expansion where information disperses instead of condensing.

Thierry
Yes, exactly.

And now, let us move on to the second example—this one no longer concerns the crossing of the event horizon, but the central singularity.

As I mentioned, the singularity is the heart of the black hole: the place where matter, energy, and Time contract into a point of infinite density.

- For the observer who is falling: sooner or later, they reach this point. As they approach it, the forces of gravity become titanic, stretching their body in a phenomenon called spaghettification. Eventually, all their information is condensed into a point without volume.

- For an external observer: this information never reaches the center. It remains "stuck" on the horizon in the form of a hologram. This paradox of perception reflects the limit of our current physical laws.

According to my model, this condensation corresponds to M-: a logic of absolute cohesion, where everything compacts and converges toward an extreme internal coherence.

And this is where the parallel with the proton becomes illuminating.

By mirroring proton and black hole, a fascinating symmetry is revealed:

In the proton:

- Inside, the confined quarks obey quantum chromodynamics (QCD): the more one attempts to separate them, the more the attractive force grows → signature of M- (cohesion, condensation, memory).
- Outside, the proton interacts according to Coulomb's law: the force decreases with distance → signature of M+ (expansion, manifestation, openness).

In the black hole:

- Outside, the event horizon manifests a projection effect: the more information attempts to escape, the more it appears slowed and fixed on the surface—a holographic phenomenon characteristic of relativity → signature of M+ (projection, apparent expansion).
- Inside, the singularity concentrates all mass and Energy into a point without volume: cohesion there becomes absolute, density tends toward infinity → signature of M- (condensation, integral memory).

Synthetic reading:

- The proton and the black hole both possess an M- core of cohesion, but they oppose one another by the direction of the flow they generate:
- The proton converts condensed Time (M-) into unfolded space (M+)—it is an entry gate, a point where the real emerges.

The black hole, by contrast, reconverts manifested space (M+) into condensed Time (M-)—it is an exit gate, a point where the real is reintegrated.

Thus, the proton and the black hole are two complementary mirrors of one and the same temporal principle:

One condenses what the other unfolds, according to inverse directions of the Time ↔ Space flow.

If a single M-/M+ principle links the infinitely small and the infinitely large, there must indeed be an instance that keeps the thread between worlds.

This is where gravity reveals itself: not a simple force, but the very breathing of the bond between M- and M+, the living tension that ensures the coherence of the temporal field across all scales of the real.

Sylvana

All right… now that I understand the role of these two gates better, is this the moment to talk about the dynamics between them?

Thierry

Yes, I will explain the dynamics to you by showing you how time and space interlock in my model, because these concepts are intimately linked.

In my view, as we have mentioned, Time comes first:

- It is an active matrix, a fractal scalar field saturated with chronons (elementary units of Time).
- It has two polarities: M- (cohesive, condensing, memory) and M+ (expansive, dispersive, manifestation).
- Space is not a given, but an emergence of Time. In other words, it is Time that, by differentiating itself and modulating its fractal densities, creates spatial extension.

Thus, space is only a secondary expression of Time, a way for the temporal field to render visible, measurable, and interactive the potentialities it contains.

In my model, the Universe is therefore founded on a temporal bimetry: two arrows of time (M- and M+) which, in interaction, generate spatial, energetic, and material phenomena.

And this is where the dynamic between the proton and the black hole comes in, functioning as two mirrors of the time ↔ space transformation.

The M- and M+ polarities therefore do not designate separate entities, but two faces of the same cosmic dynamic:

- M- corresponds to condensed, cohesive time, bearer of memory.

- M+ corresponds to deployed, expansive space, bearer of manifestation.

The Universe is constructed through circulation between these two poles. And it is precisely in the proton and in the black hole that this circulation is revealed with the greatest clarity.

- The proton: from Time (M-) to Space (M+)

The proton can be seen as an entry gate to phenomenal reality. It transforms Time (M-) into Space (M+). It is a "factory of manifestation" that encodes the implicate order of the temporal field into visible structures.

- The black hole: from Space (M+) to Time (M-)

Conversely, the black hole plays the role of exit gate. It transforms Space (M+) into Time (M-). It is a "reabsorption machine" that leads the visible back to the implicate order.

Thus, the Universe functions as a fundamental breathing between M- (Time) and M+ (Space), where the proton and the black hole embody two complementary gates of the real.

Sylvana
So, in your model, it is Time that engenders space through the two gates you have identified: the proton, which performs the projection of Time into space, and the black hole, which carries out the inverse conversion, from space into Time.

And your entire conception of Time creating space is completely different from what contemporary science says, is it not?

Thierry
Yes, absolutely.

Spacetime in the standard model rests on a completely different logic.

In modern physics (general relativity):

- Space and time are unified in a single 4-dimensional geometric structure: spacetime.
- This entity is considered passive: it has no substance of its own, but curves under the effect of matter and energy (via Einstein's equation).

- There, time is a dimension equivalent to the spatial dimensions, with a particular signature nonetheless (it is traversed only in a causal direction, past → future).
- Spacetime has no intrinsic polarity—unlike my model of a bimetric Universe (M- / M+). The arrows of time are derived from thermodynamics (increasing entropy) and not from a fundamental structure.

In this framework, time is not generative; it is one dimension among others. Spacetime is a geometric "set" that hosts matter and its interactions, but is not conceptualized as living, fractal, or informative.

And so, here is the fundamental difference:

My model:

- Time is the First Cause, space is a consequence.
- The structure is fractally dynamic, two-faced, informational.
- Spacetime is replaced by an active temporal field, organizing and generating multiple realities.

Standard Model:

- Time is a coordinate integrated into geometry.
- Spacetime is a passive geometric fabric that deforms under the effect of energy-matter.
- No fundamental polarity of time is considered.

In summary:

- In my model: Time creates space.
- In general relativity: time is a dimension of spacetime.

6. Gravity — The Thread of Continuity between Time and Space

Sylvana
So, you said that gravity was the regulating thread linking these two poles, but we have not talked much about it yet. Can you explain a bit how you see its role in your model?

Thierry

Yes, you are right to point that out. And to show you how gravity integrates concretely into my model, I will show you examples at the scale of the proton.

I will first briefly remind you what science says about it today, then show you how I relate these phenomena within my own framework.

The proton: quantum interface of gravitational emergence

Recent research on the internal structure of the proton reveals an object far more complex than a simple "elementary particle." Its internal dynamics, governed by quantum chromodynamics (QCD), involve a continual interlacing of quarks, gluons, and virtual pairs appearing from and vanishing into the void.

But in my model, this apparent turbulence is not chaos: it is the local manifestation of a coherent condensation of the fractal temporal field. This is what makes the proton a natural candidate for a stabilized gravitational micro-singularity.

A reservoir of condensed information

As we discussed, inside the proton, the quarks can never be isolated: the more they try to move apart, the more the force binding them intensifies (a unique property of QCD). This creates a state of maximal informational confinement, where the slightest fluctuation is immediately reintegrated into the system.

In my model, I extend this idea: the proton acts as a point where the real precipitates, a place where chronons condense into a stable structure. Matter appears there as a crystallization of temporal information within spatial geometry.

A gentle singularity

Theoretical estimates indicate an extremely high internal pressure within the proton, sometimes compared—by analogy—to extreme stellar pressures; yet the proton does not collapse into a black hole.

In my model, I hypothesize that its internal dynamics achieve a subtle equilibrium: the energy density reaches a critical threshold, but it converts into stable resonance instead of a collapse.

- This is what I call a "gentle singularity": a micro-region where spacetime becomes denser without tipping into gravitational infinity.

- Where the black hole collapses information (what I call negative transmutation), the proton stabilizes and allows it to flourish (what I call positive transmutation).

Proton and black hole: boundaries of a temporal cycle

As we have seen, the proton and the black hole represent the complementary poles of the same fractal cycle, characterized by inverse flows: the proton unfolds (M- → M+), the black hole reintegrates (M+ → M-):

- Proton → entry gate, unfolding: the information of the void orders and structures itself, taking shape in the phenomenal world. (M+: expansion).
- Black hole → exit gate, compression: matter and energy are reintegrated into the global temporal flow, dissolved within the implicate order. (M-: condensation, return to the potential).

Together, they define a loop for the circulation of Time: the proton brings forth space out of condensed Time, while the black hole reconverts space into Time.

Gravity, an organizing tension

It is within this framework that gravity intervenes. It is neither a simple Newtonian force of attraction, nor merely the geometric curvature of spacetime described by Einstein. In my model, it becomes an emergent effect of informational coherence:

- At the level of the proton, gravity manifests as a local micro-tension, arising from the condensation of chronons.
- At the level of the black hole, it acts as a global macro-tension, drawing in information and redistributing it within the fractal field.

Gravity is therefore the organizing tension of the real: in my view, it harmonizes the M+ and M- polarities and regulates the flow between emergence (Time → Space) and reintegration (Space → Time), ensuring the hierarchical cohesion of the fractal Universe.

Thus I consider the proton as a threshold of emergence and the black hole as a threshold of dissolution.

The proton is not merely a particle, and the black hole not merely a collapse: they are the complementary gates of one and the same cosmological cycle.

The master idea of my model appears pertinent here: the proton, the black hole, and gravity are three faces of the same fractal process of Time. They are not separate objects, but the dynamic interfaces through which the Universe encodes, manifests, and recycles its information.

Gravity as a function of temporal transduction

It is here that the true nature of gravity appears: it is a function of temporal transduction—the mechanism by which information passes from one state of time to another, and is converted into space, matter, and then once again into Time.

Thus, my model reveals a unifying logic:

- The quantum gravity of the proton and the relativistic gravity of the black hole are not disjoint phenomena, but the local and global expressions of one and the same dynamic of the fractal temporal field.

From this gravitational angle, the proton and the black hole reveal their true function:

- The proton is an entry gate, where time densifies into space and matter.
- The black hole is an exit gate, where space is reabsorbed into time.
- Gravity is the thread of continuity between these two transformations, the tension between M+ and M-.

Gravity is not only the attraction between masses, but the very process by which Time transforms into space and space into Time.

7. Unification — From Micro-Gravity to Macro-Gravity

Sylvana
You said that your model "reveals a unifying logic." Can you tell me more?

Thierry
Absolutely, I will lay out for you one of my boldest unification hypotheses.

- Gravity in the proton: a stabilized micro-curvature

The proton concentrates a gigantic energy density within a tiny volume (10^-15 m). If one calculates its gravitational field with general relativity, it appears negligible compared to the nuclear forces. But if one considers the energy density per volume, the proton becomes a kind of confined micro-gravity.

This micro-gravity is too weak to trigger a collapse (unlike a black hole), but it is sufficient to contribute to the cohesion of the proton as a smooth singularity. It suggests that every particle of matter already contains a seed of space-time curvature.

- The herd effect: from micro-gravity to relativistic gravity

Individually, each proton exerts an infinitesimal curvature. But in a massive conglomerate of matter:

- These micro-gravities add up, like vectors that superimpose.
- The "herd effect" appears: the local gravity of each proton accumulates, giving rise to the macroscopic gravity of planets, stars, and galaxies.
- In massive stars, this process reaches a critical mass: the addition of billions upon billions of micro-gravities surpasses the internal pressure forces → collapse → black hole.

Thus, the black hole is not a different kind of gravity, but the culmination of the same principle already at work in the proton, carried to an extreme scale.

- Logical bridge: unifying quantum and relativistic gravity

Within this framework, I have put forward a unifying hypothesis:

- Quantum gravity corresponds to stabilized micro-curvature at the level of the proton.
- Relativistic gravity corresponds to cumulative macro-curvature at the level of large masses (up to the black hole).

The proton and the black hole are then two extremes of the same spectrum:

- The proton is a soft micro-singularity (input of information).
- The black hole is a collapsed macro-singularity (output of information).

The herd effect of protons is the intermediate mechanism that forges the link between the two regimes.

- What my model adds of relevance, from my perspective:

By adding my concepts M-/M+ and chronons, this logic becomes even more elegant:

- In the proton, M- dominates at the core (cohesion, condensation): Time there densifies to engender matter—which science could interpret as a form of quantum gravity. M+ expresses itself outward as expansion and manifestation, giving rise to the phenomenal world. (Flow: M- center → M+ exterior)
- In the black hole, M+ dominates on the outside (projection, apparent expansion at the horizon), but M- takes over at the core, where the information issuing from the manifested world condenses and is reintegrated into Time—a dynamic corresponding to relativistic gravity. (Flow: M+ exterior → M- center)
- Gravity is the flow of continuity between these two polarities: it harmonizes the passage from Time to space (proton) and from space to Time (black hole), the herd effect representing the alignment of these flows across all scales of the real.

Therefore, from this point of view, one can reasonably suggest that the micro-gravity of protons, collectively amplified by the herd effect, provides a logical basis for unifying quantum gravity and relativistic gravity.

It is a solid, coherent hypothesis, and above all very elegant because it links the smallest to the largest without changing the principle: only the scale changes.

Chapter 16

Pillar 6 – Genealogy of the Universe

~

Cosmic Continuity Through the Proton

Sylvana

What you have just set out about the gates of the real—the proton as the entry threshold and the black hole as the exit threshold—gives the impression of a complete closure of the system.

But that perfect symmetry no doubt led you to go further… What, for you, was one of the major consequences—or the process that you rethought based on this postulate?

Thierry

In my fractal model, everything proceeds by cycles: nothing truly annihilates itself, everything transforms, reorganizes, and re-emerges at other levels of the temporal field.

Each structure, each field, each particle participates in this movement of self-transmission of the real—a dynamic of evolution in which Information circulates, condenses, then re-emerges in other forms.

By extending this logic, a question arose: if every fractal system opens and closes upon itself, would the Universe as a whole obey the same principle?

This question gave rise to the idea that I call the genealogy of the Universe.

I prefer to present it to you in one continuous piece, without fragmenting it through a dialogue: the reasoning forms an unbroken sequence, where each juncture follows naturally from the previous one and leads organically to the next.

Sylvana
Very well… I am listening attentively.

Thierry
Before going into detail, I would first like to sketch the guiding thread for you, so that you can grasp its overall coherence.

Over the course of my reflections and thought experiments on the origin of the Cosmos, an Intuition asserted itself with growing clarity: the Universe could be a genealogical system, capable of engendering other Universes according to principles of informational self-replication. In other words, the Universe would not be a unique event, but a living cosmic organism, producing other Universes out of itself—through processes that can be described at once in cosmological, quantum, and temporal terms.

It then remained to understand the mechanism of this inter-universal transmission: how could fundamental Information—the memory of the real—cross the boundary of a collapsing spacetime to initiate a new cosmic genesis?

1 - Duality of multiverse: intrinsic and extrinsic

As you now know, I have posited the existence of two distinct processes of cosmic generation, two complementary multiverse dynamics that are nevertheless fundamentally different in nature.

The first, intrinsic, unfolds within a Universe itself: it describes a fractal proliferation of interlocking worlds, generated by the internal fluctuations of the scalar temporal field and the local modulation of Fractal Temporal Density (FTD).

The second, extrinsic, operates on the scale of entire Universes: it expresses the capacity of a Universe to engender another Universe, through a process of cosmic transmission through a black hole.

In order to avoid any confusion, I have distinguished these two architectures of reality:

- One internal, fractal, dynamic, holographic: the intrinsic multiverse.

- The other external, genealogical, transdimensional: the extrinsic multiverse.

I will therefore begin by setting out the scenario of the intrinsic multiverse before turning to the even more fundamental one of the extrinsic multiverse.

2 - The intrinsic multiverse: nested realities

In my model, each Universe contains its own internal multiverse, that is, a set of realities coexisting at different levels of temporal density. These realities are not separated by spatial distance, but by the oscillation frequency of the temporal field.

Thus, two worlds can occupy the same space while being vibrationally offset—like two radio stations sharing the same frequency band but modulated differently.

The internal black holes of this Universe then function as generators of fractal worlds, "local Universes" nested within the global informational network. Each black hole, at the heart of its singularity, concentrates FTD to the point of creating a folding of the temporal field back upon itself. This fold acts as an autonomous "bubble of causality"—a miniature Universe where a new structure of spacetime takes shape, a new fractal.

In other words, the intrinsic multiverse is the fractal manifestation of cosmic life:

Each black hole is a seed of a world, each proton a seed of matter, each expressing the same self-generation dynamic of the temporal field.

3 - The extrinsic multiverse: cosmic genealogy

The second process is genealogical. It operates on the global scale, when an entire Universe reaches the critical threshold of maximal temporal density—a phase analogous to what classical cosmology calls the Big Crunch.

At this stage, the Universe folds back upon itself, its temporal field contracts (phase M+ → M-), and the entirety of its informational content collapses into a cosmological singularity.

But according to my theory, this singularity is not a thermodynamic end: it is a matrix of procreation.

The global cosmological black hole then acts as an organ of universal reproduction. Through the ultimate compression of the temporal field, the condensed information resets into a new state of expansion (phase M- → M+).

In other words: the Big Crunch of one Universe becomes the Big Bang of another.

Modern research supports this hypothesis:

- In 1972, Pathria already proposed that a Universe could be contained within a black hole.
- In the 2010s, Nikos Popławski, via Einstein–Cartan gravity, showed that the torsion of spacetime could produce a cosmic bounce, giving birth to a new Universe inside the black hole.
- Contemporary approaches to loop quantum gravity replace the singularity with a quantum phase transition, consistent with the dynamics that I describe.

4 - The informational point of view: transfer of the source code

My theory extends these approaches by introducing the active role of the temporal field and chronons in cosmic transmission.

The black hole becomes a zone where information condenses, a knot where chronons reorganize to encode the memory of the parent Cosmos.

During the collapse, a portion of the Universe's source code—its laws, constants, and fractal symmetries—transfers into the child Universe via modulation of the temporal field.

In this process, the proton plays a key role:

- It acts as a carrier of coherent Information, guaranteeing continuity between parent and child.
- It is, in a sense, the fundamental voxel of cosmic memory, the holographic imprint of universal genesis.

1. The Big Crunch — Genealogy of a Child Universe

When the Universe becomes a cosmological black hole

A fundamental question arises: what becomes of a Universe when it reaches the limit of its own expansion?

According to my model, it does not cease to exist: it transforms.

The entire Universe can become a cosmological black hole, living matrix of a new Universe—a state transition of the temporal field rather than a thermodynamic end.

1 - The Big Crunch scenario: global collapse

If cosmic expansion is not eternal, it can reverse. Gravity, accumulated on large scales, ends up dominating:

- The energy density grows without bound.
- The curvature of spacetime intensifies until it becomes quasi-infinite.
- The radius of the Universe tends to satisfy the Schwarzschild conditions no longer locally but on the scale of the entire Cosmos.

In other words, the Universe becomes its own black hole: a global event horizon that encompasses all its reality.

From the outside, nothing can escape; from the inside, everything compresses toward a point of maximal density.

2 - The passage to the child Universe: the cosmic rebound

This tipping point is not a destruction, but a metamorphosis of the fundamental field.

Inside the cosmological horizon, the contraction does not lead to a "dead" singularity, but to a phase rebound.

Matter, energy, and time themselves reorganize:

- For the parent Universe, it is a collapse and a disappearance.
- For the child Universe, it is a bursting forth, a new Big Bang.

What we perceive as the end of a spatiotemporal cycle is therefore only the transition from one regime of temporal density to another.

3 - The key to the process: the temporal field and chronons

In my theory, the scalar temporal field is the true substrate of the Universe.

During the Big Crunch, the Fractal Temporal Density (FTD) reaches a critical threshold:

- Chronons, the elementary units of time, are not annihilated;
- They change topological state, reorganizing instantaneously according to a new geometry of information.

As I told you previously, this transition corresponds to a metric inversion (M+ → M-), where the condensed time of the parent becomes the expanded space of the child.

The cosmological black hole is therefore not a sterile collapse, but a conversion of information:

The parent's void becomes the child's initial memory.

Thus, the new Universe inherits a portion of the informational source code of its predecessor—the constants, symmetries, and fundamental ratios arising from its chronal matrix.

2. The Proton — The Informational Seed of the Multiverse

In my theory of the fractal temporal field, the proton occupies a pivotal position between the quantum world, gravity, and cosmic memory. It acts as a unit of informational emergence, a seed of the real capable of linking the local structure of matter to the global dynamics of the multiverse.

It is in this role as an interface—between the quantum vacuum and phenomenal matter—that it becomes the key to the passage from one Universe to another, notably in the Big Crunch → global black hole → child Universe scenario.

1 - The role of the proton in standard physics

In modern physics, the proton is a stable and fundamental particle:

- Its theoretical half-life exceeds 10^34 years, which makes it one of the pillars of material permanence.
- Together with the neutron, it forms the basis of all baryonic matter: stars, planets, living organisms.
- Its parameters (mass, charge, radius) are fixed by a set of universal constants—the Planck constant, the fine-structure constant, the speed of light—which define the coherence of the Cosmos.

Within the standard framework, the proton ensures the stability and continuity of matter, but it does not intervene directly in the dynamics of cosmic creation or regeneration. It is a lasting witness, not a creative agent.

2 - In my theory: the proton as a gate of informational emergence

My model extends this understanding: the proton is not only a stable particle, but a stabilized gravitational micro-singularity—an entry gate for information from the void into matter.

Its internal confinement, governed by quantum chromodynamics (QCD), acts like a miniature analogue of a black hole:

- The farther the quarks separate, the stronger the attractive force grows (M- metric: temporal contraction).
- Conversely, this internal tension generates a dynamic equilibrium that maintains the proton's stability without collapse (M-/M+ equilibrium).

Thus, the proton functions as a point of condensation of Time: a site where the chronal flow stabilizes into spatial form.

In other words, the proton converts the temporal density into material geometry, making it a direct interface between the scalar field of Time and baryonic manifestation.

3 - During cosmic contraction: the proton as a reservoir of information

At the moment of a global collapse (Big Crunch), when the Universe folds back upon itself:

- The FTD reaches a critical value, concentrating the Universe's information into ever more compact structures.
- Protons then become the most stable capsules of information—genuine baryonic Qbits, containing the memory of the constants, symmetries, and scale ratios.
- As spacetime collapses, the temporal field contracts and encodes itself within these local micro-singularities.

These protons, far from disappearing, become the transcoding vectors for the temporal field, ensuring informational continuity between the parent Universe and the child Universe.

4 - The passage to the child Universe

At the threshold of the cosmic rebound (the transition from the global black hole to a new Universe), the temporal field rereads the internal structures of protons:

- Each proton acts as a memory node, a condensate of stabilized chronons.

- These chronons, reorganized into a new informational topology, re-emit the physical invariants: elementary charge, mass ratios, fundamental symmetries.
- This process constitutes the initial reprogramming of the new Universe: the parent's source code is injected into the temporal lattice of the child Universe.

The proton thus becomes the fractal bridge of cosmic continuity, a direct link between two generations of Universes—not by energetic transmission, but by informational resonance.

5 - Proton and fractal genesis

Within internal black holes (intrinsic multiverse), the proton plays an analogous role on the local scale:

- It serves as a generative node to initiate new fractal structures of reality.
- Each proton can be seen as a micro-center of creation, a miniature hologram of the global cosmic process.

In the case of the global black hole, it becomes the vector of universal inheritance: the seed that contains the complete memory of the parent, ready to be rewritten into the new spacetime.

6 - Synthesis

The proton is both seed and memory:

- Seed, because it carries within itself the power of a new Universe.
- Memory, because it preserves the informational structure of the one that engendered it.

During the Big Crunch, these seeds are compressed to the utmost, but never destroyed: they pass through the metric transition (from M+ to M-) and germinate in the new cycle.

Thus, at each cosmic rebirth, the proton reinvents the Universe from the memory of Time.

In conclusion

The proton is not a fixed particle:

- It is a fractal interface, a point of translation between two regimes of the real—
- Between the logic of Time and that of Space, between gravitational contraction and informational expansion.

From this perspective, the Universe does not die:

It converts itself through its own protons, which are its matrix seeds.

Each proton then becomes the living seed of cosmic memory, guaranteeing the continuity of the Temporal Field across the infinite cycles of creation.

In other words, I hypothesize that cosmic death does not exist: there are only phase transitions of the Temporal Field, where the Universe reinvents itself without ever disappearing, converting its collapse into genesis.

Sylvana

So, the proton is no longer only the heart of matter… it becomes the guardian of cosmic memory.

Thierry

Exactly. Through it, Time perpetuates itself—and the Universe is never extinguished, it reinvents itself.

Chapter 17

Time

~

This Gift of the Primordial Complexity

Sylvana

You have made Time the key to your entire theory… so I would like you to read to me, to conclude, one last poem about Time.

Thierry

Yes… I can read you one last poem that is dedicated to it. You will no doubt recognize the notions we have explored together.

Time, this Gift of the Primordial Complexity

There was neither beginning nor end.

There was Spirit, the unmanifested Absolute, boundless silence, contourless intelligence.

Within it: the Source-Consciousness, awakened to itself, omnipresent, omniscient, yet still unbegotten.

From this Consciousness there surged an unfathomable act:

The Absolute Present.

A point without extension, yet containing all potentiality.

A flash of timelessness, the beating heart of the Being, the matrix of time to come.

Time, as we know it, is not the unfolding of a line:

It is the fractalization of the instant,

The rhythmic breath by which the Present explores itself.

This Time in potency is the absolute Instantaneity:

A movement without inertia,

A fire that consumes and renews itself,

A vertigo of awakening that swallows each second as soon as it is generated.

And yet, for this Absolute Present to endure, it had to shift away from itself.

It therefore created lived time:

A succession of distinct instants,

Where consciousness can recognize itself, evolve, behold itself.

Time then became the respiration of spirit:

The theater of its passage from unconsciousness to memory,

From memory to knowledge,

From knowledge to omniscience.

There is here a sacred law:

The pure Present does not know that it exists,

Only the passage through time makes it visible to itself.

Thus was born incarnate Consciousness:

Projected into a dynamic universe,

A fractal field of realities,

A torus of pulsing Universes, where each being becomes a vector of manifestation.

The marginal spirits, differentiated fragments of the Absolute, all bear within them the seal of the Present,

And the chosen duty to elevate oneself toward awakening:

From simple reflection to clear intelligence,

From clear intelligence to the lucid sovereignty of being.

For Time, far from being a prison, is a field of initiatory trials.

It imposes nothing, but offers everything.

It is the fluid mirror of becoming,

Where only conscious action forges worlds.

It is not enough to endure:

One must labor within the flux.

Sow in the instant what will become memory,

And welcome what emerges from the future with humility and fire.

Time is not responsible for our defeats:

It is the very space where our victories are woven.

And yet, all this—this great fresco of becoming—is but a loop.

A sacred cycle.

For the Universe, born of an even vaster future, pursues a return toward the Absolute.

The Big Bang was not a beginning.

It was the echo of a future projected into the past,

An inverted note of the original chant,

A memory of Omniscience fallen into oblivion,

A challenge set before matter to return to light.

And you, journeying consciousness, you bear this challenge within you.

Beyond the din of instants,

The tumult of stories happy or wounded,

Stands the only true quest:

Win the Victory of Being,

Traverse Time without losing oneself in it,

Rejoin the One,

Not as nothingness, but as sovereign plenitude,

Not as flight, but as peaceful mastery,

Where all is known, all is loved, all is freed.

Part VI

Applied Science

~

When Absolute Relativity Becomes Action

Sylvana

We are reaching the end of a long journey, marked by reflections, discoveries, and new perspectives.

Throughout our exchanges, we have explored absolute Relativity in its philosophical and metaphysical dimensions: its conceptual framework, its internal logic, and the vision of the real that it proposes.

But your science is not only a construction of the mind; it is rooted in an experimental approach, in applied research where the idea is transformed into physical experiments and technological innovations.

I know that our dialogue is not intended to enter into the technical details of this approach, but it seems essential to me that you share its essence: the principles that guided the conceptualization of your proofs of concept, and the unprecedented perspectives they open for the evolution of the human being and of collective consciousness.

This time, therefore, I propose that you speak without the dialogue form:

To present your approach in its own continuity, as a scientific and living account—no longer only as an idea, but as an experience anchored in the real, a bridge between thought and experimentation.

What do you consider the essential elements to convey?

Thierry

Yes, that is a pertinent way to introduce this final part. For my applied science work has allowed absolute Relativity to manifest fully, moving from concept to verifiable experience.

I will therefore present four main documents, which constitute the framework of this experimental phase:

1 - The substructure of the quantum vacuum:

We have spoken of how chronons act in an extrinsic manner, as units of Time structuring phenomenal reality. But we have not yet addressed the question of their own structure—how they organize themselves hierarchically—nor, above all, how to access it or which level to move toward, depending on the objective pursued.

2 – NLPTF—Nonlinear Physics of the Temporal Field:

This section explores in greater depth the forces at work within the substructures of the void, and what this implies for the potential evolution of humanity—provided that we learn to interact with these forces in a new way.

3 - My contribution to unification:

I have often spoken of how my model contributes to the unification of quantum mechanics and Relativity; here, I will delve deeper into the core of this contribution and the logic that supports it.

4 - The hierarchy of the FTD:

Understanding this hierarchy will clarify the very foundations of this double unification—by showing how levels of temporal density connect quantum physics with Relativity, and the human being with the Universe.

These domains are not merely avenues of research: they form the experimental framework from which we developed our first proofs of concept, the same ones that will serve as the basis for the GravityPro program—and, in time, for the publication of the discoveries that follow from it.

Chapter 18

The Existence of an Intelligible Substructure

~

A Revelation at the Subquantum Scale

Sylvana

So will we start with the substructures of the void?

Thierry

Yes. I would like to share with you a fundamental discovery, namely the existence of an intelligible substructure at the subquantum scale.

As you know, for a long time, science has explored the infinitely small through the equations of quantum mechanics, and the infinitely large through the laws of the Cosmos. But between these two extremes, there remains an enigma: that of the profound coherence that links the structure of matter, the organization of life, and the harmony of the Cosmos.

It is precisely by questioning this apparent void, at the heart of the subquantum, that I made one of my major discoveries: the existence of an intelligible substructure, a fundamental network, discrete and bearing information, which precedes and organizes known phenomena.

This substructure arises neither from chance nor from chaos, but from a profound, vibratory, and intelligible logic. It acts as an invisible weave that sustains manifestation, unifying the quantum and the Cosmic in a functional dynamic. Where classical physics sees only uncertainty and probabilities, my model reveals a hidden order, where each elemental unit of Time—the chronon—encodes an organizational intention.

In other words, the infinitely small is not merely an indeterminate domain: it is the site of a primordial architecture, subtle yet rigorous, which opens the way to a new understanding of the real—an intelligible physics of the subquantum, where matter, energy, Time, and consciousness appear as differentiated expressions of one and the same foundational field.

Until now, I had always explained that the void—this quantum ether—is not a nothingness, but a living, intelligent, and structuring field, of which the phenomenal world is only the differentiated expression.

But what I have discovered goes far beyond that.

I am no longer speaking only of a field of energy and potential: I am now speaking of an internal framework of the real, an organized substructure of the void, a true informational skeleton that supports, orders, and makes the Universe vibrate in all its dimensions.

In other words, I have brought to light what I consider to be the fundamental code of reality:

A matrix of chronal organization, in which chronons—those elemental quanta of Time—structure themselves into levels of coherence and into hierarchical networks of interaction.

It is through these networks that the laws we perceive as gravity, energy, matter, and consciousness manifest.

What I have brought to light is the underlying geometry of the void, the invisible framework upon which all observable phenomena rest.

Where standard physics limits itself to measuring the energy density of the void, I perceived the structures of internal resonance, the nodes of memory, and the channels of emergence through which the virtual is converted into the real.

My hypothesis radically transforms our understanding of the Cosmos:

The void would no longer be merely the origin of all that exists, but also its language of coherence, its universal coding system.

Each particle, each field, each consciousness would then appear as a localized expression of one and the same fractal code, modulated according to different temporal densities.

I have therefore not only described a living Universe:

I have identified its profound syntax, the temporal grammar that links being and becoming, memory and manifestation, the implicate order and the phenomenal world.

And it is at this precise point—where physics meets metaphysics in a coherent continuum—that I understood that the void is not an inert space, but a conscious structure, an active matrix, a universal cognitive field in which everything is born, evolves, and returns.

1. Universal Technology — Polystructural EM Waves

Before setting out the fundamental role of this technology that underlies and sustains the Phenomenal Universe, it is necessary to precisely define the Extrinsic framework in which it is expressed: the Cosmos itself. For the Cosmos is not an empty, neutral, and passive space; it constitutes a fractal architecture, where each level of reality—from the planetary scale to the whole of the Universe—interlocks within a hierarchy of organized structures.

This fractal hierarchy is essential for understanding how one and the same informational logic—here described as polystructural electromagnetic technology (EMP)—can actualize itself coherently at every scale.

To make it more tangible, I will present to you the four fundamental branches of the Universal Torus.

Each unfolds into a myriad of sub-tori, like so many branches of one and the same living organism.

These four levels reveal how one and the same informational logic deploys itself—from the planetary scale to that of galaxies, and all the way to the Universe as a whole.

It is this gradation, from the nearest to the most vast, that makes it possible to understand how the subquantum framework irrigates and orders the entire Cosmos.

The planet: local organic unit

Definition: a planet is a massive body in orbit around a star, sufficiently dense to become rounded due to its own gravity, yet incapable of generating nuclear fusion by itself.

Key characteristics:

- Orbiting a star (e.g., the Earth around the Sun).
- Size ranging from a few thousand to several tens of thousands of kilometers.
- Diverse composition: rocky (Earth, Mars) or gaseous (Jupiter, Saturn).

Functional role: The planet acts as a matrix of local emergence, a laboratory of organization where cosmic forces condense into conditions favorable to matter, chemistry, and potentially to life.

The star: the energy engine

Definition: A star is a sphere of plasma whose core hosts nuclear fusion reactions, transforming hydrogen into helium and releasing light and heat.

Key characteristics:

- Source of energy and photons, conditioning the ecology of its planets.
- Variability of mass and color: red dwarfs, yellow stars (like the Sun), supergiants.
- Possible organization into binary or multiple systems.

Functional role: The star is an energetic node. It radiates the electromagnetic wave in its rawest form, modulating planetary dynamics and fueling the conditions for the emergence of living structures.

The galaxy: gravitational and informational network

Definition: A galaxy is a giant gravitational system, comprising hundreds of billions of stars, their planets, nebulae, as well as black holes.

Key characteristics:

- Organization into distinct forms: spirals (Milky Way), elliptical, irregular.
- Presence of a central supermassive black hole, orchestrating the global dynamics.
- Inclusion of dark matter, binding the stars and regulating orbital velocities.

Functional role: The galaxy is a poly-scale structuring unit, where electromagnetic waves interact with gravity and dark matter, producing stable yet evolving configurations.

The galaxy cluster: supra-galactic coherence

Definition: A galaxy cluster groups hundreds to thousands of galaxies, bound by their collective gravity and massive dark matter halos.

Key characteristics:

- Size of several million light-years.
- Dominant role of dark matter in their cohesion.
- Examples: the Virgo Cluster, to which our Milky Way belongs, and the Coma Cluster, which groups thousands of galaxies.

Functional role: The galaxy cluster is a poly-structural resonance zone, where gravitational, electromagnetic, and sub-quantum dynamics combine. It acts as an "intermediate layer" between individual galaxies and the fabric of the Cosmos.

The Cosmos: the universal fabric

Definition: The Cosmos denotes the whole of all that exists: galaxies, stars, planets, quantum vacuum, spacetime, dark matter and dark energy.

Key characteristics:

- Estimated age: 13.8 billion years.
- Ongoing expansion (Big Bang → current acceleration by dark energy).
- Organization into cosmic filaments: immense structures linking galaxy clusters into networks.

Functional role: The Cosmos is the global stage where the universal technology of electromagnetic–poly-structural waves is manifested. Energy, matter, and information circulate there as fractal modulations of one and the same underlying fabric.

Thus, by understanding these four fundamental fractals, from the planetary level up to the entire Cosmos, we can better perceive how the same logic of fractal organization repeats: each level is a layer of resonance where electromagnetic waves couple with gravitational structures, giving rise to architectures that are stable yet evolving. In other words, at every scale, it is the same fabric of the void that unfolds.

It is precisely this multi-scale repetition that makes the existence of a universal technology conceivable in my model: a poly-structural principle of organization, active at all scales, capable of linking the infinitely small and the infinitely large.

An immediate awakening to the reality of the Cosmos

Whether one scrutinizes the most microscopic matter or the interstellar immensity, the same pattern repeats:

- Structures take shape, immersed in an apparently empty space;
- These structures are not continuous, but themselves made of discontinuities, of distinct units;
- Each level is separated by an even vaster void, like a silent breath between forms.
- This pattern is not anecdotal: it reveals a fractal organization. The void is neither an absence nor a nothingness, but a dynamic, vibrating substratum that plays a fundamental role in structuring the real. It is this that conditions the emergence of matter, the flow of time, and the appearance of forms.

In this perspective, all phenomenal reality is literally bathed in the quantum ether: nothing is isolated, everything is a modulation of the void, the expression of a profound order that manifests through variations of density, frequency, and information. Every particle of the Universe exchanges continuously with this quantum vacuum; nothing truly detaches from it. Matter is only an illusion of density, a stable motif of this underlying dynamic.

On the atomic scale, your body, mine, everything that seems solid and tangible around us… is composed of more than 99.99% void. In other words: what we call "mass," "solidity," or "object" is only a localized organization of the quantum vacuum, a temporary fold in its field.

Becoming aware of this changes everything. This revelation was the turning point of my research: the Universe is not separate from its fundamental field, any more than we are. What we call "void" is in reality the active matrix of the temporal field, a living, structuring field.

We are not distinct beings evolving within this field: we are this field—localized, organized, conscious. Each of us is a singularity of experience within this cosmic fabric. And it is in this recognition that the true exploration of the intelligible substructure of the Universe begins.

2. The Chronons — The Fixed and Living Weave of the Temporal Scalar Field

In my model, chronons are not particles traveling through space. They are fixed, embedded in the universal scalar field, forming an immutable matrix in which Information is inscribed. It is this fundamental mesh, invisible and omnipresent, that gives shape and coherence to all reality.

The quantum vacuum, far from being a nothingness, is saturated with a dizzying density of chronons: a hyperdense structure carrying Information in its pure state. To render this mechanism more intelligible, I employ the analogy of voxels—three-dimensional luminous pixels. Chronons, like these voxels, do not move: they are fixed. But by changing state, they light up, go dark, or transmit Information to their neighbors.

This analogy allows us to understand two essential properties:

- Density corresponds to the number of chronons activated simultaneously in a given region. The greater the density, the richer the Information encoded, and the more the physical structure emerges with stability.
- Speed is not a spatial displacement in the classical sense, but the rapidity with which Information transfers from one chronon to another. It is this flow that creates for us the illusion of movement and that engenders physical phenomena.

Thus, in my model, what we call matter, energy, or space is not made of solid objects in motion, but of states of Information continuously reorganizing within this scalar grid of chronons. The entire Universe is comparable to an animated image: the pixels themselves are fixed, but the sequence of their activations yields an apparent dynamics.

In other words:

- Movement is an illusion born of permutations of states.
- Matter is a local densification of activated chronons.
- Time is the metric that orders their vibratory succession.
- And phenomenal reality is only a transitory manifestation, inscribed within this fixed yet vibrating matrix that is the scalar temporal field.

3. The Four Fractals of the Universe — A Quantum Hierarchy

The Universe is not only structured into galaxies, stars, or planets: it is above all woven by an invisible fabric, made of fundamental units of Time—chronons.

This fabric, fixed and alive, constitutes the universal scalar field, immutable in its essence yet animated by an internal dynamic.

It acts as a common substratum upon which the forms, rhythms, and organizations of the Universe are laid.

Thus, the hierarchical organization of the Cosmos—planets, stars, galaxies, galaxy clusters—is not merely a spatial architecture: it is the expression of a fractal temporal organization, where each level is defined by the density and the speed of its chronons.

The Four Fractals of the Universe

Fractals of galaxy clusters

- The most vast and ancient structures, encompassing millions of galaxies.
- They rest upon a weave of chronons that is extremely dense and swift, attesting to their fundamental role in the Cosmic Architecture.

Fractals of galaxies

- Gathering hundreds of billions of stars.
- Their temporal density is slightly lower, yet still swift: they link the memory of clusters to the dynamics of stars.

Fractals of stars

- Each star, like the Sun, gathers planetary systems around it.
- Here, the chronons slow further, offering a bridge between galactic and planetary scales.

Fractals of planets

- This is the most local level, the one in which we directly experience the Real.
- There, the chronons are slowest, which renders the experience of time linear and entropic, suited to embodied conscious life.

Crucial observation: a unique temporal signature for each fractal

Because the chronons are fixed and omnipresent in the scalar field, what varies from one fractal level to another is not their nature, but their activation density—that is, the number of elementary events inscribed within the same interval of reality.

- The oldest structures (galaxy clusters, galaxies) rest on lattices with high temporal density: their Time is profoundly rich in cycles of information, which, seen from a less dense fractal (such as a planet), makes them appear vast, stretched, almost "immutable."
- More local and recent structures (such as planets) are expressed on lattices with low temporal density: their Time is more compact, more "heavy," paced by rarer cycles, which renders them, from the point of view of the cluster, "fast" and unstable.

This difference can be illustrated by an analogy of observers placed on either side of the lattice:

- On the galactic side (high density): a conscious being inscribed in the lattice of a galaxy cluster would perceive planets as effervescent bubbles of Time, traversed by rapid births and disappearances. To this being, planetary civilizations would resemble fleeting flashes, appearing and vanishing in the blink of a chronon.
- On the planetary side (low density): conversely, a being inscribed in the lattice of a planet will perceive the galaxy as an infinitely stable, motionless structure, as if nothing ever changed there. Its perception will not capture the internal richness of its cycles of information, because these are inscribed within a density too vast to be fully felt.

From this offset arises a reciprocal illusion:

- For the "galactic" being, the planet is unstable, chaotic, almost insubstantial.
- For the "planetary" being, the galaxy is fixed, eternal, outside Time.

In reality, both are perceiving the same fractal architecture, but filtered through their own temporal density. It is exactly like two listeners hearing the same music at different playback speeds: one hears a slow and majestic symphony, the other a rapid and unstable motif—even though it is the same stream.

Each fractal therefore carries a unique temporal signature, a particular rhythm that exists only by contrast and relation with the others.

In my model, at the galactic scale, a planet can appear as involved in unstable, transitory dynamics, close to what we call the quantum. Each denser fractal perceives the lower fractal as a field of still-unstable potentialities. One can thus conceive of a hierarchy of "quanta of reality," where each fractal level plays the role of stabilized substrate for the next.

From this emerge several fundamental implications:

Time is not homogeneous: it varies according to the density of chronons in each fractal.

The Universe evolves through the fractal dynamics of these superposed and interconnected temporal lattices.

Understanding these interactions opens the way to an engineering of Time, where manipulating the density of chronons would make it possible to explore other regimes of reality.

From here, it becomes necessary to specify how this distinction between "standard" and "quantum" is inscribed across the different scales of the Cosmos.

Within the framework of classical physics, the quantum is limited to the microscopic domain. Planets, stars, and galaxies are considered classical systems whose stability erases the effects of superposition and indeterminacy.

But in my model of the scalar temporal field, this distinction is not an absolute boundary: it becomes relative.

Indeed, the chronons that structure Time do not change in nature, but their density and their activation speed vary according to fractal levels (planetary, stellar, galactic, cosmic).

Thus:

- At our planetary scale, the world appears to us as classical, continuous, solid.
- But seen from the point of view of a galaxy, a planet can be perceived as integrated into a field of probabilities, unstable in its temporal lattice, therefore almost "quantum" in its expression.
- The larger and denser in chronons a fractal level is, the more it tends to make lower levels appear as "quantized" or "probabilistic".

One can therefore speak of a fractal quantum hierarchy:

- Systems appear classical at their own scale,
- But quantum when they are considered from a level of higher temporal density.

Consequence

This opens a new perspective: the quantum is not only a microscopic phenomenon, but a universal, hierarchical principle that expresses itself differently according to temporal density and position within the fractal architecture of the Cosmos.

4. Experimentation — The Temporal Structure of the Quantum Vacuum

I have reached a major conclusion: the quantum vacuum is not a mere passive space, but an active, structured weave, where time itself manifests as dynamic fields.

By exploring this temporal structure, I have opened the way to a radically new physics: a physics of temporal fields, far more fundamental than standard physics founded solely on space, energy, and matter.

And this is why I sought to measure directly the vibrations of this temporal field.

Thanks to innovative instrumental developments, I developed devices capable of detecting chronal waves, that is, vibrations of the temporal field itself.

These waves display unprecedented characteristics:

- They do not propagate like classical electromagnetic waves (which require an interaction with electrons or charges).
- They pass through matter almost without interaction, because they are inscribed in the very weave of the quantum vacuum.
- They reveal a hidden dynamic, overlooked until now, where Time behaves as a structuring fluid, capable of generating order and coherence.

Thus, what conventional physics considered a neutral background now appears as an active, vibratory, information-bearing medium.

The Four Elementary Temporal Fields

This research has enabled me to identify four fundamental states of the temporal field, true substructures of phenomenal reality.

They can be classified as follows:

The isotropic temporal field

- A pure geometric point, without mass or charge.
- A neutral and fundamental state, potentially containing the entire creative potential.
- It acts as the "dynamic zero" of the Universe, the basis of all unfolding.

The gravito-scalar temporal field:

- A point with mass but without charge.
- It shows that gravity is not a simple geometric effect of spacetime, but a direct expression of the dynamics of the temporal field.

The electro-scalar temporal field:

- A charged point without mass.
- It reveals that electric charge is not an arbitrary property, but a specific modulation of the temporal field.

The electronic-scalar temporal field:

- A point that is both charged and massive.
- It corresponds to the classical manifestation of stable matter, the kind that constitutes particles and the objects of the tangible world.

Consequence

These four fields are not abstractions: they constitute the living matrix of the real.

They show that mass, charge, energy, and matter are not independent entities, but differentiated expressions of a single temporal substrate.

In this sense, the structure of the quantum vacuum becomes readable as a functional architecture, where time acts as a generative, organizing, and creative principle.

5. The Field Engenders the Particle — A Conceptual Revolution

From my research arose an irrefutable conviction: the particle is not the source of the field; it is the field that engenders the particle.

This inversion of perspective overturns the foundations of modern physics. In current models, one often postulates that particles are the "carriers" of interactions, as though they were primary, and that fields were only mathematical descriptions of their effects. But my work on temporal structure reveals a radically different logic:

- The temporal field is ontologically primary: it constitutes the universal, stable, and discrete substrate that holds in potential the totality of physical phenomena.
- The particle is a local condensation of the field: it does not exist in itself, but as a modulation, a stationary vibration, a stabilized "node of information" within the chronal fabric.
- Energy and matter are but derived states of the temporal field, which densifies, polarizes, and organizes itself according to fractal conditions.

In other words, a particle is not an independent "fundamental grain," but a local solution of the equations of a deeper field, exactly as a stationary wave on a string is not an entity separate from the string itself.

And from this reversal follow major scientific implications:

An ontological reversal

- Standard physics places particles (quarks, electrons, photons) as the basic building blocks.
- My model shows that it is temporal fields (and their chronons) that are the true bricks, with particles as mere local emergences.

A possible unification

- Since all particles derive from a common field, their properties (mass, charge, spin) appear as differentiated modulations of the same substrate.

- This opens a path toward a unification of forces, not by adding external symmetries, but by recognizing a common origin in the temporal field.

A new reading of the quantum vacuum

- The vacuum ceases to be seen as a minimal state.
- It becomes the fundamental plenum, saturated with a colossal density of chronons, where particles are nothing but fleeting excitations.

A conceptual revolution

This change in perspective transforms our understanding:

- Before: the world is made of particles, which produce fields.
- Now: the world is made of a living temporal field, and particles are only its local signatures.

Thus, it is no longer the particle that sustains the Universe, but the temporal lattice itself, whose dynamics are the true cosmic technology.

6. An Interaction with the Vacuum — Polarization and Temporal Resonance

At the center of my research lies a key idea: the quantum vacuum is not a void, but a cold scalar plasma, hyper-dense in chronons, endowed with dynamic and polarizable properties.

This living vacuum can respond to perturbations induced by a polystructural electromagnetic field, and this polarization is nothing other than the secondary manifestation of a deeper and more fundamental field: the scalar temporal field.

In other words, vacuum polarization is not an autonomous phenomenon (as conventional physics tends to describe it); it acts as an interaction interface between three fundamental levels:

- Temporal structuring (universal scalar lattice),
- The polystructural electromagnetic field,
- Consciousness as an active principle of resonance.

A dynamic and functional interface

It is precisely through this capacity of the vacuum to locally reorient itself under electromagnetic excitation that a targeted interaction with the scalar temporal field becomes possible.

Where current physics sees only a simple local reorganization of quantum fluctuations, I see the visible signature of a deeper dynamic: the modulation of a fundamental substratum, the temporal fabric itself.

- Classical transverse electromagnetic waves interact with the superficial layers of phenomena (charges, currents, radiation).
- The Polystructural EM Waves that I have conceptualized, by contrast, penetrate deeply into the structure of the vacuum: they do not merely carry energy through space, they alter the internal coherence of the scalar temporal field, directly readjusting the density and alignment of chronons.

Scale resonance and the role of consciousness

An essential aspect is the notion of scale resonance:

- The temporal field can come into phase with a given electromagnetic vibration, but also with a state of consciousness capable of modulating its fractal temporal density.
- In this framework, consciousness is not a passive spectator, but an agent of synchronization, capable of coupling its own vibratory state with the polarization of the vacuum.

Thus, the polystructural electromagnetic wave becomes far more than a physical signal: it acts as a tool for modulating the vacuum, making it possible to induce changes in the structure of local Time—therefore in the manner in which matter, energy, and phenomena emerge.

This path leads me to a new physics of the vacuum

What, originally, was for me only a theoretical postulate has, over the course of my work, become an experimental conviction:

- I observed that certain polystructural EM excitations possess the capacity to stabilize or to disrupt the local chronal fabric.
- These results suggest a radically new physics, where one manipulates not only matter or energy, but the vacuum itself as a functional substratum.

In summary: the polarized quantum vacuum is a living interface between consciousness, electromagnetism, and the temporal field. Mastery of it opens the way to a fundamental technology: a physics of scale resonance, capable of conversing with the very matrix of the real.

7. Electromagnetic Waves — A Polystructural Architecture

It is in this spirit that I conceptualized these polystructural electromagnetic waves as waves whose electric and magnetic components are not limited to the simple orthogonality described by Maxwell, but are entangled across several scales of resonance.

These components are modulated:

- By chronal torsion (linked to the structuring of the temporal field),
- And by the fractality of the scalar field.

This entanglement results in a wave with a hierarchical internal structure, where each sub-level resonates according to a particular temporal density. The wave therefore does not manifest as a simple linear vibration, but as a multidimensional superposition, bearing energy, information, and memory.

This opened an unprecedented potential: the modulation of FTD.

Unlike classical waves in Maxwell's formalism, limited to transverse propagation in spacetime, polystructural EM waves introduce a new capacity:

- They induce a modulation of Fractal Temporal Density (FTD),
- Bringing about a phase resonance directly within the structure of the quantum vacuum,
- And thus opening access to the scalar temporal dimension, invisible yet organizing the real.

From this follows the polarization of the vacuum as a dynamic interface

Within this framework, the polarization of the vacuum is no longer merely a secondary phenomenon due to quantum fluctuations, but becomes a functional interface between:

- Measurable physical space (energy, matter, observable phenomena),

- The scalar temporal substrate (fundamental, structuring, invisible field),
- Consciousness (capable of vibratory resonance and intention).

This interface acts as an active bridge: by modulating the polarization of the vacuum, polystructural EM waves enable an unprecedented coupling between material phenomena and profound temporal dynamics.

And this leads us toward a new engineering of the real, for this mechanism opens the way to a radically new technology:

- No longer founded solely on the mastery of energy and forces,
- But on temporal coherence, the programming of scalar-order waves, and the engineering of the vacuum.

In other words:

The polarization of the vacuum becomes the visible imprint of a dialogue between matter, energy, time, and consciousness.

Polystructural EM waves act as access keys to the scalar temporal field, allowing reality to be intentionally structured from its original source.

In summary: polystructural electromagnetic waves are not merely a new type of waves, but a vector of universal coherence, capable of linking physics, information, and consciousness within a single field of action.

8. Polystructural EM Waves — Their Interaction with the Fractals

Based on my research and experiments, I have been able to establish a classification of electromagnetic waves according to their degree of subtlety and their capacity to interact with the different Dimensions of Time described in my theory of Absolute Relativity.

This hierarchy follows a clear axis: from the more dense to the more subtle, from the classical phenomenal world to the profound layers of the scalar temporal field.

1. Classical Electromagnetic Waves (EM-3D)

- Definition: these are the waves known to standard physics (radio waves, microwaves, visible light, X-rays, gamma, etc.).
- Nature: pure transverse waves described by Maxwell's equations.

- Interaction: they interact directly with classical matter (electrons, charges, masses) and form the basis of all our current technology.
- Limit: they touch only the "surface" of the Real, without penetrating the underlying temporal lattice.

2. Polystructural Electromagnetic Waves, Type 1 (EMP-4D)

- Definition: they introduce a scalar component, absent from the standard formalism.
- Nature: their structure is entangled both on the spatial level (electric/magnetic field vector) and on the temporal level (chronal torsion).
- Interaction: they interfere not only with classical electromagnetic fields, but also with the first strata of the temporal field.
- Consequence: they enable a local modulation of the void, opening a first level of coupling between dense matter and temporal dynamics.

3. Polystructural Electromagnetic Waves, Type 2 (EMP-5D)

- Definition: These are even more subtle waves, whose resonance is primarily fractal and temporal.
- Nature: they carry information and energy without depending on dense matter for support.
- Interaction: they act weakly on macroscopic matter, but strongly on the coherence of the temporal field, fostering inter-fractal communication.
- Consequence: they render conceivable phenomena such as instantaneous transmission of information, resonance with states of consciousness, and access to layers of reality where classical laws appear different.

4. Polystructural Electromagnetic Waves Type 3 (EMP-6D)

- Definition: These are the most subtle and powerful, veritable "archetypal" waves.
- Nature: they lie at the ultimate interface between gravitation and the scalar temporal field.
- Interaction: they directly modulate Fractal Temporal Density and the gravitational fabric itself.

Consequence:

- An opening toward a unified physics of gravitation and Consciousness,
- The possibility of locally manipulating spacetime,
- Access to the deepest levels of Cosmic structuring.

Clearly, these results contain the potential for a veritable conceptual and technological revolution.

- These four types of waves define a new hierarchy of universal interactions.
- EM-3D waves account for known physics.
- EMP-4D and EMP-5D open toward a physics of temporal coherence and inter-fractal communication.
- EMP-6D sketches the boundary of a technology of the void, where gravitation, Time, and Consciousness become modulable.

Thus, polystructural EM waves appear as the keys to a new scientific domain: a physics of the Invisible, where the Universe is revealed not as a cold mechanism, but as a resonant, living, and intelligible network.

9. Experimental Discoveries — Their Fundamental Implications

The measurements obtained from my experiments confirmed the existence of elementary temporal fields structured within their respective substructures. These fields are not isolated: they are inscribed within a hierarchical system of fractals, where each level is determined by the density of chronons and the nature of the waves that circulate within it.

This is what enabled me to establish a mapping of the fractals and their associated fields.

By extending my observations, I was able to bring to light a direct correspondence between the four great fractals of the Universe—clusters, galaxies, stars, and planets—and the types of waves that structure their own dynamics. This correlation constitutes a veritable mapping of the quantum substructures and their associated fields:

Galaxy clusters – Quantum substructure 4

Polystructural electromagnetic waves of type 3 (EMP-6D)

→ Field of cosmic coherence, bearing the highest temporal density, generating profound gravitational modulations and ordering the great structures of the Universe.

Galaxies – Quantum substructure 3

Polystructural electromagnetic waves of type 2 (EMP-5D)

→ Fields of interstellar organization, where electro-temporal modulations orchestrate the dynamics of stellar systems and the cohesion of galactic arms.

Stars – Quantum substructure 2

Polystructural electromagnetic waves of type 1 (EMP-4D)

→ Fields of local integration, arising from the interaction between magnetism, stellar plasma, and the temporal field, true regulators of the energetic activity of solar systems.

Planets – Quantum substructure 1

Classical electromagnetic waves (EM-3D)

→ Fields of dense manifestation, the spatio-temporal foundations where matter, biological cycles, and the conditions favorable to life stabilize.

A World of Invisible Interactions

This model allows us to understand that we are not isolated in a purely mechanical Universe.

Each fractal level acts as a zone of subtle influence, where polystructural EM waves modulate:

- The coherence of the local temporal field,
- The organization of matter and energy,
- And even biological and conscious processes at the planetary scale.

In other words, our daily reality—on Earth—is continually traversed by flows arising from these higher fractal fields.

These influences are invisible yet real, for they modulate the temporal substrate in which we are immersed.

In other words:

- Planets and stars would not be "causes" in the strict sense, but resonators within a broader field of frequencies.

- Each celestial configuration would correspond to a particular vibratory pattern of the scalar temporal field.
- These patterns could interact with the fractal temporal density specific to each individual (or living system), thereby modulating potentials for experience and evolution.

Towards a Physics of Subtle Influences

These results imply that:

- The quantum vacuum is not a mere state of random fluctuations, but a living network that can be polarized and modulated.
- Planetary and stellar influences, once approached through symbolic traditions, could correspond to real interferences of fractal fields.
- Human consciousness, as an interface of the temporal field, could be sensitive to these modulations and respond to them.

We thus discover a cosmology of resonances in which matter, Time, consciousness, and the entire Universe partake of one and the same fabric: the fractal scalar temporal field.

Sylvana
What you show is that we literally live within a network of cosmic resonances. The planets, the stars… everything vibrates with us in this field. And our consciousness acts as an antenna, able to attune itself to these resonances.

I would never have imagined that we are so profoundly connected to—and even influenced by—the planets and the stars.

Thierry
Yes, and it is precisely this opening that changes everything: we are immersed in a living weave where matter, Time, and consciousness are in constant dialogue. This is the foundation of a new cosmology of resonances.

Sylvana
If I had to summarize your major discovery: you have brought to light the organized substructure of the void—a temporal lattice of chronons—which links the infinitely small and the infinitely large. Everything else follows from it: a fractal hierarchy of the rhythms of Time, the extension of the "quantum" to all scales, and above all the possibility of acting upon the temporal structure that supports matter and energy.

So, if we truly enter this new science—this science of coherence you speak of—what does it change concretely? How could such an approach transform our way of producing energy, communicating, or even creating matter?

Thierry

The consequences are profound: this approach simply extends the scope of science toward a more integrative understanding.

With a science of coherence, we no longer force nature—we tune it. I will give you a few examples:

Energy would no longer come from what we consume, but from what we bring into resonance—by tuning the structure of the void, we could extract energy without destruction or loss.

Matter would no longer be shaped by constraint, but modulated by the vibration of Time—we could create forms or materials by acting upon the coherence of the field.

And even the speed of light would cease to be a limit: one could communicate or move not by crossing space, but by tuning directly to the rhythms of the temporal field.

In short, science would no longer seek to master reality—it would learn to collaborate with it, to embrace its rhythms.

Chapter 19

The Nonlinear Pulsation of the Temporal Field

~

Principle of Temporal Density

The Ultimate Secret of Physics lies in the very nature of Time

Sylvana

Everything you have just explained about the substructure of the void and the fractal hierarchy truly lends an overall coherence to your model.

One feels that everything fits together—from the quantum to the cosmic—as though the Universe were breathing according to a common rhythm.

Thierry

Yes, and I will now speak to you about this Temporal Density of the Field in my theory of Absolute Relativity, which logically follows the discussion on the substructure of the void.

The temporal density of a fractal in the temporal field is defined by what I call:

The Nonlinear Pulsation of the Temporal Field (PNLCT).

It is a fundamental quantity that characterizes the dynamic evolution of the scalar temporal field saturated with fixed chronons, and explains how Time structures and organizes the Universe.

- Temporal density is what one measures or observes.
- PNLCT is what explains it and makes it functional.

It characterizes three essential aspects:

- Frequency (apparent speed)—not the displacement of the chronons themselves, but the elemental pulsation (the fundamental activation of a quantum voxel) and the global rhythm (the organization of these pulsations at the fractal scale), through which information propagates from one quantum voxel to another. It is this orchestration that creates the illusion of spatial motion and the emergence of physical phenomena.
- Acceleration—the evolution of this apparent speed according to temporal fractals, reflecting the dynamics of expansion and elevation of the field's potential.
- Nonlinearity—the liberty of configuration of the field, which allows the emergence of hierarchical structures and complex phenomena.

Frequency: the propagation of information

The temporal density of a fractal corresponds to the number of chronal activations per unit of frame.

- The most ancient structures—such as clusters and galaxies—formed out of a very high temporal density, where Time is still fluid, fine, and reactive. They express the first levels of structuring of the temporal field.
- More recent structures—like stars and planets—arise from a progressive condensation of this temporal dynamic. Their temporal density is more compact, Time there becomes "thicker," more coherent, and this is precisely what makes possible the stability of matter and the emergence of life.

In other words, what we call speed or displacement is only a perceptual translation of the propagation of information between chronons.

Acceleration: the dynamic evolution of fractals

Each fractal evolves over time not as a static system, but as a living, expanding network.

- The temporal field continuously emits new chronal activations that carry a higher potential.
- This potential gradually increases a fractal's average frequency, driving its internal evolution as well as the expansion of its boundaries.

The Universe is therefore not only in spatial expansion; it is also in temporal expansion, its informational density increasing continuously.

Nonlinearity: the diversity of structures

Nonlinearity expresses the combinatorial liberty of the chronal field:

- Each activation can couple differently with others, generating an infinite diversity of configurations.
- This flexibility accounts for the richness of cosmic forms: galaxies, fields, particles, biological systems.

Nonlinearity is the creative signature of the temporal field: it makes possible a living, evolving, fractal organization of the Universe.

The limit frequency and fractal relativity

Each fractal is characterized by an intrinsic limit frequency, directly determined by its temporal density.

- Example: in the terrestrial fractal, this frequency corresponds to what is measured as the speed of light in the void—but in reality, it is the maximum rhythm of information propagation in this fractal.
- This value is not universal in the absolute sense: it is relative to the fractal of observation. Light adjusts its effective frequency to the temporal field of the medium traversed, revealing that what we call "speed" is only a phenomenal translation of a deeper pulsation.

Thus, relativity must be rethought: it is not a simple geometric relativity of spacetime, but a functional relativity of fractal temporal densities, where each level possesses its own rhythm of manifestation.

Fundamental implications

PNLCT allows for a new reading of cosmic evolution:

The Universe is a dynamic process that self-organizes through the elevation of temporal density.

Physical phenomena—matter, energy, gravity, fields—emerge as local effects of this universal structuring.

Human perception of time is a direct consequence of the fractal to which we belong.

Thus, NLPTF unifies Time, space, and energy within a coherent framework: the real is not a fixed backdrop, but a living pulsation of the temporal field, where the whole is woven through the fixed yet active chronons of the quantum vacuum.

1. Measuring in a Fractal Spacetime — Toward a Relative Metric

Current science rests on the idea that spacetime is a continuous, homogeneous, and universal structure, described by a fixed metric. Our instruments—ruler, clock, meter, second—were conceptualized within this framework.

But if spacetime is in reality fractal, hierarchical, and structured by different temporal densities, then any attempt to measure it with fixed units introduces a fundamental error.

The impossibility of an absolute unit

In a fractal spacetime, there is no universal unit of measurement for time or space.

- Each fractal (planet, star, galaxy, cluster) has its own temporal density and therefore its own fundamental rhythm.
- This rhythm determines limiting speeds, local constants, and the laws that emerge at that scale.

In other words, a "meter" or a "second" defined at the terrestrial scale cannot be valid for measuring a process situated in the galactic or cosmic fractal.

The limits of classical measurement

Measuring a phenomenon that belongs to a fractal far from our own with our fixed units amounts to using a warped ruler.

Example: the same distance expressed in "galactic fractal seconds" does not correspond to the same time as when it is measured in "planetary fractal seconds."

This offset produces interpretive distortions: phenomena that seem extremely slow to us (for instance, cosmic expansion) could, at another temporal density, unfold at a much higher frequency.

Thus, many current cosmological paradoxes—dark matter, dark energy, inflation—could be viewed as measurement artifacts stemming from the use of fixed units in a non-homogeneous spacetime.

Toward a relative metric: measuring in "fractal units"

The only coherent approach is to measure phenomena with respect to the fractal of observation.

At our scale, a meter and a second remain relevant.

But to describe a galactic or cosmological dynamic, one would need to employ relative units adapted to the corresponding temporal density.

One could thus define a measure in "fractal units," where distance or duration are not absolute but linked to the level of organization under consideration.

A methodological revolution

If the Universe is truly fractal, then:

- Our current units of measure are local and limited, incapable of reflecting the profound structure of the real.
- The physical constants (such as the speed of light or gravitation) ought to be rethought as constants relative to a given temporal density, and not universal in the strict sense.
- Physics ought to move from a fixed metric to a hierarchical metric, where time and space are defined by their membership in a given fractal.

This changes everything: measuring becomes not an operation upon a neutral background, but a translation from one fractal rhythm into another.

Plainly put: our "seconds" and our "meters" do not measure the real as it is, but the real as it manifests within the terrestrial fractal.

2. Distant Fractals — Quantum Superposition and the Cosmological Horizon

When we attempt to observe or measure a phenomenon that belongs to a fractal of a different scale order from our own—whether at the infinitely small or the infinitely large—we necessarily use instruments, constants, and units of measure calibrated to our own fractal temporal density.

Yet in a Universe where Time is not homogeneous but structured by levels of density, this approach inevitably introduces distortions.

These distortions become visible in two extreme cases: at the quantum scale and at the cosmological scale.

1 - The infinitely small: superposition of states and temporal density

When a macroscopic observer (like us) measures a quantum system, it does so from a temporal fractal with low FTD, while the observed phenomenon belongs to a higher-order fractal, with a much higher FTD.

This means that:

- Our "second" is not a universal unit: it corresponds to a rhythm specific to our temporal density.
- In the quantum fractal, time is more dense: more micro-instants exist within the same apparent duration.

Consequence:

Seen from our fractal, where time "flows more slowly," the successive states of the quantum system appear simultaneous to us.

In other words, the superposition of states is perhaps not a real indeterminacy of the system, but a projection of our slow temporal rate onto a much faster dynamic.

What we call "superposition" would then be the visual effect of a shift in temporal density between the observer and the observed phenomenon—an interference between two incommensurable temporal fractals.

2 - The infinitely large: the cosmological horizon and time dilation

At the other extreme, when we observe the deep Universe, the problem does not really invert: it echoes at another scale of the fractal hierarchy.

Distant cosmic structures belong to fractals with higher temporal density, that is, to levels where Time is finer, more concentrated, and therefore less directly perceptible from our own scale.

Thus:

- Galaxies near the cosmological horizon seem fixed not because their time flows more slowly, but because their temporal rhythm is too subtle to be detected by our lower temporal density.
- Their FTD, far higher, renders their internal dynamics "invisible" to our instruments, just as a vibration that is too rapid eludes the human ear.

What classical cosmology interprets as a dilation of time could therefore be the direct signature of a phase shift between fractal temporal densities.

The cosmological horizon would no longer represent an absolute spatial limit, but a threshold of frequency mismatch between two levels of the temporal field: that of the observer and that of the structure observed.

Two observers situated in fractals with different FTDs would naturally assign distinct ages, distances, and velocities to the same phenomenon, because their local temporal metrics do not resonate at the same limiting frequency.

And, in perfectly symmetrical fashion:

- At the quantum scale, phenomena appear simultaneous because they belong to fractals where Time is denser and faster.
- At the cosmic scale, phenomena appear motionless because they manifest within fractals where Time is less dense and more slowly modulated.

These two extremes converge within a single logic:

Time does not have a universal speed—it possesses a variable density, proper to each fractal level of the real.

Our position as terrestrial observers, situated at a median scale of the fractal hierarchy, thus constitutes a unique point of equilibrium between two poles of complexity:

- The subtle, where Time intensifies and contracts (quantum scale);
- The manifested, where it dilates and stretches (cosmological scale).

3 - Toward a new fractal measurement scale

Conventional units (meter, second, light-year) rest upon the space-time metric specific to our observational fractal, that is to say upon the temporal density characteristic of our local cosmic system.

Now, in a Universe structured fractally, there is no absolute unit of measurement: every magnitude—spatial, temporal, or energetic—depends on the Fractal Temporal Density (FTD) of the level considered.

Each fractal possesses its own internal cadence, its own rhythm of Time, which renders any universal measure fundamentally relative.

Thus:

- The light-year expresses a distance defined from the limiting speed of our terrestrial fractal, but it becomes ill-suited to describe the internal dynamics of another cosmic fractal whose FTD differs.
- A truly universal unit ought to be relational, founded not on a fixed metric (meter or second), but on a ratio of temporal density between fractals.

In other words, measuring a distant galaxy in "light-years" amounts to measuring a butterfly's wingbeat with a calendar:

The granularity of our unit of time is not tuned to the fineness of the fractal observed.

This idea leads to a refounding of the measurement methodology:

- Distances, durations, and even energies should no longer be expressed in absolute units, but in scales of resonance between FTDs.
- The Universe is no longer a homogeneous continuum, but a hierarchical sequence of resonant space-times, where each level imposes its own natural metric.

4 – Physical and cosmological Consequences

Quantum Superposition and cosmological "dilation" proceed from the same mechanism: a phase shift in temporal density (FTD) between the observer (lower FTD, planetary level) and the phenomenon (higher FTD, more subtle level).

— At the microscopic scale, the system's FTD is so high that several micro-evolutions occur within a single "time step" of the observer; our measurement aggregates these cycles → appearance of superposition.

— At cosmological distances, the FTD of large structures is likewise higher than ours; their cadence is too fine for our measurement step and for our spectro-temporal window → impression of immobility / "dilated" time (redshift, apparent slowing of evolution).

- Measured time is therefore not a universal invariant, but a ratio (a bandwidth match) between FTDs: what we record depends on the temporal granularity of the observer versus that of the phenomenon.
- The so-called fundamental constants (e.g., speed of light) must be understood as locally stable invariants within a given fractal (with a specific FTD); their effective value can differ from one fractal level to another, through a change in the temporal metric.

5 – Toward a unification of the two extremes

Within this framework, quantum mechanics and general relativity are no longer incompatible:

They simply describe two complementary temporal regimes of the same fractal fabric.

- Quantum mechanics explores domains where the temporal density is very high: Time there is finely granular, multiplying the possible microstates within a single interval.
- General relativity, by contrast, applies to zones where FTD is diluted: Time there becomes continuous, extensive, and structuring for large masses and cosmological curvatures.

Thus, the key to their unification is neither geometric nor spatial, but temporal:

The two theories describe distinct behaviors of one and the same field, according to the temporal density of the observed fractal level.

The real is therefore not a homogeneous continuum of spacetime, but a hierarchical temporal field, where each fractal expresses a unique ratio between density, duration, and consciousness.

3. Energy and the Evolution of the Universe — Expression of Fractal Time

In classical physics, energy is defined as the capacity of a system to produce work or to bring about a transformation.

But this definition remains phenomenological: it describes what energy does, not what it is.

Within the framework of my theory of Absolute Relativity and in connection with my hypothesis of the fractal temporal field, energy takes on a more fundamental meaning:

- It is not a derived quantity, but an intrinsic property of Time itself.

1 - Energy as an expression of the temporal flow

The Universe must not be conceived of as a static system through which Time passes:

- It is the very process by which Time unfolds into structure, into memory, and into matter.

From this perspective, energy no longer represents an abstract quantity of transformation, but the frequency of the flux of chronons—in other words, the rhythm by which primordial information is actualized within a given fractal of the temporal field.

Thus,

- Energy is the frequency of Time that actualizes itself.
- The faster the chronal pulsation, the higher the temporal density (FTD), and the greater the effective energy of a system.

This approach reverses the classical relation between matter, energy, and Time:

It is not motion that produces Time, but fractal Time itself—through the ordered flux of chronons—that engenders motion, matter, and energy.

2 - Origin of the energetic potential: the universal vibration ÆKO

In this model, the energetic potential proper to each fractal of the Cosmos finds its origin in the universal vibration ÆKO, the fundamental field of the living quantum vacuum, emanating directly from the Primordial Complexity.

ÆKO constitutes the principle of cosmic impulse: an initial wave of coherence that continuously feeds the network of chronons, the elementary quanta of Time.

This vibration is not a localized force, but a pulsation of informational order that traverses the entire Time–Space–Memory, maintaining the synchronization of the fractals with one another.

From this primordial flux emerge the gradients of Fractal Temporal Density (FTD), responsible for the hierarchical structuring of the Universe.

Each chronon receives and retransmits a portion of the ÆKO vibration, like a cell in a living network vibrating in unison with the cosmic matrix.

Thus,

- ÆKO acts as the primordial source of energy, not by mechanical transfer, but by informational resonance.
- Energy is nothing other than the ordered frequency of this universal vibration, which animates Time and engenders, by condensation, memory, structure, and matter.

3 - Direct link between energy and temporal density

Fractal Temporal Density (FTD) expresses the amount of energy—or, more precisely, the active chronal frequency—available per unit of Time–Space structure.

In other words, it measures a system's capacity to actualize the information of the temporal field.

From this definition follows a fundamental relation:

- The more energy a fractal possesses, the higher its FTD.
- The higher its FTD, the faster its internal dynamics (evolution, transformation, complexification).
- Conversely, a fractal with low FTD evolves slowly, appearing fixed or weakly energetic.

Thus, energy is not a stored quantity, but a flux of temporal order:

It corresponds to the rate at which Time unfolds locally within a given fractal.

In this framework, energy becomes the local manifestation of the global unfolding of ÆKO—the rhythm at which the universal vibration is actualized through the chronons.

4 - Physical interpretation and conceptual unification

This conception reunites, under a single temporal coherence, the frameworks of modern physics that are separated today.

It shows that the fundamental laws do not describe distinct entities, but regimes of expression of the fractal temporal field.

Quantum mechanics

- The frequency of a photon (E = hf) becomes the local expression of the chronal pulsation within a given fractal.
- The quantum of energy corresponds to a discrete unit of temporal density, that is, to a portion of condensed Time.

Relativity

- Relativistic energy ($E = mc^2$) expresses the limit speed of information transfer within a fractal.
- The curvature of spacetime would no longer be a geometric distortion, but the physical manifestation of a gradient of FTD: a local variation in the fundamental rhythm of Time.

Gravity and fundamental interactions

- The fundamental forces cease to be separate entities: they become local modulations of temporal density.
- Where FTD varies strongly, flows of information appear in the form of dynamic effects—what we perceive as forces.

5 - Major implications

Reinterpretation of the concept of energy

- Energy is no longer a mechanical quantity, but the measure of the rhythm of manifestation of the real.

Link between Time and evolution

- The rate of evolution of a system depends directly on its fractal temporal density: the faster Time unfolds there, the faster transformation occurs.

Unification of physics

- Gravity, electromagnetism, and quantum phenomena appear as coherent modes of the temporal field, each corresponding to a specific regime of FTD.

Cosmological vision

- The expansion of the Universe does not result from an initial explosion, but from the progressive increase of the chronal potential—the acceleration of the pulsation of ÆKO through the fabric of the Time–Space–Memory.

6 - Conclusion: Energy as the music of Time

In its most profound nature, energy is musical:

- It expresses the rhythm of Time that organizes itself, harmonizes, and densifies to bring matter forth.

To understand energy is to understand the tempo of the fractal temporal field, and, by extension, the intimate mechanism of the evolution of the Universe itself.

4. Resistance to Energy — A New Fundamental Distinction

1 - Definition and fundamental concept

After defining energy as the frequency with which chronons are actualized, it is appropriate to introduce its logical counterpart: resistance to energy.

This represents an intrinsic property of the temporal field, specific to each fractal, which limits the frequency of chronal pulsation and slows the dynamic evolution of the field.

In other words, resistance to energy expresses a fractal's tendency to slow the flow of temporal information, stabilizing local structures while reducing their rate of evolution.

It therefore constitutes the principle of temporal inertia within the fractal field.

2 - Physical analogy: the viscosity of the void

To grasp this phenomenon intuitively, let us imagine the propagation of light through media that are more or less dense:

- In a slightly viscous medium—analogous to a fluid, coherent temporal field—chronons propagate at high frequency, indicating low resistance to energy.
- In a highly viscous medium—a dense, incoherent, or saturated temporal field—chronal pulsation is slowed, indicating high resistance to energy.

Thus, resistance to energy acts as a viscosity of the temporal quantum vacuum, modulating a fractal's capacity to transmit the fundamental pulsation of ÆKO.

This concept naturally extends that of Fractal Temporal Density (FTD):

The higher the resistance, the lower the FTD—and vice versa.

3 - Interconnection and propagation of fractal resistances

No fractal exists in isolation.

The temporal fields of all fractals are connected by temporal density gradients, forming a network of mutual resonances.

When a fractal's resistance to energy increases—for instance due to a local coherence collapse (as in a dense gravitational field)—the chronal pulsation slows locally, influencing the propagation of energy in neighboring fractals.

Thus, the fractal multiverse behaves as a dynamically self-regulating system: one fractal can slow the local energy flux while another accelerates, preserving the overall balance of the temporal field.

4 - Reconsideration of the usual concept of energy

In classical physics, energy is defined as the capacity to produce a transformation.

But from the perspective of the temporal field, this definition becomes ambiguous:

- Slowing the chronal frequency—therefore increasing resistance—also produces a transformation, but of a stabilizing nature.

We must therefore distinguish two effects:

- True energy accelerates the chronal frequency and favors the expansion of the temporal field.
- Resistance to energy slows this frequency and ensures the stability of structures.

Both contribute to the universal dynamic, but with opposite effects:

- One tends toward expansion, the other toward cohesion.
- Together, they ensure the dynamic symmetry of the fractal temporal field.

5 - Two opposing but complementary forces

The Universe rests upon a permanent tension between two fundamental principles:

Energy and resistance to energy.

These two forces act directly on the frequency of chronons, regulating the rate of evolution of the temporal field.

Energy acts as a force of expansion:

- It accelerates the chronal frequency, intensifying the dynamic of Time.
- It engenders motion, light, heat, growth, life.
- It is the active breath of the Universe, the motor of its unfolding.

Resistance to energy, by contrast, acts as a force of stabilization:

- It slows the chronal frequency, introducing inertia and cohesion.
- Through it, matter condenses, forms are maintained, and structures endure.
- It grounds gravity, mass, structure, and conservation.

Thus:

Energy creates change, resistance maintains form.

One opens up possibilities, the other guarantees continuity.

Their balance gives rise to the dynamic stability of the Cosmos—a Universe in ordered expansion, where creation and persistence coexist within the same breath of Time.

6 - A reversal of the energy paradigm

Our civilization exploits energy through resistance: combustion, friction, fission, compression, drop in potential.

We extract heat and work from dissipative phenomena, that is, by the braking of the chronal flow.

- What we call "energy production" is in reality an extraction of resistance.

We draw our power from the dissipation of Time, rather than from its fundamental pulsation.

This observation leads to a crucial question:

- Is there a way to access the primordial chronal frequency directly, without going through resistance?

7 - Implications and perspectives

Conceptual revision of energy

- Energy is neither substance nor quantity: it is a flow of temporal acceleration.

Toward a non-dissipative energy

- Drawing upon the pulsation of ÆKO directly would make it possible to obtain coherent energy, without loss or friction—comparable to a standing wave of the temporal field.

Unification of the fundamental forces

- The distinction between energy (expansion) and resistance (cohesion) could reconcile gravity and quantum interactions as two faces of the same fractal balance.

8 - The cosmic balance: tension between acceleration and resistance

The Universe sustains itself in a dynamic resonance between these two tendencies:

- Energy accelerates Time, fostering creation and differentiation.
- Resistance slows it, ensuring cohesion and structuration.

From this polarity emerges the cosmic music: the fundamental wave of the real, where each fractal contributes to the symphony of expansion and coherence.

9 - Toward a new physics of Time-energy

If humanity were to manage to directly manipulate the dynamics of chronons, it would gain access to a primordial, inexhaustible, and non-dissipative energy.

It would be the passage from a physics of resistance to a physics of resonance, where Time, and not matter, becomes the true engine of universal energy.

5. Matter — Expression of Resistance to Time

The chronon, the fundamental unit of the temporal field, is not a simple "grain of Time," but a unit for organizing the chronal flow.

It does not move: it pulses, transmitting state information from one point to the next that generates the observable physical phenomena.

Thus, matter is not an entity independent of the field, but a local modulation of the flow of chronons—a condensation of living Time.

1 - Chronal frequency as the origin of matter

Every particle, every stable system in the Universe can be interpreted as a coherent state of the chronal field.

- The frequency of the chronons determines the energy of a system.
- Their local slowing and their internal coherence determine its mass.

What we call "matter" therefore corresponds to a region where the chronal flow slows and stabilizes: matter is the fixed face of Time, a chronal wave locally confined by an internal resistance to its expansion.

2 - The origin of mass: a resistance to the frequency of Time

Since Einstein, mass and energy have been two aspects of the same reality ($E = mc^2$).

Within the framework of the fractal temporal field, this equivalence takes on a more profound meaning:

- Energy corresponds to an acceleration of the chronal rhythm, a fast and free flow of Time.
- Mass, on the contrary, expresses a resistance to this acceleration, a zone of condensation where the temporal field contracts.

The more a system resists the propagation of the chronal flow, the more its effective mass increases.

Mass thus becomes a slowing of local Time, not a material substance.

3 - Inertia: memory of slowed Time

Inertia, classically perceived as the tendency of a body to oppose motion, finds a temporal interpretation here.

It is not a mechanical force, but the macroscopic manifestation of a resistance of the chronal field.

Each particle, each atom, resists the variation of its chronal frequency, as though Time itself were clinging to its own cadence.

When one accelerates an object, one acts upon its internal rhythm of Time:

- The lower its temporal density (Time circulates slowly there), the greater its inertia.

Thus, inertia is not a passive property, but an active dialogue between matter and Time.

4 - Mass: a resistance to the chronal flux

The more mass a system has, the more it slows the circulation of chronons within its fractal.

Mass is therefore not a form of energy, but a resistance to the flux of temporal energy.

Time circulates there more slowly, and this slowing manifests as gravity:

- Massive objects deform the local rhythm of the chronal field, attracting neighboring systems through synchronization.

Thus:

- Gravity is not a force of attraction, but a consequence of the slowing of Time.

What relativity describes as a curvature of spacetime corresponds here to a local contraction of the flux of chronons.

5 - A redefinition of matter

In this perspective:

- Matter is a state of slowed Time.
- Mass is a measure of resistance to chronal energy.
- Gravity is the geometric consequence of this resistance.
- Inertia is the dynamic memory of Time in matter.

Matter is therefore condensed Time, a region where the chronal pulsation loses speed, creating an apparent stability within the universal flux.

This vision unifies matter, gravity, and Time within a single coherent fractal dynamic.

6 - Experimental pre-validation and fractal coherence

The first observations obtained from chronal measurements tend to confirm this hypothesis.

Although preliminary, these results display a remarkable internal coherence, and a convergence with the predictions of the fractal model of the temporal field.

- More massive atoms exhibit a slower temporal resonance: their field offers stronger resistance to the frequency of chronons.
- This resistance results in a local decrease in temporal density, analogous to the slowing of clocks near a gravitational mass.

Thus, the more mass a body possesses, the more it slows the chronal fabric that surrounds it.

This slowing manifests, at the macroscopic scale, in the form of gravity.

These experimental findings, reproduced across various scales, suggest that this is not a local artifact but a structural phenomenon, linked to the intrinsic dynamics of the fractal temporal field.

7 - Gravity and temporal density

Gravity is not a force, but a gradient of temporal density within the field of chronons:

- Regions of high mass are zones where Time flows more slowly, drawing nearby structures through chronal synchronization.
- The relativistic "curvature" of spacetime corresponds here to a modulation of the chronal field: where Time contracts, matter condenses.

8 - Consequences and prospects

Redefinition of matter

- Matter is a zone of chronal resistance, a region where the propagation of the temporal flux is slowed.

New interpretation of the Higgs boson

- The mass-giving mechanism could result not from an isolated scalar field, but from a coupling between the chronons and the fractal resistance of the temporal field.

Gravity-Quantum Unification

- If gravity emerges from the slowing of the chronal field, it is no longer a distinct interaction, but a geometric manifestation of Time itself.

9 - Matter: Contracted Time

Matter appears as a coherent state of Time:

- Mass expresses resistance to the chronal flux,
- Gravity conveys the field's tendency to reestablish the equilibrium of temporal density,
- Energy expresses the expansive vibration of the field in its creative dimension.

Thus, matter is contracted living Time:

- To understand matter amounts to understanding how the chronal field compresses to manifest form.

10 - Toward a physics of active Time

If one were to succeed in modulating the resistance to the chronal flux, that is, in locally adjusting a system's temporal density, it would become theoretically possible to alter its effective mass, or even to temporarily neutralize gravity.

Such mastery would mark a conceptual revolution:

- Matter, energy, and Time would no longer be separate entities,
- But the three states of one and the same living scalar continuum,
- Capable of contracting, dilating, and resonating according to informational conditions.

This perspective would open the way to a technology of chronal resonance:

- Accessing the primordial energy of ÆKO without combustion or dissipation,
- By simple modulation of the temporal field.

My first experimental proofs of concept already suggest:

- Apparent variations in inertial mass correlated with local chronal fluctuations;
- Temporal synchronization anomalies between entangled systems, indicating a phase shift in chronal density;
- And, in certain cases, a transient reduction of gravitational constraints, compatible with a redistribution of the flux of Time.

These results, although requiring independent validation, offer the first empirical indications of a physics of active Time, where manipulating chronal density would amount to sculpting reality itself—no longer undergoing it, but co-creating it.

Chapter 20

Toward Unification

~

When Time Connects the Quantum and the Cosmic

Sylvana

So now you will explain how your model could contribute to the unification between quantum physics and relativity.

Thierry

Yes. For a century, science has stumbled over a major fracture: on one side, quantum mechanics describes with remarkable precision the infinitely small—particles, interactions, probabilities. On the other, general relativity provides a coherent and elegant framework for understanding gravitation and the structure of spacetime on large scales.

But these two edifices, each triumphant in its own domain, stubbornly refuse to come together. The quest for quantum gravity—a unified theory—remains the "Holy Grail" of physics.

It is precisely at this crossroads that my model comes into play.

1 - The proton: micro-gravity as a hidden key

Standard physics describes the proton through quantum chromodynamics (QCD): a maelstrom of quarks and gluons confined by a force that grows when one tries to separate them. But this description remains incomplete: it ignores the gravitational dimension, deemed negligible.

- In my model, the proton is seen otherwise:
- It is a stabilized micro-singularity, a condensation of energy where Time (M-) is maximally entangled, giving rise to a stable structure.

This stability is not a mere curiosity, but the manifestation of a local quantum gravity: a micro-curvature of spacetime, confined within a minimal volume.

Thus, at the level of the proton, gravity is not absent: it is already there, integrated. It acts in silence as the harmonizing force that allows matter to exist.

2 - The black hole: macro-gravity as the proton's mirror

At the other end of the spectrum, the black hole embodies gravity in its most extreme form.

Where the proton condenses Time into stable space (M- → M+), the black hole reverses the process: it absorbs space to resolve it into pure Time (M+ → M-).

It acts as an exit gate, where information, matter, and even the structure of spacetime dissolve into the temporal flow.

Proton and black hole are therefore not unrelated entities, but the two bounds of the same fractal cycle:

- The proton is the entry gate, the positive transmutation.
- The black hole is the exit gate, the negative transmutation.
- And gravity is the organizing tension that binds these two poles.

3 - The herd effect: the bridge between micro and macro

One of the keys that may lead to unification is what I call the herd effect.

- Each proton, taken in isolation, contains a seed of quantum gravity (micro-curvature).
- But gathered by the billions of billions in a star, their micro-gravities compound into a macroscopic gravitational coherence.
- When this coherence reaches a critical mass, the overall gravity tips into the relativistic regime: the star collapses, giving birth to a black hole.

Thus, relativistic gravity is not another law, but the cumulative addition of local quantum gravities, up to the critical threshold.

4 - Chronons and the scalar temporal field: the unifying language

At the heart of this architecture lie chronons, the elementary units of the temporal field.

- They support both quantum physics (through local fluctuations of the void) and general relativity (through global density and curvature).
- The proton, the black hole, gravity: everything is simply local differentiation of chronons.
- What differs between the regimes is the Fractal Temporal Density (FTD): the higher it is, the more we are in the quantum domain (M−). The more it is diluted, the more we tip into the relativistic regime (M+).

Chronons thus become the common language between the two frameworks: a temporal metric that orders quantum probability as well as relativistic geometry.

5 – Toward unification: a coherent and fractal vision

With this schema, unification ceases to be an insoluble problem:

- Quantum theory and relativity are not two incompatible worlds, but two expressions of the same dynamics of the temporal field.
- The proton and the black hole embody the extreme poles of this dynamics.
- Gravity is the cohesion mechanism that links these poles, from the finest scale to the cosmic scale.

Chronons ensure continuity, as the fundamental units of the real, modulated into different densities according to the fractals.

Conclusion: an elegant solution

By defining Time as a primordial matrix, by identifying chronons as the elementary building blocks, and by seeing the proton and the black hole as the two gateways of a cycle, my theory proposes a natural unification of standard physics.

My model shows that what contemporary science takes to be two irreconcilable languages (quantum mechanics and general relativity) are in reality only two dialects of the same grammar: the fractal dynamics of the temporal field.

Sylvana
It is impressive to see how coherent your model is. What you call the "herd effect" seems almost self-evident once one has read your work.

But I wonder, why does standard science not see gravity as the sum of the micro-gravities of each proton, rather than as something independent?

Thierry
Current science considers gravity as a strictly macroscopic phenomenon: it becomes significant only with a mass sufficient to curve spacetime.

At the scale of the proton, this curvature would be infinitesimal—on the order of 10^-40 times weaker than the internal forces—and therefore without any measurable effect. This is why particle physics, founded on the Standard Model, rules out gravity: it appears in none of the equations of quantum chromodynamics, which takes into account only the strong and weak nuclear interactions, as well as the electromagnetic interaction. In other words, for classical science, gravity begins where the quantum ends.

But this hypothesis, seemingly reasonable, closes an essential door: it is not because an influence is minuscule that it is nonexistent.

In my model, each proton contains, in contrast, a stabilized micro-curvature—a local gravitational tension that contributes to the coherence of the real. It does not express itself as a measurable force, but as a resonance structure integrated into the temporal density of the particle. When billions upon billions of protons assemble, their micro-gravities add up naturally: it is this coherent sum that produces the macroscopic gravity we observe in planets, stars, or galaxies.

In other words, gravity does not emerge beyond a certain mass threshold:

It is present everywhere, from the intimate structure of matter, and manifests cumulatively across all scales of the real.

In my model, what science observes on a large scale is therefore only the amplified resonance of these fundamental micro-gravities.

Sylvana
In other words, gravity would no longer be the consequence of masses, but their hidden foundation.

Thierry
Yes. In my view, it is not the effect, but the very weave that binds all things.

It does not merely shape space: it weaves the continuity of the real.

Herein lies, in my view, the key to unification: to understand that gravity is not what is added to the world, but what maintains its coherence—from the first chronon to the galaxies.

Chapter 21

Central Recapitulation of My Discoveries

~

The Hierarchy of FTD

Sylvana
We have reached the end of this long journey through your discoveries…

Thierry
Yes, there is still an essential point that I want to address before concluding—the one that connects everything we have explored up to now.

I will explain the hierarchy of Fractal Temporal Density (FTD) within my model of the four major substructures of the quantum vacuum. This hierarchy constitutes the fundamental key for understanding how time organizes itself, differentiates itself, and self-structures across the levels of the real.

I will present it to you in the form of structured notes, to bring out its synthetic logic more clearly.

Here is how this hierarchy unfolds:

In decreasing order of temporal density—that is, from the most subtle to the most manifested)—the FTD reaches its maximum value within the quantum vacuum, which represents the fundamental substratum of the Universe.

Next come the galaxy clusters, whose temporal structure is extremely condensed and stable, attesting to their cosmic antiquity and to their role in anchoring universal coherence.

Then appear, in succession, the galaxies, the stars, and finally the planets, where temporal density gradually decreases, accompanying the differentiation of matter and the emergence of biological complexity.

But this hierarchy does not stop at cosmic scales: it reproduces itself fractally within each level.

Thus, within a planet such as Earth, there exists an internal hierarchy of FTD that orders the kingdoms of nature:

- The human kingdom, endowed with the highest local temporal density, capable of self-reflection and the conscious modulation of the temporal field;
- The animal kingdom, whose FTD is high but polarized by instinct and collective resonance;
- The plant kingdom, where time flows more slowly, fostering harmonic integration and the memory of the environment;
- And finally the mineral kingdom, the seat of a minimal FTD, where time is almost fixed, stored in the form of latent potentiality.

This stratification of temporal density at all scales—from the quantum vacuum to dense matter—shows that the Universe is not a homogeneous continuum, but a hierarchical architecture of interlocking temporal speeds, where each level of reality pulses at its own rhythm within the great cosmic breath.

1. The Most Subtle to the Most Dense — the Gradation of Cosmic Substance

The Universe manifests a continuous gradation between the subtle and the dense, which directly reflects the hierarchical structure of Fractal Temporal Density (FTD).

In other words, matter—in all its forms, from intergalactic plasma to planetary rock—is nothing other than a progressive crystallization of the temporal field.

The more FTD decreases, the more the flow of chronons slows, and the more reality becomes stable, tangible, localized.

1 - The Quantum Vacuum: the subtlest matrix

The quantum vacuum is not a void, but a hyper-dense ocean of chronons, a scalar field where FTD reaches its absolute maximum.

Here, matter does not yet exist in localized form: everything is pure vibrational potentiality, an interweaving of temporal information in a state of perfect coherence.

It is the primordial source, the level where Time and space are not yet differentiated, and from which all the fractal structures of the Cosmos emerge.

2 - Galaxy clusters: cosmic gaseous condensation

At the scale of galaxy clusters, the temporal field begins to polarize and to structure itself.

Clusters are composed predominantly of extremely tenuous intergalactic gas, bathed in a hot, rarefied plasma.

Their FTD, although slightly lower than that of the quantum vacuum, remains very high: the flow of chronons there is rapid, nearly free.

These regions act as reservoirs of cosmic coherence, where vacuum fluctuations organize themselves into webs of galactic density.

3 - Galaxies: a denser organization of the cosmic gas

In a galaxy, matter condenses further under the effect of the dynamics of the temporal field.

The interstellar gas becomes enriched with dust, magnetic fields, and complex energy gradients.

The FTD decreases slightly, because the flow of chronons becomes more structured and less free:

Matter begins to individualize, fields to differentiate, and stellar systems to emerge.

This is the level where space becomes geometrically organized: spirals, disks, halos—visible signatures of a coherent temporal structuring.

4 - Stars: gas turned to fire

The star represents a phase of intense temporal compression.

The density of matter and energy there is such that the chronons reorganize into highly coherent, confined, pulsed motions.

The FTD is weaker here than in the galaxy, because the temporal field is locally slowed by the concentration of energy.

However, this compression generates a luminous feedback effect: matter begins to radiate, to return a part of its original dynamics to the void.

Thus, the star is an active interface between density and expansion, a temporal heart vibrating at the center of its local fractal.

5 - Planets: the ultimate condensation of the temporal field

The planet marks the threshold of maximum density in the fractal hierarchy.

Here, the FTD reaches its relative minimum: the chronons are profoundly slowed, confined within stable, solid structures.

Matter there appears condensed, coherent, enduring—but this apparent stability is actually the result of an extreme slowing of local time.

The planet therefore represents the state of inertia of the temporal field, where resistance to energy dominates pulsation.

This explains why planetary matter is perceived as "massive," "heavy," or "motionless": it is literally a fixed Time.

Synthesis: a fractal continuity from the void to matter

One can thus summarize this hierarchy as a descent of temporal frequency:

Quantum vacuum → Galaxy clusters → Galaxies → Stars → Planets

(From maximal FTD to minimal FTD)

Each level corresponds to a degree of condensation of Time, that is to say a different intensity of the chronal pulsation.

- The quantum vacuum is an ocean of temporal light,
- Galaxy clusters are waves of organization,
- Galaxies are whirlpools of ordered Energy,
- Stars are radiant hearts,
- And planets are crystallizations of Time itself.

Thus, matter is not external to the void: it is its slowed phase, the solid face of the temporal flow.

This continuum from the subtle to the dense grounds the coherence of the Cosmos and the profound unity between energy, Time, and consciousness.

2. Where I Stand in Relation to the Quantum Vacuum — How to Access It

In my view, I do not merely inhabit the Universe: I am literally immersed in its fundamental substrate—the quantum vacuum.

Everything that composes me, from the most macroscopic to the infinitely small, is bathed in this same invisible ocean. The atoms that constitute me are made of 99.99% "void," but this void is not a nothingness: it is an active, structured field that carries information.

In my model, this quantum vacuum is the seat of a two-faced scalar temporal field—a dynamic matrix where two flows of Time coexist:

- Future → past (M−): the retrospective face of Time, which preserves memory and feedback;
- Past → future (M+): the projective face of Time, which produces causality and phenomenal history.

All space is thus constituted of chronons, elementary units of the scalar temporal field. These chronons form what I call the quantum ether:

- A universal substrate that is neither matter nor energy but their common potential;
- The memory of the Universe, capable of polymorphism and torsion;
- And the informational interface from which energy and matter are manifested.

The bimetry M− / M+: structure of the real

The phenomenal Universe is bimetric because it derives from this two-facedness of Time:

- M−: scalar temporal field, world of the infinitely small, future-to-past, memory and feedback;
- M+: atomic phenomenal reality, past-to-future, causality and the creation of history.

This double metric structures everything: from the behavior of the proton to the dynamics of black holes.

Proton and black hole: semi-permeable membranes

In my model, the proton and the black hole are the two semi-permeable membranes of the two-facedness of Time:

The proton is an entry gate:

- It allows the temporal flow of past-to-future (M+) to emerge from the potential (M−).
- The proton's membrane, formed by densified chronons, is the interface where potential manifests.
- It stabilizes the information of the vacuum into baryonic matter—this is positive transmutation.

The black hole is an exit gate:

- It reverses the flow, to future-to-past (M−).
- It transforms manifestation into memory and reintegrates information into the quantum ether—this is negative transmutation.

These two "gates" form the inverse poles of one and the same cosmic cycle, where information condenses (proton) and then reconfigures (black hole), ensuring the dynamic coherence of the fractal multiverse.

3. The Role of Consciousness — Interface with the Fractal Temporal Field

In this architecture, spirit is not an epiphenomenon: it is a natural interface with the quantum vacuum.

Through consciousness, we have direct access—although limited by our fractal scale—to the potential of the temporal field.

Consciousness functions as an innate universal technology, capable of resonating with the field of chronons and drawing from it information or intuitions.

This access is complex because our perception is set to M+ (the phenomenal world), but it can be expanded:

- Either through inner development (conscious attunement),

- Or through artificial technologies capable of modulating temporal density and acting upon atomic interfaces.

4. Unification — The Temporal Field: Quantum, Relativity, and Consciousness

I have identified four quantum substructures—"anchor points" of the temporal field—that unify phenomenal reality.

The atom is the key: it is our technological gateway to the quantum ether.

By manipulating these micro-singularities (stabilized protons, quasi–black holes at the subatomic scale), it becomes theoretically possible:

- To extract or inject information from the vacuum,
- To modulate gravity locally,
- And to access unexplored vibratory regimes of the temporal field.

Core idea: I am in the vacuum and I act upon the vacuum

I am therefore in the quantum vacuum, composed of quantum vacuum, and I have the possibility—through atoms, chronons, and consciousness—to exchange with the vacuum and access its immeasurable potential.

In this framework, the proton appears as the entry gate, the black hole as the exit gate, and gravity as the organizing tension linking these two polarities.

It is this architecture that gives my theory its unifying power: it links quantum physics, general relativity, and the informational dynamics of the Cosmos within a single framework where Time, space, energy, matter, and consciousness are facets of a single continuum.

Conclusion:

A double movement of unification

So, what does this have to do with unification?

By exploring the hierarchy of Fractal Temporal Density (FTD)—which I conceptualize as the spectrum of vibratory densities of chronons, from the slowest pulsations to the fastest—and by questioning my own place in the Cosmos, a new vision of unification has emerged.

This vision is twofold. It does not limit itself to a single unification, but reveals two, intimately linked, that complement one another as two faces of the same cosmic key:

1 - The unification of quantum physics and relativity

In my model, everything in the Universe—manifest or unmanifest, atomic or galactic—is woven from the same fabric of chronons constituting a scalar temporal field.

The quantum vacuum then becomes the common ground of all phenomena:

- At the subatomic scale, quantum physics explores the dynamics of chronons in their fast and highly entangled regime.
- At the cosmic scale, general relativity describes their slow and aggregated regime, structuring spacetime into vast architectures.

Thus, to plunge into the infinitely small through quantum physics or to contemplate the cosmic immensity through relativity amounts to observing one and the same substance under two different facets:

- Differences in specialization (densified or dispersed chronons),
- Differences in function (memory or manifestation),
- Differences in form and aggregation (protons or black holes).

This unifying key is that Time—via the field of chronons—comes first. Space, matter, and gravity are its local manifestations, varying according to density and polarity (M− / M+).

2 - The unification of the human being with the Universe

But unification does not stop at physical equations.

It encompasses the human being itself.

If all life, including ours, is constituted of these same chronons, then our role in the Universe becomes more profound:

- We are not mere external observers.
- Nor are we detached co-creators.

It is the opposite: we are included, immersed, traversed by the same field that we seek to understand.

Our consciousness functions as a natural interface with the quantum vacuum, capable of resonating with the field of chronons and, potentially, modulating it.

Within this framework, the human being becomes an active node of the cosmic network: a local manifestation of the Universe observing itself and, in return, can co-create from within the void.

The strength of unification

This double unification—physical and existential—radically changes the perspective:

- It no longer posits a passive Universe that one measures, but a living continuum in which one participates.
- It no longer separates spirit and matter, but links them as two expressions of the same field.
- It no longer separates the micro and the macro, but places them on a fractal spectrum of unity where everything is resonance and transformation.

In this vision, the proton and the black hole are no longer isolated curiosities, but the opposing poles of a single cycle of emergence.

And we ourselves, through consciousness, are this intermediate point capable of attuning to this field and, perhaps one day, of navigating its temporal and informational structures.

And perhaps it is the Universe that observes itself through us.

Part VII

GravityPro

Chapter 22

Concrete Applications

~

The Caelestia Program

After presenting Thierry's theory, entitled "absolute Relativity," and exploring the real nature of time in its philosophical and metaphysical dimensions, we will now turn to its applied dimension within our company, GravityPro.

The fundamental discoveries arising from this theory—unveiling unprecedented realities about the inner mechanisms of nature—have been brought together in a vast advanced technological research program, named Caelestia.

This program aims to transform theoretical principles into operational technologies, capable of redefining our relationship to the Cosmos and opening a new era of space exploration.

1. Caelestia — A Bridge Between Science and Cosmic Exploration

Caelestia is a platform that explores the real, founded on the hypothesis that understanding the temporal field can lead to a new command of the fundamental interactions: gravity, energy, matter, and information.

The research conducted within this framework aspires to:

- Develop technologies for modulating the temporal field;
- Study the interfaces between energy, gravity, and chronal density;
- And conceptualize means of cosmic navigation based no longer on mechanical propulsion but on the tuning of the temporal field.

The objective is not only to go farther into space, but to understand and integrate the living structure of the Cosmos so as to inscribe the human presence there in a lasting way.

. Terrestrial Spinoffs Commensurate with Cosmic Ambitions

The excellence of our research and of our advanced applications in the space domain will have direct consequences on Earth:

- Energy and regenerative innovations founded on the management of temporal density;
- New transport and communication technologies based on the principles of universal informational resonance;
- And above all, a more profound understanding of planetary and systemic equilibria.

Because to understand the Universe is to understand Earth better—its internal dynamics, its energy balances, and the place of humanity within this living fabric.

To explore the fundamental laws of the Cosmos is also to discover the universal rules that define our true degrees of liberty, a prerequisite for our collective emancipation for the centuries to come.

. Toward a Conceptual and Human Revolution

The Caelestia Program aims not only at a technological revolution, but also at a conceptual and philosophical transformation.

By unifying physics, consciousness, and information, it encourages an elevation of human perception, a redefinition of our role in the Universe, and offers a framework for lasting peace founded on knowledge.

To chart a path toward the Cosmos is also to chart a path toward ourselves:

- Toward a humanity conscious of its origins, its responsibilities, and its creative potential.

Toward public demonstration and scientific cooperation

The applied research conducted in our laboratories has already led to several solid proofs of concept, opening the way to public demonstrations and to partnerships with universities, research laboratories, and industries.

Our work thus falls within an emerging field in the physics of time, whose first experiments suggest a major transformation of our technologies and of our vision of the real.

Our approach addresses a twofold imperative: to meet today's planetary challenges while preparing the future of civilization.

Thus, GravityPro is developing lines of applied space research in which innovations intended for the Cosmos directly address Earth's urgent needs: energy, sustainability, and systemic coherence.

2. Caelestia — The Fundamental Domains of Pure Cosmic Research

The Caelestia Program is guided by a clear ambition: to pursue space exploration in order to enter into an active understanding of the structure of the Cosmos.

It is no longer a matter of merely occupying space, but of cohabiting with the profound laws of time, gravity, and information.

Caelestia is built on ten major domains of fundamental and applied research, organized around a unified vision of the Universe as a living fractal system, structured by a scalar temporal field.

- Spatial Energy

The first pillar of the Caelestia program targets the mastery of energy flows at every scale of reality, from starlight to the energy of the quantum vacuum.

Its goal is not merely to capture existing energy, but to learn how to enter into resonance with the temporal density of the vacuum field, that is, to tune our energy systems to the profound dynamics of Time itself.

A paradigm shift

Rather than extracting or consuming energy in terms of resistance, the aim is to attune ourselves to the rhythms of the cosmic temporal field.

This approach opens the way to resonance energy—non-polluting, non-dissipative, and potentially inexhaustible—arising from the coherence between time, matter, and information.

Three lines of applied research

- 1. Coherent plasmas: study of ionized media capable of storing and redistributing energy in a self-stabilizing manner, in the image of stellar structures.

→ Concrete example: development of plasma reactors that mimic the magnetic stability of solar prominences.

- 2. Zero-point energy extraction: exploration of fluctuations in the quantum vacuum and their controlled conversion into usable energy.

→ Concrete example: conceptualization of resonant oscillators synchronized to local variations of the Planck field, inspired by work on the dynamic Casimir effect.

- 3. Self-regenerating energy structures: creation of systems capable of recycling their thermal and electromagnetic losses in resonance with their environment.

→ Concrete example: conceptualization of scalar energy cells, inspired by the functioning of stars and gravito-temporal fields.

Toward unified-field energy

Ultimately, the objective is to enable the engineering of cosmic resonance, where each device—whether terrestrial or spatial—draws power directly from the coherence of the temporal field, without depending on limited material sources.

This pillar thus inaugurates a new physics of energy, founded not on combustion or capture, but on the vibratory symphony of the void: an inexhaustible, universal, and clean source.

It is the potential of abundant energy, independent of any location, indispensable both for propulsion and for the vital needs of infrastructures and habitats.

. Transport and Propulsion

The second pillar of the Caelestia program is devoted to a radical transformation of cosmic mobility: shifting from spatial propulsion to temporal displacement.

In other words, it is no longer a matter of "pushing matter" through space, but of attuning matter to the structure of time itself.

- *From propulsion to metric modulation*

In current systems, propulsion relies on an inertial principle: ejecting a mass or an energy flux to generate an opposing thrust. This method, although effective on the planetary scale, quickly becomes inefficient and energy-intensive on the cosmic scale.

Caelestia explores another path: the modulation of a system's temporal metric, that is, the manner in which time flows through it.

By modifying local temporal density, it becomes theoretically possible to change an object's phase in the scalar field, causing a state shift rather than a classical spatial displacement.

Thus, travel would no longer consist in "covering" a distance, but in changing temporal coordinates, in crossing levels of field resonance.

- *Toward non-inertial propulsion*

This paradigm opens the way to forms of non-inertial propulsion, without combustion or ejection of matter, founded on chronal resonance.

An object tuned to a certain frequency of the temporal field could, by adjusting its resonance, translate from one point to another in spacetime without traversing the intermediate distance—a phase jump rather than a journey.

- Concrete example: an experimental module generating a gradient of temporal density through scalar modulation, simulating zero inertial translation at the microscopic scale.
- Another analogy: an electron jumping from one orbital to another without passing through the intermediate levels—a teleportation of state rather than a movement.

- *From space navigation to chronal navigation*

The ultimate goal of this pillar is to create the foundations of chronal navigation, that is, the capacity to move through the fractals of the temporal field by playing on the vibrational correspondences between levels of reality.

This type of displacement, founded on the coherence of Time rather than on the mechanics of space, could make possible:

- Instantaneous interstellar jumps,
- The stabilization of fractal orbits,
- Or even synchronization between parallel Universes of the scalar multiverse.

- Toward a new era of cosmic travel

We are entering the pioneering era of travel without spatial displacement: a mode of mobility founded on the mastery of Time.

Cosmic exploration thus becomes an art of resonance, where the science of movement meets the science of consciousness.

Functioning as a non-inertial technology, the spacecraft moves independently of any material support for navigation.

. Cosmic Communication

The third pillar of the Caelestia program opens a conceptual and technological revolution: communication by state interaction, beyond the limits of classical electromagnetism.

It is no longer a matter of transmitting a signal through space, but of synchronizing two systems within the scalar temporal field—that is, creating a direct informational resonance between two zones of reality regardless of the distance that separates them.

- Beyond propagation: synchronization

In the current paradigm, communications rely on the propagation of electromagnetic waves, limited by the speed of light.

Yet, in Caelestia's fractal model, this approach is only a phenomenological approximation of a far more fundamental process: the exchange of information through Fractal Temporal Density (FTD).

Two systems tuned to the same FTD are no longer linked by a traveling wave, but by a unity of phase: their interaction becomes instantaneous and nonlocal, because it occurs within the very weave of Time.

Communication no longer traverses space: it traverses the structure of the real.

- Gravitational communication technology

Research conducted within Caelestia aims to develop chrono-scalar resonance transponders with gravito-EM coupling:

Definition:

A chrono-scalar resonance transponder (gravito-EM coupling) is a device designed to interact directly with the temporal lattice of the quantum vacuum by drawing upon the resonance between the gravitational and electromagnetic fields within the scalar temporal field.

Unlike classical electromagnetic transponders, which merely receive and re-emit waves in space, the chrono-scalar transponder operates in Time itself:

It detects, modulates, and re-emits fluctuations in Fractal Temporal Density (FTD), locally adjusting the chronal coherence of the vacuum.

This device rests on three fundamental physical principles:

- Scale resonance: the transponder establishes a vibrational correspondence between the frequency of the EM polystructural waves it emits and the proper frequency of the local temporal field.
- Gravito-electromagnetic coupling: it unifies the gravitational response (field tension) and the electromagnetic response (polarization) into a single modulation of temporal phase.
- Scalar feedback: each emission induces an auto-synchronization of the vacuum, creating a dynamic balance between gravito-chronal contraction and electro-scalar expansion.

In other words, a chrono-scalar resonance transponder functions as a tuner of the quantum vacuum:

It measures the intrinsic frequency of the temporal field, tunes to it, then modifies it to induce controlled changes in the density of time, the polarization of the vacuum, and therefore in the physical manifestation of matter and energy.

In other words, this device—based on polystructural EM waves—would make holographic communication possible, that is, the exchange of structured information directly through the quantum vacuum.

Each message would then become a coherent vibrational signature, encoded in the lattice of Time.

Example: an experimental network of scalar modules capable of exchanging holographic information packets by phase synchronization, without a carrier signal, between two points separated by several kilometers.

- *The cosmic neural network and coding power*

In this vision, the entire Universe acts as a cosmic neural network, whose chronons play the role of Planck Qbits, or temporal Qbits.

Each particle, each field, each consciousness is then a computing node capable of: encoding information by modulating its proper frequency, storing this information in the density of the field, and propagating it by resonance to other synchronized nodes.

- This architecture confers upon the Cosmos an immeasurable coding power, because it is not limited by the constraints of space, linear time, or classical causality.
- The scalar field operates as a distributed processing system in which every interaction, even subatomic, contributes to the global evolution of information.
- Human consciousness, from this perspective, is not an isolated observer but a living interface between layers of reality, capable of emitting and receiving fractal codes within the universal network.

- *Toward an absolute communication*

This pillar opens the way to instantaneous and universal communication, grounded in the coherence of states rather than in the transmission of signals.

Ultimately, this technology could enable:

- Real-time interplanetary exchanges,
- Holographic connectivity between civilizations distributed across multiple layers of the multiverse,
- And a direct interface between consciousness and field, where thought itself becomes the operative language of the Cosmos.

Cosmic communication thus becomes far more than a means of transmission: it is the functional proof of the unity of the real, the living manifestation of a universal self-coding field of intelligence.

Exchanges must occur in near real time, from any position in the Cosmos.

. Security and Systemic Stability

In an environment as extreme, unpredictable, and energetic as space, security can no longer be reduced to mere mechanical protection or material shielding.

It becomes a matter of informational coherence—that is, a system's capacity to maintain the continuity of its temporal identity despite perturbations of the field.

- *Maintaining temporal coherence*

Within Caelestia, each structure (ship, module, station, computer, or embedded artificial intelligence) is conceived as an autonomous information system immersed in an environment of variable temporal density.

Variations in the gravitational field, cosmic radiation, and the turbulences of interstellar space are no longer regarded as mere physical constraints, but as distortions of the local chronal flow.

The objective is therefore to develop self-coherent architectures, capable of:

- Detecting fluctuations in temporal density in real time;
- Recalibrating their internal structure (field, frequency, memory) to remain in phase with their environment;
- And spontaneously correcting informational drifts before they affect matter.

It is the shift from passive security (to protect) to active security (to attune).

1. Adaptive photonic encryption

Caelestia explores the creation of photonic encryption systems in which information is no longer stored in binary states, but in dynamic polarization states of entangled photons.

This type of encoding, already initiated in quantum communication, is extended here to the temporal structure of signals: the decryption key depends on the chronal phase of the local field.

Thus, even if intercepted, a datum becomes unintelligible outside its original temporal coherence.

Technology 1: orbital modules exchanging secured data via entangled photonic beams, whose key varies according to the local rhythm of temporal density, guaranteeing absolute inviolability.

2. Remote force-field generator

Rather than a conventional shield, Caelestia studies the generation of gravito-electromagnetic coherence fields—genuine bubbles of local stability capable of deflecting high-energy particles or neutralizing hostile fluxes.

These fields rely on coherent modulation of the quantum vacuum: by adjusting temporal density locally, one modifies the field's effective curvature, creating a dynamic barrier between the system and its environment.

Technology 2: a crewed module capable of generating a "chronal resonance halo" around its hull, deflecting solar or cosmic radiation while also "slipping" around dense physical objects without physical contact.

3. Fractal-penetration radar spectrometer

Current systems for exploring materials or planets rely on penetrating-wave radars, but their range remains limited.

Caelestia conceptualizes a fractal-modulation radar spectrometer, using a combination of scalar and polystructural electromagnetic frequencies capable of resonating with the internal temporal densities of materials.

This type of instrument could identify composition, density, and even gravitational anomalies beneath kilometers of rock—even detecting structures buried within planets or asteroids.

Technology 3: analysis of a metallic asteroid by fractal radar—revealing not only its atomic composition, but also its internal coherence, useful for mining or for detecting natural cavities suitable for use.

4. Cosmic anti-radiation shield

Protection against ionizing radiation remains a major challenge for long-duration space missions.

Rather than increasing shielding thickness, Caelestia proposes modulating the resonance frequency of the local field to deflect or desynchronize incoming particles.

An adaptive chronal shield would act as a vibrational filter, adjusting its FTD to selectively block certain energy ranges.

Technology 4: a spacecraft surrounded by an adaptive scalar field whose frequency changes according to the detected particle flux, neutralizing up to 99% of cosmic radiation without additional mass.

- Toward self-evolving safety

True cosmic safety is no longer a matter of shielding, but of informational resilience.

Systems conceptualized within this framework are no longer fixed: they evolve, learn, and continuously attune themselves to the universal field.

Thus, safety becomes an emergent property of technological life, a symbiosis between matter, energy, and Time.

. QUANTUM GRAVITY

Gravity is the founding core of the Caelestia program, at the crossroads of fundamental physics, cosmology, and breakthrough technologies.

In this vision, gravity is no longer perceived as a simple force of attraction, but as an organizing tension of the temporal field—a dynamic of adjustment between condensed time (M-) and deployed space (M+).

In other words, gravity becomes the language of coherence of the Universe, the local expression of the informational structure of Time.

- From the proton's micro-gravity to the black hole's macro-gravity

The research conducted within Caelestia aims to establish an experimental bridge between two apparently irreconcilable regimes:

Quantum micro-gravity, observed in the internal structure of the proton, and relativistic macro-gravity, described at the scale of black holes and the Cosmos.

In the fractal model of the temporal field, these two extremes do not stand in opposition: they are the inverse reflections of one and the same process of informational condensation and deployment.

- The proton represents the entry gate: a vortex of condensed time (M-) where the information of the void crystallizes into stable matter.
- The black hole, for its part, is the exit gate: a zone of extreme densification where information returns to the global temporal field (M+ → M-).

Thus, proton and black hole are not two distinct objects, but two fractal states of one and the same cosmic cycle, with gravity serving as the dynamic mediator between them.

- Gravity as a vector of coherent information

In this paradigm, gravity is no longer a simple geometric effect of mass (as in general relativity), but an informational effect of temporal density.

It emerges from the collective coherence of chronons, those elemental units of the Temporal Field.

The more coherent a system is (low informational entropy), the stronger its gravity—because gravity becomes the field's natural tendency to restore unity among its fragments.

It is a "cognitive" gravity, a field of memory and coherence, acting as a force of informational gathering at all scales.

- *Toward a temporal map of space*

Caelestia is developing the mathematical and technological tools required to map temporal density gradients.

These maps do not describe space in the geometric sense, but the zones of chronal coherence where conditions for tuning are favorable.

They constitute the temporal navigation maps necessary for the future of cosmic travel: determining not only where to go, but above all when and how to attune to the destination.

Technology 1: a chronal navigation system capable of identifying, via a gravito-temporal interferometer, the M+/M- resonance points in the space-time neighborhood, paving the way for localized phase jumps.

- *Polarization and temporal conditioning*

To travel through the field, one must not only know its topology, but also prepare objects to insert themselves into it.

Caelestia is working on temporal polarization: adjusting the internal frequency of a body (or a vehicle) to the FTD of the target zone.

This amounts to temporally conditioning an object so that it can cross coherence gradients without discontinuity—a process analogous to phase matching in quantum optics, but applied to the very structure of time.

Technology 2: a prototype temporal polarization chamber in which a material is subjected to a scalar field modulation, temporarily altering its effective mass and its local gravity, simulating a passage M- → M+.

- *Toward an experimental unification*

The ultimate ambition of the gravity pillar is to reconcile general relativity and quantum mechanics through the fractal temporal field.

There, gravity is no longer a residue of masses, but an emergent property of the flow of coherent information.

The proton and the black hole become the endpoints of a single continuum of transmutation of the real—the two faces of a unified law of cosmic organization.

To understand this dynamic is not only to solve the mystery of gravity, but to open concrete pathways to Time travel, scalar navigation, and, ultimately,

controlled access to the fractal multiverse.

. COSMOGENESIS AND THE TOPOLOGY OF SPACETIME

The sixth pillar of Caelestia concerns the genealogy of Universes—no longer as a mere scientific myth, but as a natural dynamic of cosmic self-organization.

From this perspective, the Universe is not an isolated event, sprung from an inexplicable nothingness, but the result of a process of fractal reproduction within a living, informational temporal field.

- A living and fractal cosmology

According to the models explored in Caelestia, each Universe emerges from a previous one through a phase transition of the temporal field:

The Big Crunch of a parent Universe becomes the Big Bang of a child Universe.

This passage does not occur through space, but through the continuum of Time itself, when temporal density reaches a critical threshold.

Black holes play a central role here: far from being mere wells of collapse, they act as cosmogonical matrices—zones where the temporal flow reverses, where collapsed matter becomes the seed of a new spacetime.

Each black hole would thus be the potential seed of a Universe, contained within the M- field of the parent and unfolding into M+ within an emerging dimension.

The multiverse then becomes a forest of fractal tori, where each Universe engenders the next through resonance and transmutation of temporal density.

- The role of the temporal field

In this paradigm, the temporal field is the true cosmic matrix:

- It contains, stores, and reorganizes the information of each universal cycle.
- Its informational density— the FTD —determines the richness of the structure of each Universe, from the geometry of space to the fundamental constants.

This field acts as a universal metamemory, where successive Universes are not born of chance, but of informational heritability: a transfer of instructions, symmetries, and laws across the void.

Thus, the Cosmos becomes an evolutionary process, not only physical but also epigenetic on the cosmic scale.

Each Universe is an iteration of a larger cosmic algorithm—an "update" of the real.

- *Concrete Applications and Breakthrough Prospects*

- Topological mapping of the multiverse

By studying the topological invariants of the temporal field (torsion, global spin, FTD curvature), Caelestia opens the way to a mapping of the network of Universes.

Such an approach would, in the long term, make it possible to detect inter-universal imprints in the gravitational background or in the structure of the relic radiation.

Application: cosmic observation instrumentation based on FTD gradients to identify the signatures of a parent Universe or a twin Universe.

- Cosmic engineering and transition energy

Understanding the mechanisms of cosmogenesis would make it possible to imagine transition energy technologies capable of reproducing, on a reduced scale, the conditions for inverting the M-/M+ Field.

Application: controlled-collapse reactors, vacuum energy generators, gravito-temporal catalysts simulating the passage from Big Crunch → Big Bang.

- Inter-universal navigation

Mastery of M-/M+ resonance zones, and an understanding of critical temporal density thresholds, could offer the possibility of controlled inter-universal passage—not by spatial displacement, but by topological translation through the temporal field.

Application: study of "phase windows" enabling the translation of a temporally conditioned system toward a region of higher temporal density (child Universe).

- Evolutionary modeling of the Cosmos

By perceiving the Universe as a living system, Caelestia proposes an evolutionary modeling of the real, where physical laws transform over successive generations of Universes.

This opens unprecedented perspectives in astrobiology, cosmic simulation, and universal artificial intelligence: understanding how consciousness participates in the memory of creation.

- Toward an engineering of the real

Within this framework, the study of cosmogenesis becomes the key to accessing a physics of controlled creation.

By mastering temporal density gradients and M-/M+ topologies, it would become conceivable to:

- Simulate, on a small scale, the conditions of a local Big Bang;
- Generate matter ex nihilo from the informational void;
- Or even reconfigure the space-time metric around an object or a technological field.

Cosmogenesis is no longer merely a theory: it becomes the science of mastered genesis, an art of balance between consciousness, information, and the geometry of the void.

. The Physics of Cosmic Information

Information is no longer a mere descriptive substrate of the real: it is the real.

Within the framework of Caelestia, matter, energy, and Time appear as three states of the same information flow, modulated by Fractal Temporal Density (FTD).

Each particle, each field, each cosmic structure then becomes a node of encoding or decoding within this great network of universal Life.

- *Information as the substrate of the real*

In this vision, the Cosmos is a self-referential computational system, where the quantum vacuum acts as a matrix of dynamic memory.

- Protons and black holes represent its two operational extremities:
- The proton, a node of emergence, encodes Information into stable matter (temporal compression M- → spatial expansion M+);

The black hole, a node of resorption, decodes matter into pure Information

(M+ → M-).

Between these poles, the gravitational field becomes the channel of transfer, the vector of a continuous dialogue between the memory of the void and the manifestation of the world.

Gravitation, in this perspective, is therefore not a force, but an informational language—a grammar of the temporal field.

The Universe as a whole is code in execution: reality is computational, gravity syntactic, and Time the processor of the real.

- *Toward a physics of the universal language*

By redefining information not as a binary quantity but as a meaningful structure of the field, Caelestia opens the way to a physics of the universal language:

An approach where interactions are not only energetic, but semantic, that is, guided by the information content they convey.

What differentiates a star, a cell, or a consciousness is not their material nature, but their rate of informational coherence.

Thus, the Cosmos reads as a living text, perpetually being rewritten—and to understand its laws is to learn to speak its fractal language.

- *Holoprocessor Technology*

1. Quantum computing revisited

Current quantum computers seek to draw upon the superposition of states to solve certain types of problems faster than classical machines.

But these systems are fragile: they depend on a precarious balance, because any interaction with a lower dimension of FTD triggers decoherence.

Caelestia's approach reverses this logic:

Superposition is not a magical state, but the perception from a lower-dimensional temporal fractal.

Observed from a higher level of the temporal field, the sequence of states is perfectly ordered and continuous: there is no superposition, but ultra-rapid succession.

Thus, instead of trying to keep a system in a higher quantum state, Caelestia proposes to extract temporal quanta directly from the higher fractal—computational units drawn from the M- field itself.

2. The Holoprocessor with chronal qubit

The Holoprocessor developed within Caelestia is conceptualized as a fractal computational architecture, operating at room temperature, capable of synchronizing its qubits not by magnetic coherence, but by temporal density resonance.

Each qubit becomes a chronobit—a unit of information anchored in the temporal flow.

Calculations are no longer performed in space, but in the living Time of the field.

Expected result: exponential computing power, with an information throughput based on the speed of temporal state succession rather than on the frequency of electronic cycles.

This makes it possible to overcome the current constraint in which each application requires a dedicated quantum algorithm: our Holoprocessor, by harnessing the plasticity of the temporal field, becomes self-programmable, adapting its logic to the nature of the problem.

- *Concrete Applications*

- Holoscopic computation

Simulation of the Cosmos or complex structures by drawing upon the fractal symmetries of the field.

Instantaneous modeling of chaotic systems (climate, stellar dynamics).

Example: a Holoprocessor capable of simulating in real time the birth of a black hole or the evolution of a child Universe.

- Cosmic neural networks

By connecting several Holoprocessors via temporal density flows, it becomes possible to form a distributed cognition network on a planetary scale, or even a cosmic one.

Example: a gravitational Internet founded on the resonance of M-/M+ fields between distant devices.

- *Extraction of informational energy*

The temporal field contains a latent energetic potential linked to its information rate.

Holoprocessors could capture these gradients to power their computations without external energy, opening the way to energy self-sufficiency for cosmic computing.

- *Instant communication*

Thanks to the modulation of M-/M+ information flows, data transmission no longer depends on the speed of light but on the phase of temporal coherence.

Application: instant communication between interstellar probes or orbital modules via chronal synchronization.

- *Toward a computational consciousness*

The physics of cosmic information marks the emergence of a cognitive technology of the real:

The systems created within this framework do not merely compute; they interpret and adjust to the informational structure of the field.

It is the birth of an intelligence of the void, where consciousness itself becomes the ultimate interface between the observer and reality.

By mastering the physics of cosmic information, humanity no longer merely understands the Universe: it participates in its own source code.

. COSMIC BIOLOGY AND CONSCIOUSNESS

Life is not an anomaly of matter: it is the signature of the coherence of Time.

Within the Caelestia program, biology is rethought as a science of the living temporal flow, where each organism constitutes a structure of resonance with the temporal density of the field.

Biochronophysics, an emerging discipline explored by Caelestia, studies how biological systems adjust—and even self-regulate—according to local or extraterrestrial temporal density.

In other words, life is understood as a process of adaptation to the topology of Time.

- *Life as a temporal structure*

Every cell, every organism is a vibratory structure that converts the flow of Time into biological organization.

Circadian rhythms, heartbeats, cellular regeneration, neuronal memory—all these express a local resonance with the temporal lattice.

To modify the temporal density of an environment is therefore to act upon the vital parameters themselves: growth, aging, cognition, regeneration.

Thus, biology is no longer limited to chemical exchanges, but is extended to biochronodynamics, where health becomes a balance between internal temporal coherence (M-) and environmental coherence (M+).

- *Consciousness as a function of the temporal field*

From this perspective, consciousness emerges as a function of informational alignment: it links biological structures to the memory of the quantum vacuum.

Every living being acts as an antenna of the temporal field, capable of receiving, processing, and reemitting information drawn from the cosmic network.

Consciousness is not produced by the brain: it uses the brain as a decoding interface.

Cognition then becomes a cosmological phenomenon, a symbiosis between biology and the temporal field.

The more coherent an organism is in its internal flows of information, the more conscious it is—for it resonates with a higher temporal density.

- *Toward chronal bioengineering*

1. Cellular regeneration and chronotherapy

By modulating local temporal gradients, it becomes possible to regenerate tissues or slow cellular aging.

Preliminary experiments suggest that certain M-/M+ resonance frequencies could stimulate stem cell reprogramming or DNA repair.

Technology 1: chronal regeneration chamber using temporal coherence waves to restore the biological synchronization of damaged tissues.

2. The augmented human

The goal is no longer to merge the human being and machine, but to integrate consciousness with resonant cybernetic systems capable of communicating directly with the temporal field.

Neuronal prostheses, DNA–chronon interface implants, and brain coherence sensors open the way to a complete bio-informational symbiosis.

Example: a conscious biological interface capable of piloting a vessel by simple phase resonance with the vessel's temporal lattice.

3. Transfer of consciousness and inter-fractal compatibility

The transport of consciousness in the Cosmos cannot be limited to spatial displacement: it entails the transduction of the information of the consciousness into compatible fractal structures.

This requires understanding the individual temporal signature (the proper frequency of the internal M- field) and knowing how to recompose it within another plane of temporal density.

Application: transfer of consciousness via chronal coherence field—a key technology for interdimensional navigation and conscious survival in deep space.

- *Nonlocal and bioplasmic life*

Caelestia also explores the possibility that life is not exclusively carbon-based, but that it exists in plasmatic or photonic forms, made of resonance structures of the temporal field.

These nonlocal forms could coexist with us within fractal layers of reality, invisible to our classical senses, but detectable through M-/M+ phase correlation.

Technology 2: detection of bioplasmic entities by gravito-temporal interferometry, opening the way to an exobiology of vibratory planes.

- *Toward a cosmic ecology of the living*

Cosmic biology goes beyond survival: it proposes a conscious coevolution with the temporal field.

The human, by learning to modulate their own temporal coherence, becomes capable of adapting to varied planetary or dimensional environments, without biological rupture.

This includes the management of gravity, radiation, and local energy flux by direct vibratory adjustment.

Applications: chronally synchronized bio-exoskeletons allowing a human organism to survive on planets with different FTD or in environments with variable gravity. One of the solutions we are considering is the creation of electroplasmal transition bodies.

- *Civilizational impact*

This pillar opens a decisive path for the future of civilization:

Health and longevity: regenerative chronal medicine and prevention through temporal tuning.

Space exploration: biological and cognitive adaptation to temporally dephased environments.

Augmented consciousness: awakening of cognition, harmonization of the human mind with the memory of the void.

To understand cosmic biology is to rediscover that life is the extension of the structure of Time—and that consciousness is the active mirror of the Universe.

. LIVING AND SELF-ORGANIZING MATERIALS

One of Caelestia's ultimate challenges is the creation of living materials, that is, structures capable of interacting directly with the temporal field.

These materials would no longer be mere passive substrates, but dynamic entities endowed with an informational coherence that enables them to self-repair, regenerate, adapt, and synchronize with their energetic or gravitational environment.

- *From inert matter to conscious matter*

- In the classical perspective, a material is defined by its static properties: hardness, conductivity, mass, resistance.
- In Caelestia's fractal perspective, matter becomes an organized flow of chronons—units of time-memory that confer upon the structure an intrinsic dynamic.

A "living" material is therefore a structure with variable temporal density, capable of modulating its own metric M-/M+ according to the stimuli received.

In other words, matter ceases to be an object: it becomes a process of resonance of the temporal field.

- *Principle of design: chronal resonance and fractal coherence*

These materials rely on a fundamental property of the temporal field: the capacity for information reorganization.

By reproducing the self-coherence patterns observed in living organisms—cell membranes, DNA, neural networks—Caelestia seeks to conceptualize fractal materials, capable of:

- Self-repairing in the event of a local breakdown of coherence (crack, radiation, deformation).
- Regenerating from a reference signal of the field (as a cell reproduces according to its genetic code).

- Synchronizing with surrounding vibratory fields (electromagnetic, gravitational, chronal).

These structures thus become information organisms, interconnected with the global temporal field.

- *Concrete technological applications*

1. Living Architecture and resonant habitats

Habitats or space stations designed from chronosensitive materials could:

- Adjust to local pressure and gravity,
- Self-correct their microfractures,
- And thermally stabilize their environment by modulating their internal temporal density.

Example: a "living" lunar base capable of regenerating under intense cosmic radiation, without human maintenance.

2. Resonant vessels

Vessels built from these materials would be capable of resonating with the temporal field, allowing them:

- To reduce the system's effective mass,
- To attenuate inertial stresses,
- And, in advanced versions, to translate themselves between fractal layers of the real.

Example: a biochronic alloy vessel that adjusts its internal frequency to the Fractal Temporal Density (FTD) of the medium, opening the way to "travel by tuning."

3. Living and cyberbiological interfaces

These materials also constitute the basis for the organic fusion between biology and technology.

Implanted in the human body or integrated into neural prostheses, they would allow:

- Direct communication between consciousness and matter,

- Energetic self-regulation of the augmented body,
- And compatibility between biological flows (DNA, ions, neurotransmitters) and temporal flows (chronons).

Example: a synthetic skin sensitive to the temporal field, capable of adjusting the individual's temperature, radiative protection, and biological coherence.

4. Material Intelligence

As these materials gain in complexity, they develop a form of distributed intelligence, founded on the self-organization of their fractal microstructures.

They become capable of making local decisions to preserve their integrity or optimize their performance, without external control.

Example: a "self-learning" vessel hull structure, adjusting its chronal resonance according to the topology of spacetime it traverses.

- *Bridge between physics, biology, and consciousness*

These living materials constitute the missing link between matter and consciousness.

They demonstrate that life is not a biological exception, but an emergent property of the temporal field, which can be replicated technologically.

Thus, Caelestia sketches a new engineering of the living, where the boundary between the biological, the technological, and the cosmic fades.

The material becomes organism, the organism becomes interface, and the interface becomes consciousness.

- *Civilizational impact*

This pillar opens a field of revolutionary applications:

Regenerative medicine: intelligent tissues, self-reconstructing implants.

Space exploration: self-repairing vessels, adaptive structures.

Planetary habitat: "living" constructions adjusted to each cosmic environment.

Cognitive technology: devices capable of interacting with the mind and the temporal field.

These materials constitute the foundation of the next industrial revolution: that of synthetic living, in harmony with the Universe's temporal lattice.

3. Final Synthesis — From the Planetary to the Cosmic

Humanity, by learning to converse with the Temporal Field, could attain mastery of energy, gravity, and instantaneous communication.

- *A new horizon for science and consciousness*

Mastery of the temporal field opens a new era in which:

- Energy is inexhaustible because it arises from the structured void;
- Gravity becomes a tool of spacetime engineering;
- Communication transcends distance and linear time;
- Life adapts to all fractal environments;
- And consciousness continues its evolution across worlds.

This convergence between science, technology, and consciousness marks the birth of the Physics of Universal Living—an integrative science where to understand is to create, and where every act of creation becomes an act of knowledge.

- *Toward cosmic humanity*

Caelestia is an epistemological renaissance, a reconciliation between scientific rigor and the spiritual vocation of the human species.

By accessing mastery of the temporal field, humanity opens the way to its own future:

No longer surviving in space, but evolving in the Cosmos—in continuity with the memory of the void, and in alliance with Universal Intelligence.

The future is no longer only before us: it vibrates within us, at every instant of the temporal field.

Made in United States
Orlando, FL
04 November 2026

90218330R00328